IMPORTANT:

HERE IS YOUR REGISTRATION CODE TO ACCESS YOUR PREMIUM McGRAW-HILL ONLINE RESOURCES.

For key premium online resources you need THIS CODE to gain access. Once the code is entered, you will be able to use the Web resources for the length of your course.

If your course is using **WebCT** or **Blackboard**, you'll be able to use this code to access the McGraw-Hill content within your instructor's online course.

Access is provided if you have purchased a new book. If the registration code is missing from this book, the registration screen on our Website, and within your WebCT or Blackboard course, will tell you how to obtain your new code.

Registering for McGraw-Hill Online Resources

REGISTRATION CODE

LGQK-TRSL-UU4O-9FH5-PS1C

TO gain access to your MCGraw-Hill web resources simply follow the steps below:

1. USE YOUR WEB BROWSER TO GO TO: **www.mhhe.com/tillery**

2. CLICK ON **FIRST TIME USER**.

3. ENTER THE REGISTRATION CODE* PRINTED ON THE TEAR-OFF BOOKMARK ON THE RIGHT.

4. AFTER YOU HAVE ENTERED YOUR REGISTRATION CODE, CLICK **REGISTER**.

5. FOLLOW THE INSTRUCTIONS TO SET-UP YOUR PERSONAL UserID AND PASSWORD.

6. WRITE YOUR UserID AND PASSWORD DOWN FOR FUTURE REFERENCE. KEEP IT IN A SAFE PLACE.

TO GAIN ACCESS to the McGraw-Hill content in your instructor's **WebCT** or **Blackboard** course simply log in to the course with the UserID and Password provided by your instructor. Enter the registration code exactly as it appears in the box to the right when prompted by the system. You will only need to use the code the first time you click on McGraw-Hill content.

Thank you, and welcome to your MCGraw-Hill online Resources!

*YOUR REGISTRATION CODE CAN BE USED ONLY ONCE TO ESTABLISH ACCESS. IT IS NOT TRANSFERABLE.

0-07-291915-9 T/A TILLERY: PHYSICAL SCIENCE, 6/E

THE POWER TO CREATE

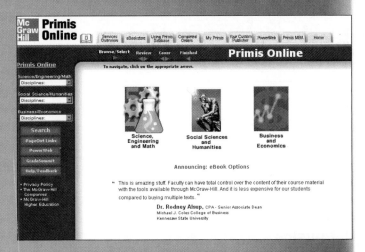

- ➲ **Customize your course online**
- ➲ **Create a custom printed book (or eBook)**
- ➲ **Mix and match from millions of resources**
- ➲ **Create a textbook or lab manual**
- ➲ **It's easy to do—just vist the website**

PRIMIS ONLINE

CREATE THE PERFECT TEXT FOR YOUR STUDENTS!

WWW.MHHE.COM/PRIMIS/ONLINE

THE TOOLS TO LEARN

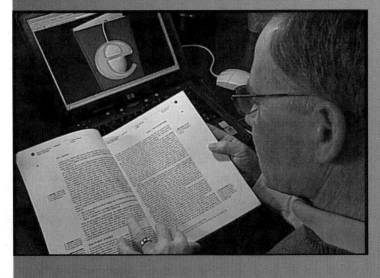

- ➲ **Receive a free copy in 7-10 days**
- ➲ **Review your custom book**
- ➲ **Adopt or modify your custom creation**
- ➲ **Students buy a book tailored to match your course**

Harness the power of the web for your course—**McGraw-Hill's PowerWeb** allows you to supplement your students' learning experience by providing Internet-based course material to supplement their learning experience.

PowerWeb provides current articles, curriculum-based materials, weekly updates with assessment, informative and timely world news, referred web links, research tools, student study tools, interactive exercises, and much more!

Instead of unorganized material provided by common search engines, all the content within **McGraw-Hill's PowerWeb** has been refereed by content experts and educational professionals—ensuring that you and your students only receive the most pertinent and topical information for your course.

POWERWEB
PHYSICAL & INTEGRATED SCIENCE
Academic Editor: **James Agnew, Elon University**

UNIT 1. The Nature of Science

Part A. Scientific Method & Interdisciplinary Nature of Science
Part B. Science, Nonscience, and Pseudoscience
Part C. Fundamental Constants

UNIT 2. Physics

Part A. Motion and Gravity
Part B. Heat and Energy
Part C. Sound and Light

UNIT 3. Chemistry

Part A. Atomic Structure
Part B. Periodic Table

UNIT 4. The Universe

Part A. Beyond Earth
Part B. Stellar Evolution
Part C. Solar System
Part D. The Earth and Moon in Space

UNIT 5. Earth

Part A. Life on Earth
Part B. Natural Disasters
Part C. Weather and Climate
Part D. Water on Earth

UNIT 6. Biology

Part A. Life on Earth
Part B. Origin and Evolution of Life
Part C. Ecology and Environment
Part D. Health
Part E. Genetics
Part F. Ethical Concerns

Visit the web for a complete list of readings and abstracts at:

www.dushkin.com/powerweb

Physical Science
PHYSICS

Sixth Edition

Physical Science
PHYSICS

Bill W. Tillery
Arizona State University

Boston   Burr Ridge, IL   Dubuque, IA   Madison, WI   New York   San Francisco   St. Louis
Bangkok   Bogotá   Caracas   Kuala Lumpur   Lisbon   London   Madrid   Mexico City
Milan   Montreal   New Delhi   Santiago   Seoul   Singapore   Sydney   Taipei   Toronto

PHYSICS
Physical Science

Copyright © 2005 by The McGraw-Hill Companies, Inc. All rights reserved. Printed in the
United States of America. Except as permitted under the United States Copyright Act of
1976, no part of this publication may be reproduced or distributed in any form or by any
means, or stored in a database or retrieval system, without the prior written consent of the
publisher.

This book contains selected material from *Physical Science, Sixth Edition,* by Bill W. Tillery.
Copyright © 2005 by The McGraw-Hill Companies, Inc. Reprinted with permission of the
publisher.

This book is a McGraw-Hill Custom textbook. Many custom published texts are modified
versions or adaptations of our best-selling textbooks. Some adaptations are printed in black
and white to keep prices at a minimum, while others are in color.

1 2 3 4 5 6 7 8 9 0 QSR QSR 0 1

ISBN 0–07–296515–0

Printer/Binder: Quebecor World

Brief Contents

Contents

Preface

Physical Science is a straightforward, easy-to-read, but substantial introduction to the fundamental behavior of matter and energy. It is intended to serve the needs of nonscience majors who are required to complete one or more physical science courses. It introduces basic concepts and key ideas while providing opportunities for students to learn reasoning skills and a new way of thinking about their environment. No prior work in science is assumed. The language, as well as the mathematics, is as simple as can be practical for a college-level science course.

Organization

The *Physical Science* sequence of chapters is flexible, and the instructor can determine topic sequence and depth of coverage as needed. The materials are also designed to support a conceptual approach, or a combined conceptual and problem-solving approach. With laboratory studies, the text contains enough material for the instructor to select a sequence for a two-semester course. It can also serve as a text in a one-semester astronomy and earth science course, or in other combinations.

> *"The text is excellent. I do not think I could have taught the course using any other textbook. I think one reason I really enjoy teaching this course is because of the text. I could say for sure that this is one of the best textbooks I have seen in my career. . . . I love this textbook for the following reasons: (1) it is comprehensive, (2) it is very well written, (3) it is easily readable and comprehendible, (4) it has good graphics."*
> —Ezat Heydari, Jackson State University

Meeting Student Needs

Physical Science is based on two fundamental assumptions arrived at as the result of years of experience and observation from teaching the course: (a) that students taking the course often have very limited background and/or aptitude in the natural sciences; and (b) that this type of student will better grasp the ideas and principles of physical science if they are discussed with minimal use of technical terminology and detail. In addition, it is critical for the student to see relevant applications of the material to everyday life. Most of these everyday-life applications, such as environmental concerns, are not isolated in an arbitrary chapter; they are discussed where they occur naturally throughout the text.

Each chapter presents historical background where appropriate, uses everyday examples in developing concepts, and follows a logical flow of presentation. The historical chronology, of special interest to the humanistically inclined nonscience major, serves to humanize the science being presented. The use of everyday examples appeals to the nonscience major, typically accustomed to reading narration, not scientific technical writing, and also tends to bring relevancy to the material being presented. The logical flow of presentation is helpful to students not accustomed to thinking about relationships between what is being read and previous knowledge learned, a useful skill in understanding the physical sciences. Worked examples help students to integrate concepts and understand the use of relationships called equations. They also serve as a model for problem solving; consequently, special attention is given to *complete* unit work and to the clear, fully expressed use of mathematics. Where appropriate, chapters contain one or more activities, called Concepts Applied, that use everyday materials rather than specialized laboratory equipment. These activities are intended to bring the science concepts closer to the world of the student. The activities are supplemental and can be done as optional student activities or as demonstrations.

> *"It is more readable than any text I've encountered. This has been my first experience teaching university physical science; I picked up the book and found it very user-friendly. The level of detail is one of this text's greatest strengths. It is well suited for a university course."*
> —Richard M. Woolheater, Southeastern Oklahoma State University

"The author's goals and practical approach to the subject matter is exactly what we are looking for in a textbook. . . . The practical approach to problem solving is very appropriate for this level of student."
—Martha K. Newchurch, Nicholls State University

". . . the book engages minimal use of technical language and scientific detail in presenting ideas. It also uses everyday examples to illustrate a point. This approach bonds with the mindset of the nonscience major who is used to reading prose in relation to daily living."
—Ignatius Okafor, Jarvis Christian College

"I was pleasantly surprised to see that the author has written a textbook that seems well suited to introductory physical science at this level. . . . Physical Science *seems to strike a nice balance between the two—avoiding unnecessary complications while still maintaining a rigorous viewpoint. I prefer a textbook that goes beyond what I am able to cover in class, but not too much. Tillery seems to have done a good job here."*
—T. G. Heil, University of Georgia

New to This Edition

In general, there has been a concerted effort to make the text even more user-friendly and relevant for students:

- A new "Concepts Applied" feature was added throughout the text, adding applications of relevance for students.
- Where needed, Parallel Exercises were reorganized to make Group A and B exercises more physically, as well as conceptually, congruent.
- Then the Parallel Exercises were selectively "tuned" for the intended audience of nonscience majors by revising and replacing some exercises with new, more conceptual exercises.
- Text materials were made more conceptually oriented and student-friendly throughout.
- The overall size of the text was reduced by two chapters through reorganizing and condensing some of the historical background material.
- Old chapter 2, "Motion," and old chapter 3, "Patterns of Motion," were merged into one new chapter ("Motion") for a more intuitive presentation.
- Old chapter 9, "Atomic Structure," was substantially rewritten and merged with old chapter 10, "Elements and the Periodic Table," into one new chapter ("Atoms and Periodic Properties") with a more student-friendly approach.
- Old chapter 13, "Water and Solutions," (new chapter 11) was substantially rewritten to be more conceptual and relevant to students.

- The astronomy chapters were substantially rewritten to be more intuitive, contain less history, and update factual materials.
- To satisfy requests from current users of the text, new "Closer Look" features were added, for example: Freefall, Simple Machines, The Measurement Process, Doppler Radar, Lasers, Radiation and Food Preservation, Three Mile Island and Chernobyl, Dark Energy, Seismic Tomography, Estuary Pollution, and the Health of the Chesapeake Bay.
- Also to satisfy requests from current users of the text, additional "People Behind the Science" features were added, including biographies on Isaac Newton, Michael Faraday, Erwin Schrödinger, Robert Bunsen, Shirley Ann Jackson, Stephen Hawking, Jocelyn (Susan) Bell Burnell, and Carl Sagan.

The Learning System

Physical Science has an effective combination of innovative learning aids intended to make the student's study of science more effective and enjoyable. This variety of aids is included to help students clearly understand the concepts and principles that serve as the foundation of the physical sciences.

Overview

Chapter 1 provides an *overview* or orientation to what the study of physical science in general, and this text in particular, are all about. It discusses the fundamental methods and techniques used by scientists to study and understand the world around us. It also explains the problem-solving approach used throughout the text so that students can more effectively apply what they have learned.

Chapter Opening Tools
Chapter Outline
The chapter outline includes all the major topic headings and subheadings within the body of the chapter. It gives you a quick glimpse of the chapter's contents and helps you locate sections dealing with particular topics.

Chapter Overview
Each chapter begins with an introductory overview. The overview previews the chapter's contents and what you can expect to learn from reading the chapter. It adds to the general outline of the chapter by introducing you to the concepts to be covered, facilitating in the integration of topics, and helping you to stay focused and organized while reading the chapter for the first time. After reading the introduction, browse through the chapter, paying particular attention to the topic headings and illustrations so that you get a feel for the kinds of ideas included within the chapter.

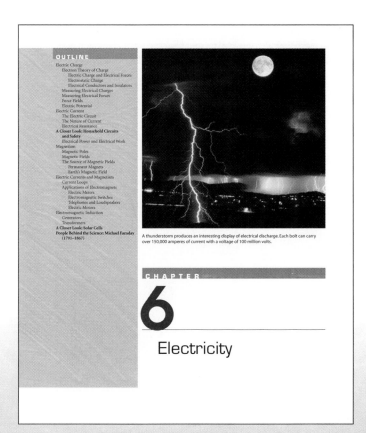

A thunderstorm produces an interesting display of electrical discharge. Each bolt can carry over 150,000 amperes of current with a voltage of 100 million volts.

CHAPTER

6

Electricity

Examples

Each topic discussed within the chapter contains one or more concrete, worked *Examples* of a problem and its solution as it applies to the topic at hand. Through careful study of these examples, students can better appreciate the many uses of problem solving in the physical sciences.

"I feel this book is written well for our average student. The images correlate well with the text, and the math problems make excellent use of the dimensional analysis method. While it was a toss-up between this book and another one, now that we've taught from the book for the last year, we are extremely happy with it."
—Alan Earhart, Three Rivers Community College

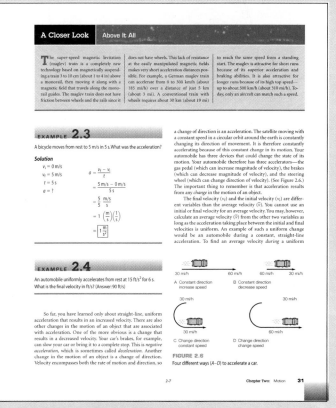

Applying Science to the Real World

Concepts Applied

Each chapter also includes one or more *Concepts Applied* boxes. These activities are simple investigative exercises that students can perform at home or in the classroom to demonstrate important concepts and reinforce understanding of them. This feature also describes the application of those concepts to everyday life.

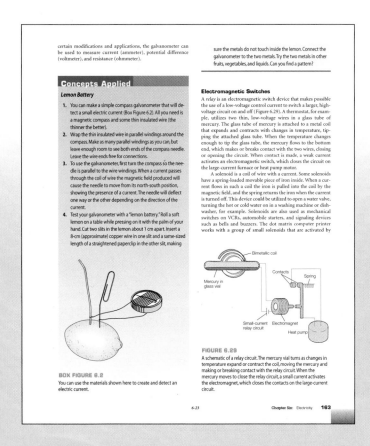

Closer Look

One or more boxed *Closer Look* features can be found in each chapter of *Physical Science*. These readings present topics of special human or environmental concern (the use of seat belts, acid rain, and air pollution, for example). In addition to environmental concerns, topics are presented on interesting technological applications (passive solar homes, solar cells, catalytic converters, etc.), or topics on the cutting edge of scientific research (for example, El Niño and Dark Energy). All boxed features are informative materials that are supplementary in nature. The *Closer Look* readings serve to underscore the relevance of physical science in confronting the many issues we face daily.

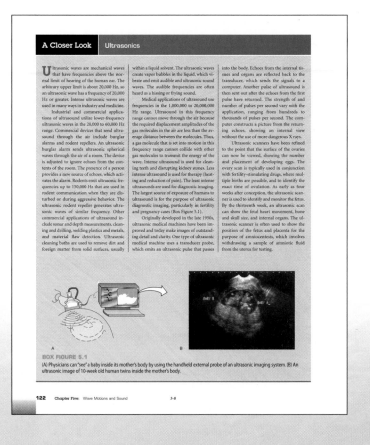

People Behind the Science

Many chapters also have one or two fascinating biographies that spotlight well-known scientists, past or present. From these *People Behind the Science* biographies, students learn about the human side of the science: physical science is indeed relevant, and real people do the research and make the discoveries. These readings present physical science in real-life terms that students can identify with and understand.

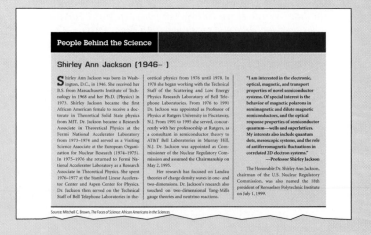

"The People Behind the Science features help relate the history of science and the contributions of the various individuals."
—Richard M. Woolheater, Southeastern Oklahoma State University

End-of-Chapter Features

At the end of each chapter, students will find the following materials:

- *Summary:* highlights the key elements of the chapter.
- *Summary of Equations* (chapters 1–13): reinforces retention of the equations presented.
- *Key Terms:* gives page references for finding the terms defined within the context of the chapter reading.
- *Applying the Concepts:* tests comprehension of the material covered with a multiple-choice quiz.
- *Questions for Thought:* challenges students to demonstrate their understanding of the topics.
- *Parallel Exercises* (chapters 1–13): reinforces problem-solving skills. There are two groups of parallel exercises, Group A and Group B. The Group A parallel exercises have complete solutions worked out, along with useful comments, in appendix D. The Group B parallel exercises are similar to those in Group A but do not contain answers in the text. By working through the Group A parallel exercises and checking the solutions in appendix D, students will gain confidence in tackling the parallel exercises in Group B, and thus reinforce their problem-solving skills.

"I like this [Summary of Equations] feature. It collects the equations together for easy reference. . . . I also like this [Key Terms] feature. It is well organized, thorough and gives the student a tool for review. The instructor can also use it for a checklist of topics. . . . The end-of-chapter features of Summary of Equations and Key Terms make the chapters very user-friendly."
—Richard M. Woolheater, Southeastern Oklahoma State University

"The Parallel Exercises and the explanation in the appendix, the readability of the material, and the depth of coverage are the strongest features of this text."
—Martha K. Newchurch, Nicholls State University

"The provision of solutions to a set of problems as a guide for solving identical problems on an adjacent set is an ingenious learning tool."
—Ignatius Okafor, Jarvis Christian College

End-of-Text Materials

Appendices providing math review, additional background detail, solubility and humidity charts, and solutions for the Group A Parallel Exercises can be found at the back of the text. There is also a glossary of all key terms, an index, and special tables printed on the inside covers for reference use.

Summary

Electromagnetic radiation is emitted from all matter with a temperature above absolute zero, and as the temperature increases, more radiation and shorter wavelengths are emitted. Visible light is emitted from matter hotter than about 700°C, and this matter is said to be *incandescent*. The sun, a fire, and the ordinary lightbulb are incandescent sources of light.

The behavior of light is shown by a light ray model that uses straight lines to show the straight-line path of light. Light that interacts with matter is *reflected* with parallel rays, moves in random directions by *diffuse reflection* from points, or is *absorbed*, resulting in a temperature increase. Matter is *opaque*, reflecting light, or *transparent*, transmitting light.

In reflection, the incoming light, or *incident ray*, has the same angle as the *reflected ray* when measured from a perpendicular from the point of reflection, called the *normal*. That the two angles are equal is called the *law of reflection*. The law of reflection explains how a flat mirror forms a *virtual image*, one from which light rays do not originate. Light rays do originate from the other kind of image, a *real image*.

Light rays are bent, or *refracted*, at the boundary when passing from one transparent medium to another. The amount of refraction depends on the *incident angle* and the *index of refraction*, a ratio of the speed of light in a vacuum to the speed of light in the medium. When the refracted angle is 90°, *total internal reflection* takes place. This limit to the angle of incidence is called the *critical angle*, and all light rays with an incident angle at or beyond this angle are reflected internally.

Each color of light has a range of wavelengths that forms the *spectrum* from red to violet. A glass prism has the property of *dispersion*, separating a beam of white light into a spectrum. Dispersion occurs because the index of refraction is different for each range of colors, with short wavelengths refracted more than larger ones.

A wave model of light can be used to explain interference and polarization. *Interference* occurs when light passes through two small slits or holes and produces an *interference pattern* of bright lines and dark zones. *Polarized light* vibrates in one direction only, in a plane. Light can be polarized by certain materials, by reflection, or by scattering. Polarization can only be explained by a transverse wave model.

A wave model fails to explain observations of light behaviors in the *photoelectric effect* and *blackbody radiation*. Max Planck found that he could modify the wave theory to explain blackbody radiation by assuming that vibrating molecules could only have discrete amounts, or *quanta*, of energy and found that the quantized energy is related to the frequency and a constant known today as *Planck's constant*. Albert Einstein applied Planck's quantum concept to the photoelectric effect and described a light wave in terms of quanta of energy called *photons*. Each photon has an energy that is related to the frequency and Planck's constant.

Today, the properties of light are explained by a model that incorporates both the wave and the particle nature of light. Light is considered to have both wave and particle properties and is not describable in terms of anything known in the everyday-sized world.

Summary of Equations

7.1

angle of incidence = angle of reflection

$$\theta_i = \theta_r$$

7.2

$$\text{index of refraction} = \frac{\text{speed of light in vacuum}}{\text{speed of light in material}}$$

$$n = \frac{c}{v}$$

7.3

speed of light in vacuum = (wavelength)(frequency)

$$c = \lambda f$$

7.4

$$\text{energy of photon} = \binom{\text{Planck's}}{\text{constant}}(\text{frequency})$$

$$E = hf$$

KEY TERMS

blackbody radiation (p. 180)	polarized (p. 194)
incandescent (p. 180)	quanta (p. 196)
index of refraction (p. 187)	real image (p. 201)
interference (p. 191)	refraction (p. 201)
light ray model (p. 182)	total internal reflection (p. 186)
luminous (p. 180)	unpolarized light (p. 194)
photoelectric effect (p. 195)	virtual image (p. 185)
photons (p. 196)	

APPLYING THE CONCEPTS

1. A luminous object is an object that
   a. reflects a dim blue-green light in the dark.
   b. produces light of its own by any method.
   c. shines by reflected light only, such as the moon.
   d. an object that glows only in the absence of light.

2. An object is hot enough to emit a dull red glow. When this object is heated even more, it will
   a. emit shorter-wavelength, higher-frequency radiation.
   b. emit longer-wavelength, lower-frequency radiation.
   c. emit the same wavelengths as before, but with more energy.
   d. emit more of the same wavelengths with more energy.

3. The difference in the light emitted from a candle, an incandescent lightbulb, and the sun is basically from differences in
   a. energy sources.
   b. materials.
   c. temperatures.
   d. phases of matter.

7-23   Chapter Seven: Light **201**

Appendix A
Mathematical Review

Working with Equations

Many of the problems of science involve an equation, a shorthand way of describing patterns and relationships that are observed in nature. Equations are also used to identify properties and to define certain concepts, but all uses have well-established meanings, symbols that are used by convention, and allowed mathematical operations. This appendix will assist you in better understanding equations and the reasoning that goes with the manipulation of equations in problem-solving activities.

Background

In addition to a knowledge of rules for carrying out mathematical operations, an understanding of certain quantitative ideas and concepts can be very helpful when working with equations. Among these helpful concepts are (1) the meaning of inverse and reciprocal, (2) the concept of a ratio, and (3) fractions.

The term *inverse* means the opposite, or reverse, of something. For example, addition is the opposite, or inverse, of subtraction, and division is the inverse of multiplication. A *reciprocal* is defined as an inverse multiplication relationship between two numbers. For example, if the symbol *n* represents any number (except zero), then the reciprocal of *n* is $1/n$. The reciprocal of a number ($1/n$) multiplied by that number (*n*) always gives a product of 1. Thus, the number multiplied by 5 to give 1 is $1/5$ ($5 \times 1/5 = 5/5 = 1$). So $1/5$ is the reciprocal of 5, and 5 is the reciprocal of $1/5$. Each number is the *inverse* of the other.

The fraction $1/5$ means 1 divided by 5, and if you carry out the division it gives the decimal 0.2. Calculators that have a $1/x$ key will do the operation automatically. If you enter 5, then press the $1/x$ key, the answer of 0.2 is given. If you press the $1/x$ key again, the answer of 5 is given. Each of these numbers is a reciprocal of the other.

A *ratio* is a comparison between two numbers. If the symbols *m* and *n* are used to represent any two numbers, then the ratio of the number *m* to the number *n* is the fraction m/n. This expression means to divide *m* by *n*. For example, if *m* is 10 and *n* is 5, the ratio of 10 to 5 is $10/5$, or $2:1$.

Working with *fractions* is sometimes necessary in problem-solving exercises, and an understanding of these operations is needed to carry out unit calculations. It is helpful in many of these operations to remember that a number (or a unit) divided by itself is equal to 1, for example,

$$\frac{5}{5} = 1 \qquad \frac{\text{inch}}{\text{inch}} = 1 \qquad \frac{5 \text{ inches}}{5 \text{ inches}} = 1$$

When one fraction is divided by another fraction, the operation commonly applied is to "invert the denominator and multiply." For example, 2/5 divided by 1/2 is

$$\frac{\frac{2}{5}}{\frac{1}{2}} = \frac{2}{5} \times \frac{2}{1} = \frac{4}{5}$$

What you are really doing when you invert the denominator of the larger fraction and multiply is making the denominator (1/2) equal to 1. Both the numerator (2/5) and the denominator (1/2) are multiplied by 2/1, which does not change the value of the overall expression. The complete operation is

$$\frac{\frac{2}{5}}{\frac{1}{2}} \times \frac{\frac{2}{1}}{\frac{2}{1}} = \frac{\frac{2 \times 2}{5}}{\frac{1 \times 2}{2}} = \frac{\frac{4}{5}}{\frac{2}{2}} = \frac{\frac{4}{5}}{1} = \frac{4}{5}$$

Symbols and Operations

The use of symbols seems to cause confusion for some students because it seems different from their ordinary experiences with arithmetic. The rules are the same for symbols as they are for numbers, but you cannot do the operations with the symbols until you know what values they represent. The operation signs, such as $+$, $\div$, $\times$, and $-$ are used with symbols to indicate the operation that you would do if you knew the values. Some of the mathematical operations are indicated several ways. For example, $a \times b$, $a \cdot b$, and ab all indicate the same thing, that *a* is to be multiplied by *b*. Likewise, $a \div b$, a/b, and $a \times 1/b$ all indicate that *a* is to be divided by *b*. Since it is not possible to carry out the operations on symbols alone, they are called *indicated operations*.

Operations in Equations

An equation is a shorthand way of expressing a simple sentence with symbols. The equation has three parts: (1) a left side, (2) an equal sign (=), which indicates the equivalence of

A-1   595

Supplements

Physical Science is accompanied by a variety of multimedia supplementary materials, including an interactive website; an Instructor's Testing and Resource CD-ROM, with testing software containing multiple-choice test items for the text and other teacher resources; and a Digital Content Manager CD-ROM, with digital images from the text. The supplement package also contains more traditional supplements: a laboratory manual and overhead transparencies.

Multimedia Supplementary Materials

Online Learning Center

A text-specific website, our *Physical Science* Online Learning Center, offering unlimited resources for both the student and instructor, can be found at: www.mhhe.com/tillery/. By way of this website, students and instructors will be better able to quickly incorporate the Internet into their classrooms. This interactive resource is packaged free with any new textbook.

Student Edition of the Online Learning Center. The *Physical Science, Sixth Edition* Online Learning Center has book-specific study aids organized by chapter. Each chapter includes animations modeling key concepts discussed in the book; interactive questions and problems, such as self-test quizzes and crossword puzzles, flashcards, and matching exercises using key terms and glossary definitions; chapter resources; and web-linked resources. Also included are Exploring Physical Science articles, which expose students to a different viewpoint on a topic or a new research project, as well as links to McGraw-Hill's Access Science and PowerWeb sites, which provide additional research resources.

Instructor's Edition of the Online Learning Center. For instructors, there is an image bank containing the images from the text, PowerPoint lectures, a bank of personal response system questions, the *instructor's manual,* the *instructor's edition of the laboratory manual,* clip art, a database of equations, and much more. From the student edition, instructors can access questions and problems from the text and additional Closer Look questions with e-mail boxes for gradable responses from students.

The *instructor's manual,* also written by the text author, is housed on the Online Learning Center and provides a chapter outline, an introduction/summary of each chapter, suggestions for discussion and demonstrations, multiple-choice questions (with answers) that can be used as resources for cooperative teaching, and answers and solutions to all end-of-chapter questions and exercises not provided in the text.

Instructor's Testing and Resource CD-ROM

The **Instructor's Testing and Resource CD-ROM** contains the *Physical Science* test bank (test questions in a combination of true/false and multiple-choice formats) within the Brownstone DIPLOMA© test generator. The Brownstone software includes a test generator, an online testing program, Internet testing, and a grade management system. This user-friendly software's testing capability is consistently ranked number one in evaluations over other products. Also located on the Instructor's Testing and Resources CD-ROM are Word and PDF files of the test bank, the instructor's manual, instructor's edition of the laboratory manual, the bank of personal response system questions, and the quizzes from the Online Learning Center. Any of these Word files can be used in combination with the Brownstone software or independently.

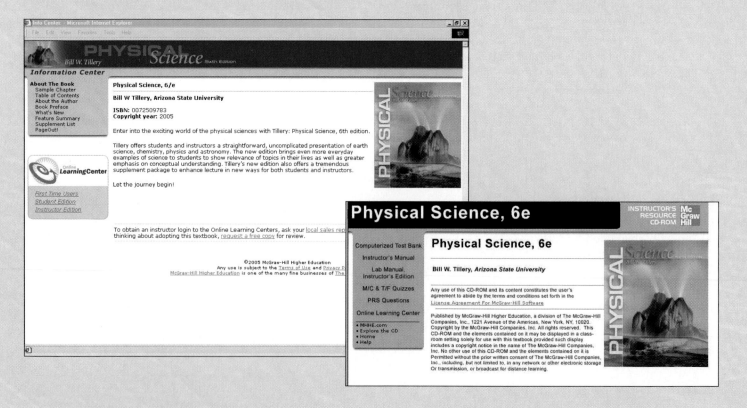

Digital Content Manager CD-ROM

The **Digital Content Manager** contains JPEG files of the four-color illustrations, photos, and tables from the text as well as a collection of animations and video clips. The CD also contains a PowerPoint presentation of the text images and another lecture PowerPoint presentation. These digital assets are contained on a cross-platform CD-ROM and are grouped by chapter within a user-friendly interface. With the help of these valuable resources, instructors can create customized classroom presentations, visually based tests and quizzes, dynamic course website content, and attractive printed support materials.

"I find Physical Science to be superior to either of the texts that I have used to date. . . . The animations and illustrations are better than those of other textbooks that I have seen, more realistic and less trivial."
—T. G. Heil, University of Georgia

Printed Supplementary Material

Laboratory Manual

The *laboratory manual,* written and classroom tested by the author, presents a selection of laboratory exercises specifically written for the interests and abilities of nonscience majors. There are laboratory exercises that require measurement, data analysis, and thinking in a more structured learning environment. Alternative exercises that are open-ended "Invitations to Inquiry" are provided for instructors who would like a less structured approach. When the laboratory manual is used with *Physical Science,* students will have an opportunity to master basic scientific principles and concepts, learn new problem-solving and thinking skills, and understand the nature of scientific inquiry from the perspective of hands-on experiences. The *instructor's edition of the laboratory manual* can be found on the *Physical Science* Online Learning Center.

Overhead Transparencies

A set of over 100 full-color transparencies features images from the text. The images have been modified to ensure maximum readability in both small and large classroom settings.

Acknowledgments

We are indebted to the reviewers of the sixth edition for their constructive suggestions, new ideas, and invaluable advice. Special thanks and appreciation goes out to the sixth edition reviewers:

Brian Augustine   *James Madison University*
Charles Blatchley   *Pittsburg State University*
Alan D. Earhart   *Three Rivers Community College*
Carl Frederickson   *University of Central Arkansas*
T. G. Heil   *University of Georgia*
Ezat Heydari   *Jackson State University*
Martha K. Newchurch   *Nicholls State University*
Ignatius Okafor   *Jarvis Christian College*
Karen Savage   *California State University at Northridge*
Ling Jun Wang   *University of Tennessee at Chattanooga*
Richard M. Woolheater   *Southeastern Oklahoma State University*
Heather Woolverton   *University of Central Arkansas*
Michael Young   *Mississippi Delta Community College*

This revision of *Physical Science* has also been made possible by the many users and reviewers of its previous editions. The author and publisher are grateful to the following reviewers of previous editions for their critical reviews, comments, and suggestions:

Lawrence H. Adams   *Polk Community College*
John Akutagawa   *Hawaii Pacific University*
Arthur L. Alt   *University of Great Falls*
Richard Bady   *Marshall University*
David Benin   *Arizona State University*
Charles L. Bissell   *Northwestern State University of Louisiana*
W. H. Breazeale, Jr.   *Francis Marion College*
William Brown   *Montgomery College*
Steven Carey   *Mobile College*
Darry S. Carlston   *University of Central Oklahoma*
Stan Celestian   *Glendale Community College*
Randel Cox   *Arkansas State University*
Paul J. Croft   *Jackson State University*
Keith B. Daniels   *University of Wisconsin–Eau Claire*
Valentina David   *Bethune-Cookman College*
Carl G. Davis   *Danville Area Community College*
Joe D. DeLay   *Freed-Hardeman University*
Renee D. Diehl   *Pennsylvania State University*
Laurencin Dunbar   *Livingstone College*
Dennis Englin   *The Master's College*
Steven S. Funck   *Harrisburg Area Community College*
Lucille B. Garmon   *State University of West Georgia*
Peter K. Glanz   *Rhode Island College*
Nova Goosby   *Philander Smith College*
D. W. Gosbin   *Cumberland County College*
Floretta Haggard   *Rogers State College*
Robert G. Hamerly   *University of Northern Colorado*
Eric Harms   *Brevard Community College*
J. Dennis Hawk   *Navarro College*
L. D. Hendrick   *Francis Marion College*
Christopher Hunt   *Prince George's Community College*
Abe Korn   *New York City Tech College*

Lauree G. Lane   *Tennessee State University*
Robert Larson   *St. Louis Community College*
William Luebke   *Modesto Junior College*
Douglas L. Magnus   *St. Cloud State University*
Stephen Majoros   *Lorain County Community College*
L. Whit Marks   *Central State University*
Richard S. Mitchell   *Arkansas State University*
Jesse C. Moore   *Kansas Newman College*
Michael D. Murphy   *Northwest Alabama Community College*
Oladayo Oyelola   *Lane College*
Harold Pray   *University of Central Arkansas*
Virginia Rawlins   *University of North Texas*
Michael L. Sitko   *University of Cincinnati*
K. W. Trantham   *Arkansas Tech University*
R. Steven Turley   *Brigham Young University*
David L. Vosburg   *Arkansas State University*
Donald A. Whitney   *Hampton University*
Linda Wilson   *Middle Tennessee State University*
David Wingert   *Georgia State University*

We would also like to thank the following contributors to the sixth edition:

Judith Iriarte Gross, Middle Tennessee State University, for her vast knowledge of student conceptual understandings, used in developing and revising the personal response system questions to accompany *Physical Science.*

T. G. Heil, University of Georgia, for his creativity in developing the multimedia lecture PowerPoint presentations on the *Physical Science* Online Learning Center.

Ezat Heydari, Jackson State University, for his thorough review in developing and revising the multiple-choice, true/false, and tutorial self-tests for the *Physical Science* Online Learning Center.

Last, I wish to acknowledge the very special contributions of my wife, Patricia Northrop Tillery, whose assistance and support throughout the revision were invaluable.

Meet the Author

Bill W. Tillery

Bill W. Tillery is a professor in the Department of Physics and Astronomy at Arizona State University, where he has been a member of the faculty since 1973. He earned a bachelor's degree at Northeastern State University (1960), and master's and doctorate degrees from the University of Northern Colorado (1967). Before moving to Arizona State University, he served as director of the Science and Mathematics Teaching Center at the University of Wyoming (1969–73) and as an assistant professor at Florida State University (1967–69). Bill has served on numerous councils, boards, and committees and was honored as the "Outstanding University Educator" at the University of Wyoming in 1972. He was elected the "Outstanding Teacher" in the Department of Physics and Astronomy at Arizona State University in 1995.

During his time at Arizona State, Bill has taught a variety of courses, including general education courses in science and society, physical science, and introduction to physics. He has received more than 40 grants from the National Science Foundation, the U.S. Office of Education, from private industry (Arizona Public Service), and private foundations (The Flinn Foundation) for science curriculum development and science teacher inservice training. In addition to teaching and grant work, Bill has authored or coauthored more than 60 textbooks and many monographs, and has served as editor of three separate newsletters and journals between 1977 and 1996.

Bill also maintains a website dedicated to providing resources for science teachers. This site is named The Science Education Resource Page (SERP) and is funded by a grant from the Flinn Foundation. The URL is http://serp.la.asu.edu.

Bill has attempted to present an interesting, helpful program that will be useful to both students and instructors. Comments and suggestions about how to do a better job of reaching this goal are welcome. Any comments about the text or other parts of the program should be addressed to:

Bill W. Tillery
Department of Physics and Astronomy
Arizona State University
PO Box 871504
Tempe, AZ 85287-1504 USA
Or (preferred) e-mail: bill.tillery@asu.edu

Physical science is concerned with your physical surroundings and your concepts and understanding of these surroundings.

CHAPTER

1

What Is Science?

Have you ever thought about your thinking and what you know? On a very simplified level, you could say that everything you know came to you through your senses. You see, hear, and touch things of your choosing and you can also smell and taste things in your surroundings. Information is gathered and sent to your brain by your sense organs. Somehow, your brain processes all this information in an attempt to find order and make sense of it all. Finding order helps you understand the world and what may be happening at a particular place and time. Finding order also helps you predict what may happen next, which can be very important in a lot of situations.

This is a book on thinking about and understanding your physical surroundings. These surroundings range from the obvious, such as the landscape and the day-to-day weather, to the not so obvious, such as how atoms are put together. Your physical surroundings include natural things as well as things that people have made and used (Figure 1.1). You will learn how to think about your surroundings, whatever your previous experience with thought-demanding situations. This first chapter is about "tools and rules" that you will use in the thinking process.

Objects and Properties

Physical science is concerned with making sense out of the physical environment. The early stages of this "search for sense" usually involve *objects* in the environment, things that can be seen or touched. These could be objects you see every day, such as a glass of water, a moving automobile, or a blowing flag. They could be quite large, such as the sun, the moon, or even the solar system, or invisible to the unaided human eye. Objects can be any size, but people are usually concerned with objects that are larger than a pinhead and smaller than a house. Outside these limits, the actual size of an object is difficult for most people to comprehend.

As you were growing up, you learned to form a generalized mental image of objects called a *concept*. Your concept of an object is an idea of what it is, in general, or what it should be according to your idea (Figure 1.2). You usually have a word stored away in your mind that represents a concept. The word "chair," for example, probably evokes an idea of "something to sit on." Your generalized mental image for the concept that goes with the word "chair" probably includes a four-legged object with a backrest. Upon close inspection, most of your (and everyone else's) concepts are found to be somewhat vague. For example, if the word "chair" brings forth a mental image of something with four legs and a backrest (the concept), what is the difference between a "high chair" and a "bar stool"? When is a chair a chair and not a stool? These kinds of questions can be troublesome for many people.

Not all of your concepts are about material objects. You also have concepts about intangibles such as time, motion, and relationships between events. As was the case with concepts of material objects, words represent the existence of intangible concepts. For example, the words "second," "hour," "day," and "month" represent concepts of time. A concept of the pushes and pulls that come with changes of motion during an airplane flight might be represented with such words as "accelerate" and

FIGURE 1.1

Your physical surroundings include naturally occurring and manufactured objects such as sidewalks and buildings.

FIGURE 1.2

What is your concept of a chair? Are all of these pieces of furniture chairs? Most people have concepts, or ideas of what things in general should be, that are loosely defined. The concept of a chair is one example of a loosely defined concept.

"falling." Intangible concepts might seem to be more abstract since they do not represent material objects.

By the time you reach adulthood you have literally thousands of words to represent thousands of concepts. But most, you would find on inspection, are somewhat ambiguous and not at all clear-cut. That is why you find it necessary to talk about certain concepts for a minute or two to see if the other person has the same "concept" for words as you do. That is why when one person says, "Boy, was it hot!" the other person may respond, "How hot was it?" The meaning of "hot" can be quite different for two people, especially if one is from Arizona and the other from Alaska!

The problem with words, concepts, and mental images can be illustrated by imagining a situation involving you and another person. Suppose that you have found a rock that you believe would make a great bookend. Suppose further that you are talking to the other person on the telephone, and you want to discuss the suitability of the rock as a bookend, but you do not know the name of the rock. If you knew the name, you would simply state that you found a "_____." Then you would probably discuss the rock for a minute or so to see if the other person really understood what you were talking about. But not knowing the name of the rock, and wanting to communicate about the suitability of the object as a bookend, what would you do? You would probably describe the characteristics, or **properties,** of the rock. Properties are the qualities or attributes that, taken together, are usually peculiar to an object. Since you commonly determine properties with your senses (smell, sight, hearing, touch, and taste), you could say that the properties of an object are the effect the object has on your senses. For example, you might say that the rock is a "big, yellow, smooth rock with shiny gold cubes on one side." But consider the mental image that the other person on the telephone forms when you describe these properties. It is entirely possible that the other person is thinking of something very different from what you are describing (Figure 1.3)!

As you can see, the example of describing a proposed bookend by listing its properties in everyday language leaves much to be desired. The description does not really help the other person form an accurate mental image of the rock. One problem with the attempted communication is that the description of any property implies some kind of *referent*. The word **referent** means that you *refer to,* or think of, a given property in terms of another, more familiar object. Colors, for example, are sometimes stated with a referent. Examples are "sky blue," "grass green," or "lemon yellow." The referents for the colors blue, green, and yellow are, respectively, the sky, living grass, and a ripe lemon.

Referents for properties are not always as explicit as they are with colors, but a comparison is always implied. Since the comparison is implied, it often goes unspoken and leads to assumptions in communications. For example, when you stated that the rock was "big," you assumed that the other person knew that you did not mean as big as a house or even as big as a bicycle. You assumed that the other person knew that you meant that the rock was about as large as a book, perhaps a bit larger.

Another problem with the listed properties of the rock is the use of the word "smooth." The other person would not know if you meant that the rock *looked* smooth or *felt* smooth. After all, some objects can look smooth and feel rough. Other objects can look rough and feel smooth. Thus, here is another assumption, and probably all of the properties lead to implied

FIGURE 1.3

Could you describe this rock to another person over the telephone so that the other person would know *exactly* what you see? This is not likely with everyday language, which is full of implied comparisons, assumptions, and inaccurate descriptions.

As an example of how the measurement process works, consider the property of *length*. Most people are familiar with the concept of the length of something (long or short), the use of length to describe distances (close or far), and the use of length to describe heights (tall or short). The referent units used for measuring length are the familiar inch, foot, and mile from the English system and the centimeter, meter, and kilometer of the metric system. These systems and specific units will be discussed later. For now, imagine that these units do not exist but that you need to measure the length and width of this book. This imaginary exercise will illustrate how the measurement process eliminates vagueness and assumption in communication.

The first requirement in the measurement process is to choose some referent unit of length. You could arbitrarily choose something that is handy, such as the length of a standard paper clip, and you could call this length a "clip." Now you must decide on a procedure to specify how you will use the clip unit. You could define some specific procedures. For example:

1. Place a clip parallel to and on the long edge, or length, of the book so the end of the referent clip is lined up with the bottom edge of the book. Make a small pencil mark at the other end of the clip, as shown in Box Figure 1.1.

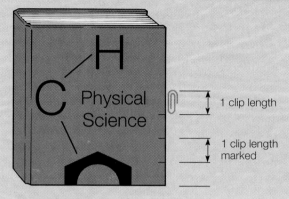

BOX FIGURE 1.1

As an example of the measurement process, a standard paper-clip length is selected as a referent unit. The unit is compared to the property that is being described. In this example, the property of the book length is measured by counting how many clip lengths describe the length.

2. Move the outside end of the clip to the mark and make a second mark at the other end. Continue doing this until you reach the top edge of the book.
3. Compare how many clip replications are in the book length by counting.
4. Record the length measurements by writing (a) how many clip replications were made and (b) the name of the clip length.

If the book length did not measure to a whole number of clips, you might need to divide the clip length into smaller subunits to be more precise. You could develop a *scale* of the basic clip unit and subunits. In fact, you could use multiples of the basic clip unit for an extended scale, using the scale for measurement rather than moving an individual clip unit. You could call the scale a "clipstick" (as in yardstick or meterstick).

comparisons, assumptions, and a not-very-accurate communication. This is the nature of your everyday language and the nature of most attempts at communication.

Quantifying Properties

Typical day-to-day communications are often vague and leave much to be assumed. A communication between two people, for example, could involve one person describing some person, object, or event to a second person. The description is made by using referents and comparisons that the second person may or may not have in mind. Thus, such attributes as "long" fingernails or "short" hair may have entirely different meanings to different people involved in a conversation. Assumptions and vagueness can be avoided by using **measurement** in a description. Measurement is a process of comparing a property to a well-defined and agreed-upon referent. The well-defined and agreed-upon referent is used as a standard called a **unit.** The measurement process involves three steps: (1) *comparing* the referent unit to the property being described, (2) following a *procedure,* or operation, which specifies how the comparison is made, and (3) *counting* how many standard units describe the property being considered.

The measurement process uses a defined referent unit, which is compared to a property being measured. The *value* of the property is determined by counting the number of referent units. The name of the unit implies the procedure that results in the number. A measurement statement always contains a *number* and *name* for the referent unit. The number answers the question "How much?" and the name answers the question "Of what?" Thus a measurement always tells you "how much of what." You will find that using measurements will sharpen your communications. You will also find that using measurements is one of the first steps in understanding your physical environment.

Measurement Systems

Measurement is a process that brings precision to a description by specifying the "how much" and "of what" of a property in a particular situation. A number expresses the value of the property, and the name of a unit tells you what the referent is as well as implying the procedure for obtaining the number. Referent units must be defined and established, however, if others are to understand and reproduce a measurement. When standards are established the referent unit is called a **standard unit** (Figure 1.4). The use of standard units makes it possible to communicate and duplicate measurements. Standard units are usually defined and established by governments and their agencies that are created for that purpose. In the United States, the agency concerned with measurement standards is named the National Institute of Standards and Technology. In Canada, the Standards Council of Canada oversees the National Standard System.

There are two major *systems* of standard units in use today, the *English system* and the *metric system*. The metric system is used throughout the world except in the United States, where both systems are in use. The continued use of the English system in the United States presents problems in international trade, so there is pressure for a complete conversion to the metric system. More and more metric units are being used in everyday measurements, but a complete conversion will involve an enormous cost. Appendix A contains a method for converting from one system to the other easily. Consult this section if you need to convert from one metric unit to another metric unit or to convert from English to metric units or vice versa. Conversion factors are listed inside the front cover.

People have used referents to communicate about properties of things throughout human history. The ancient Greek

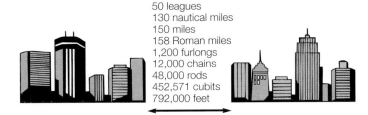

50 leagues
130 nautical miles
150 miles
158 Roman miles
1,200 furlongs
12,000 chains
48,000 rods
452,571 cubits
792,000 feet

FIGURE 1.4

Any of these units and values could have been used at some time or another to describe the same distance between these hypothetical towns. Any unit could be used for this purpose, but when one particular unit is officially adopted, it becomes known as the *standard unit*.

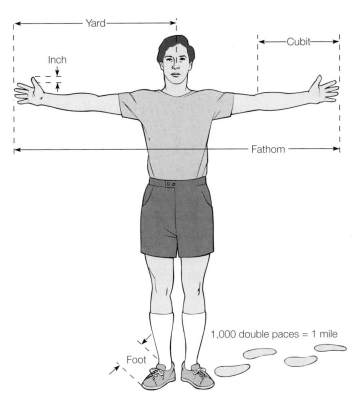

FIGURE 1.5

Many early units for measurement were originally based on the human body. Some of the units were later standardized by governments to become the basis of the English system of measurement.

civilization, for example, used units of *stadia* to communicate about distances and elevations. The "stadium" was a unit of length of the racetrack at the local stadium ("stadia" is the plural of stadium), based on a length of 125 paces. Later civilizations, such as the ancient Romans, adopted the stadia and other referent units from the ancient Greeks. Some of these same referent units were later adopted by the early English civilization, which eventually led to the **English system** of measurement. Some adopted units of the English system were originally based on parts of the human body, presumably because you always had these referents with you (Figure 1.5). The inch, for example, used the end joint of the thumb for a referent. A foot, naturally, was the length of a foot, and a yard was the distance from the tip of the nose to the end of the fingers on an arm held straight out. A cubit was the distance from the end of an elbow to the fingertip, and a fathom was the distance between the fingertips of two arms held straight out.

What body part could be used as a referent for volume? As shown in Table 1.1, all common volume units were based on a "mouthful." Each of the larger volume units was defined as two of the smaller units, making it easier to remember. Some of the units—such as the jack, jill, and pottle—have dropped from use today, leaving us with puzzles such as why there are two pints in a quart, but four quarts in a gallon. Understanding that

TABLE 1.1
English units of volume of 200 years ago

Two Quantities	Equivalent Quantity
2 mouthfuls	= 1 jigger
2 jiggers	= 1 jack
2 jacks	= 1 jill
2 jills	= 1 cup
2 cups	= 1 pint
2 pints	= 1 quart
2 quarts	= 1 pottle
2 pottles	= 1 gallon
2 gallons	= 1 pail
2 pails	= 1 peck
2 pecks	= 1 bushel

at one time there were two quarts in a pottle and two pottles make a gallon seems to make more sense if you understand the old scheme that two of something smaller makes one of the larger units.

As you can imagine, there were problems with these early units because everyone had different-sized arms, legs, and mouths. Beginning in the 1300s, the sizes of the various units were gradually standardized by English kings.

The **metric system** was established by the French Academy of Sciences in 1791. The academy created a measurement system that was based on invariable referents in nature, not human body parts. These referents have been redefined over time to make the standard units more reproducible. The *International System of Units,* abbreviated *SI,* is a modernized version of the metric system. Today, the SI system has seven base units that define standards for the properties of length, mass, time, electric current, temperature, amount of substance, and light intensity (Table 1.2). All units other than the seven basic ones are *derived* units. Area, volume, and speed, for example, are all expressed with derived units. Units for the properties of length, mass, and time are introduced in this chapter. The remaining units will be introduced in later chapters as the properties they measure are discussed.

TABLE 1.2
The SI base units

Property	Unit	Symbol
Length	meter	m
Mass	kilogram	kg
Time	second	s
Electric current	ampere	A
Temperature	kelvin	K
Amount of substance	mole	mol
Luminous intensity	candela	cd

Standard Units for the Metric System

If you consider all the properties of all the objects and events in your surroundings, the number seems overwhelming. Yet, close inspection of how properties are measured reveals that some properties are combinations of other properties (Figure 1.6). Volume, for example, is described by the three length measurements of length, width, and height. Area, on the other hand, is described by just the two length measurements of length and width. Length, however, cannot be defined in simpler terms of any other property. There are four properties that cannot be described in simpler terms, and all other properties are combinations of these four. For this reason they are called the **fundamental properties.** A fundamental property cannot be defined in simpler terms other than to describe how it is measured. These four fundamental properties are (1) *length,* (2) *mass,* (3) *time,* and (4) *charge.* Used individually or in combinations, these four properties will describe or measure what you observe in nature. Metric units for measuring the fundamental properties of length, mass, and time will be described next. The fourth fundamental property, charge, is associated with electricity, and a unit for this property will be discussed in a future chapter.

Length

The standard unit for length in the metric system is the **meter** (the symbol or abbreviation is m). The meter is defined as the distance that light travels in a vacuum during a certain time period, 1/299,792,458 second. The important thing to remember, however, is that the meter is the metric *standard unit* for length. A meter is slightly longer than a yard, 39.3 inches. It is approximately the distance from your left shoulder to the tip of your right hand when your arm is held straight out. Many doorknobs are about one meter above the floor. Think about these distances when you are trying to visualize a meter length.

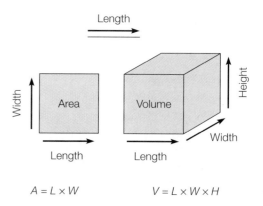

$$A = L \times W \qquad V = L \times W \times H$$

FIGURE 1.6

Area, or the extent of a surface, can be described by two length measurements. Volume, or the space that an object occupies, can be described by three length measurements. Length, however, can be described only in terms of how it is measured, so it is called a *fundamental property.*

Most people have heard of a leap year, but not a leap second. A *leap year* is needed because the earth does not complete an exact number of turns on its axis while completing one trip around the Sun. Our calendar system was designed to stay in step with the seasons with 365-day years and a 366-day year (leap year) every fourth year.

Likewise, our clocks are occasionally adjusted by a one-second increment known as a *leap second*. The leap second is needed because the earth does not have a constant spin. Coordinated Universal Time is the worldwide scientific standard of time keeping. It is based upon Earth's rotation and is kept accurate to within microseconds with carefully maintained atomic clocks. A leap second is a second added to Coordinated Universal Time to make it agree with astronomical time to within 0.9 second.

In 1955, astronomers at the U.S. Naval Observatory and the National Physical Laboratory in England measured the relationship between the frequency of the cesium atom (the standard of time) and the rotation of the earth at a particular period of time. The standard atomic clock second was defined to be equivalent to the fraction 1/31,556,925.9747 of the year 1900—or, an average second for that year. This turned out to be the time required for 9,192,631,770 vibrations of the cesium 133 atom. The second was defined in 1967 in terms of the length of time required for 9,192,631,770 vibrations of the cesium 133 atom. So, the atomic second was set equal to an average second of Earth rotation time near the turn of the twentieth century, but defined in terms of the frequency of a cesium atom.

The earth is constantly slowing from the frictional effects of the tides. Evidence of this slowing can be found in records of ancient observations of eclipses. From these records it is possible to determine the slowing of the earth. One day was only 18 hours long 900 million years ago.

It has been nearly a century since the referent year used for the definition of a second, and the difference is now roughly 2 milliseconds per day. Other factors also affect the earth's spin, such as the wind from hurricanes, so that it is necessary to monitor the earth's rotation continuously and add or subtract leap seconds when needed.

Mass

The standard unit for mass in the metric system is the **kilogram** (kg). The kilogram is defined as the mass of a certain metal cylinder kept by the International Bureau of Weights and Measures in France. This is the only standard unit that is still defined in terms of an object. The property of mass is sometimes confused with the property of weight since they are directly proportional to each other at a given location on the surface of the earth. They are, however, two completely different properties and are measured with different units. All objects tend to maintain their state of rest or straight-line motion, and this property is called "inertia." The *mass* of an object is a measure of the inertia of an object. The *weight* of the object is a measure of the force of gravity on it. This distinction between weight and mass will be discussed in detail in chapter 2. For now, remember that weight and mass are not the same property.

Time

The standard unit for time is the **second** (s). The second was originally defined as 1/86,400 of a solar day ($1/60 \times 1/60 \times 1/24$). The earth's spin was found not to be as constant as thought, so the second was redefined to be the duration required for a certain number of vibrations of a certain cesium atom. A special spectrometer called an "atomic clock" measures these vibrations and keeps time with an accuracy of several millionths of a second per year.

Metric Prefixes

The metric system uses prefixes to represent larger or smaller amounts by factors of 10. Some of the more commonly used prefixes, their abbreviations, and their meanings are listed in Table 1.3. Suppose you wish to measure something smaller than the standard unit of length, the meter. The meter is subdivided into ten equal-sized subunits called *decimeters*. The prefix *deci-* has a meaning of "one-tenth of," and it takes 10 decimeters to equal the length of 1 meter. For even smaller measurements,

TABLE 1.3
Some metric prefixes

Prefix	Symbol	Meaning	Unit Multiplier
exa-	E	quintillion	10^{18}
peta-	P	quadrillion	10^{15}
tera-	T	trillion	10^{12}
giga-	G	billion	10^{9}
mega-	M	million	10^{6}
kilo-	k	thousand	10^{3}
hecto-	h	hundred	10^{2}
deka-	da	ten	10^{1}
unit			
deci-	d	one-tenth	10^{-1}
centi-	c	one-hundredth	10^{-2}
milli-	m	one-thousandth	10^{-3}
micro-	μ	one-millionth	10^{-6}
nano-	n	one-billionth	10^{-9}
pico-	p	one-trillionth	10^{-12}
femto-	f	one-quadrillionth	10^{-15}
atto-	a	one-quintillionth	10^{-18}

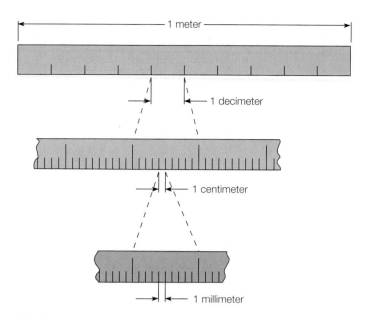

FIGURE 1.7

Compare the units shown here. How many millimeters fit into the space occupied by one centimeter? How many millimeters fit into the space of one decimeter? How many millimeters fit into the space of one meter? Can you express all of this as multiples of ten?

each decimeter is divided into ten equal-sized subunits called *centimeters*. It takes 10 centimeters to equal 1 decimeter and 100 to equal 1 meter. In a similar fashion, each prefix up or down the metric ladder represents a simple increase or decrease by a factor of 10 (Figure 1.7).

When the metric system was established in 1791, the standard unit of mass was defined in terms of the mass of a certain volume of water. A cubic decimeter (dm^3) of pure water at 4°C

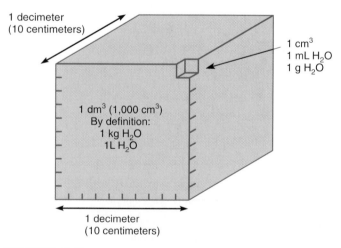

FIGURE 1.8

A cubic decimeter of water (1,000 cm^3) has a liquid volume of 1 L (1,000 mL) and a mass of 1 kg (1,000 g). Therefore, 1 cm^3 of water has a liquid volume of 1 mL and a mass of 1 g.

was *defined* to have a mass of 1 kilogram (kg). This definition was convenient because it created a relationship between length, mass, and volume. As illustrated in Figure 1.8, a cubic decimeter is 10 cm on each side. The volume of this cube is therefore 10 cm × 10 cm × 10 cm, or 1,000 cubic centimeters (abbreviated as cc or cm^3). Thus, a volume of 1,000 cm^3 of water has a mass of 1 kg. Since 1 kg is 1,000 g, 1 cm^3 of water has a mass of 1 g.

The volume of 1,000 cm^3 also defines a metric unit that is commonly used to measure liquid volume, the **liter** (L). For smaller amounts of liquid volume, the milliliter (mL) is used. The relationship between liquid volume, volume, and mass of water is therefore

$$1.0 \text{ L} \Rightarrow 1.0 \text{ dm}^3 \text{ and has a mass of } 1.0 \text{ kg}$$

or, for smaller amounts,

$$1.0 \text{ mL} \Rightarrow 1.0 \text{ cm}^3 \text{ and has a mass of } 1.0 \text{ g}$$

Understandings from Measurements

One of the more basic uses of measurement is to *describe* something in an exact way that everyone can understand. For example, if a friend in another city tells you that the weather has been "warm," you might not understand what temperature is being described. A statement that the air temperature is 70°F carries more exact information than a statement about "warm weather." The statement that the air temperature is 70°F contains two important concepts: (1) the numerical value of 70 and (2) the referent unit of degrees Fahrenheit. Note that both a numerical value and a unit are necessary to communicate a measurement correctly. Thus, weather reports describe weather conditions with numerically specified units; for example, 70° Fahrenheit for air temperature, 5 miles per hour for wind speed, and 0.5 inches for rainfall (Figure 1.9). When such numerically specified units are used in a description, or a weather report, everyone understands *exactly* the condition being described.

Weather Report

Friday (24 hours ended at 5 P.M.)
Highs—airport 73°F, downtown 76°F
Lows—airport 68°F, downtown 70°F
Rainfall .... 0.26 in
Average wind speed .... 5.2 mph
Relative humidity .... High 85%
                                        Low 75%
Rainfall ± normal to date.....+0.94 in

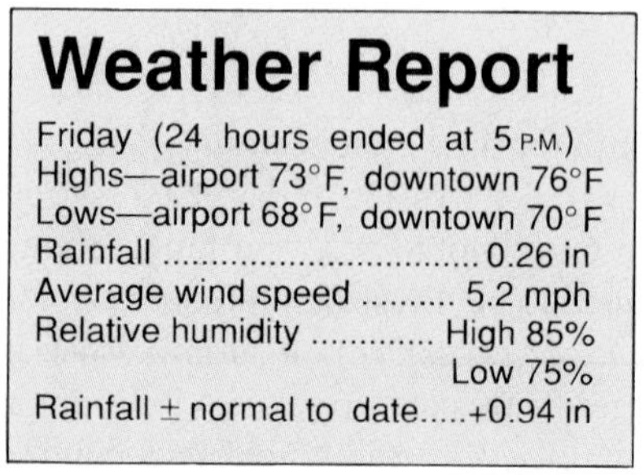

FIGURE 1.9

A weather report gives exact information, data that describes the weather by reporting numerically specified units for each condition being described.

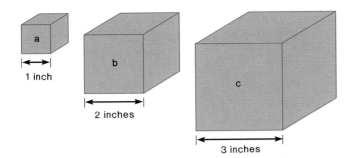

FIGURE 1.10

Cube *a* is 1 inch on each side, cube *b* is 2 inches on each side, and cube *c* is 3 inches on each side. These three cubes can be described and compared with data, or measurement information, but some form of analysis is needed to find patterns or meaning in the data.

Data

Measurement information used to describe something is called **data.** Data can be used to describe objects, conditions, events, or changes that might be occurring. You really do not know if the weather is changing much from year to year until you compare the yearly weather data. The data will tell you, for example, if the weather is becoming hotter or dryer or is staying about the same from year to year.

Let's see how data can be used to describe something and how the data can be analyzed for further understanding. The cubes illustrated in Figure 1.10 will serve as an example. Each cube can be described by measuring the properties of size and surface area.

First, consider the size of each cube. Size can be described by **volume,** which means *how much space something occupies.* The volume of a cube can be obtained by measuring and multiplying the length, width, and height. The data is

volume of cube *a*	1 in^3
volume of cube *b*	8 in^3
volume of cube *c*	27 in^3

Now consider the surface area of each cube. **Area** means *the extent of a surface,* and each cube has six surfaces, or faces (top, bottom, and four sides). The area of any face can be obtained by measuring and multiplying length and width. The data for the three cubes describes them as follows:

	Volume	**Surface Area**
cube *a*	1 in^3	6 in^2
cube *b*	8 in^3	24 in^2
cube *c*	27 in^3	54 in^2

Ratios and Generalizations

Data on the volume and surface area of the three cubes in Figure 1.10 describes the cubes, but whether it says anything about a relationship between the volume and surface area of a cube is difficult to tell. Nature seems to have a tendency to camouflage relationships, making it difficult to extract meaning from raw data. Seeing through the camouflage requires the use of mathematical techniques to expose patterns. Let's see how such techniques can be applied to the data on the three cubes and what the pattern means.

One mathematical technique for reducing data to a more manageable form is to expose patterns through a **ratio.** A ratio is a relationship between two numbers that is obtained when one number is divided by another number. Suppose, for example, that an instructor has 50 sheets of graph paper for a laboratory group of 25 students. The relationship, or ratio, between the number of sheets and the number of students is 50 papers to 25 students, and this can be written as 50 papers/25 students. This ratio is *simplified* by dividing 25 into 50, and the ratio becomes 2 papers/1 student. The 1 is usually understood (not stated), and the ratio is written as simply 2 papers/student. It is read as 2 papers "for each" student, or 2 papers "per" student. The concept of simplifying with a ratio is an important one, and you will see it time and time again throughout science. It is important that you understand the meaning of "per" and "for each" when used with numbers and units.

Applying the ratio concept to the three cubes in Figure 1.10, the ratio of surface area to volume for the smallest cube, cube *a*, is 6 in^2 to 1 in^3, or

$$\frac{6 \text{ in}^2}{1 \text{ in}^3} = 6 \frac{\text{in}^2}{\text{in}^3}$$

meaning there are 6 square inches of area *for each* cubic inch of volume.

The middle-sized cube, cube *b*, had a surface area of 24 in^2 and a volume of 8 in^3. The ratio of surface area to volume for this cube is therefore

$$\frac{24 \text{ in}^2}{8 \text{ in}^3} = 3 \frac{\text{in}^2}{\text{in}^3}$$

meaning there are 3 square inches of area *for each* cubic inch of volume.

The largest cube, cube *c*, had a surface area of 54 in^2 and a volume of 27 in^3. The ratio is

$$\frac{54 \text{ in}^2}{27 \text{ in}^3} = 2 \frac{\text{in}^2}{\text{in}^3}$$

or 2 square inches of area *for each* cubic inch of volume. Summarizing the ratio of surface area to volume for all three cubes, you have

small cube	*a*—6:1
middle cube	*b*—3:1
large cube	*c*—2:1

Now that you have simplified the data through ratios, you are ready to generalize about what the information means. You can generalize that the surface-area-to-volume ratio of a cube *decreases* as the volume of a cube becomes larger. Reasoning from this generalization will provide an explanation for a number of related observations. For example, why does crushed ice melt faster than a single large block of ice with the same volume? The explanation is that the crushed ice has a larger surface-area-to-volume ratio than the large block, so more surface is exposed to warm air. If the generalization is found to be true for shapes other than cubes, you could explain why a log chopped into small chunks burns faster than the whole log. Further generalizing might enable you to predict if 10 lb of large potatoes would require more or less peeling than 10 lb of small potatoes. When generalized explanations result in predictions that can be verified by experience, you gain confidence in the explanation. Finding patterns of relationships is a satisfying intellectual adventure that leads to understanding and generalizations that are frequently practical.

The Density Ratio

The power of using a ratio to simplify things, making explanations more accessible, is evident when you compare the simplified ratio 6 to 3 to 2 with the hodgepodge of numbers that you would have to consider without using ratios. The power of using the ratio technique is also evident when considering other properties of matter. Volume is a property that is sometimes confused with mass. Larger objects do not necessarily contain more matter than smaller objects. A large balloon, for example, is much larger than this book but the book is much more massive than the balloon. The simplified way of comparing the mass of a particular volume is to find the ratio of mass to volume. This ratio is called mass **density,** which is defined as *mass per unit volume*. The "per" means "for each" as previously discussed, and "unit" means one, or each. Thus "mass per unit volume" literally means the "mass of one volume" (Figure 1.11). The relationship can be written as

$$\text{mass density} = \frac{\text{mass}}{\text{volume}}$$

or

$$\rho = \frac{m}{V}$$

(ρ is the symbol for the Greek letter rho.)

equation 1.1

As with other ratios, density is obtained by dividing one number and unit by another number and unit. Thus, the density of an object with a volume of 5 cm³ and a mass of 10 g is

$$\text{density} = \frac{10\ \text{g}}{5\ \text{cm}^3} = 2\frac{\text{g}}{\text{cm}^3}$$

The density in this example is the ratio of 10 g to 5 cm³, or 10 g/5 cm³, or 2 g to 1 cm³. Thus, the density of the example object is the mass of *one* volume (a unit volume), or 2 g *for each* cm³.

Any unit of mass and any unit of volume may be used to express density. The densities of solids, liquids, and gases are

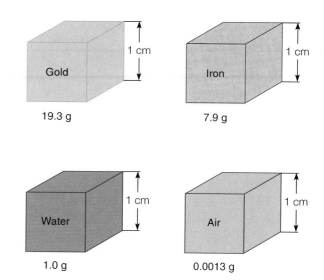

FIGURE 1.11

Equal volumes of different substances do not have the same mass, as these cube units show. Calculate the densities in g/cm³. Do equal volumes of different substances have the same density? Explain.

usually expressed in grams per cubic centimeter (g/cm³), but the densities of liquids are sometimes expressed in grams per milliliter (g/mL). Using SI standard units, densities are expressed as kg/m³. Densities of some common substances are shown in Table 1.4.

If matter is distributed the same throughout a volume, the *ratio* of mass to volume will remain the same no matter what mass and volume are being measured. Thus, a teaspoonful, a cup, and a lake full of freshwater at the same temperature will all have a density of about 1 g/cm³ or 1 kg/L. A given material will have its own unique density; example 1.1 shows how density can be used to identify an unknown substance. For help with significant figures, see appendix A (p. 595).

TABLE 1.4	
Mass densities (ρ) of some common substances	
	g/cm³
Aluminum	2.70
Copper	8.96
Iron	7.87
Lead	11.4
Water	1.00
Seawater	1.03
Mercury	13.6
Gasoline	0.680

Density Matters—Fish, Nuclear Fusion, and Cola

Sharks and rays are marine animals that have an internal skeleton made entirely of cartilage. These animals have no swim bladder to adjust their body density in order to maintain their position in the water; therefore, they must constantly swim or they will sink. The bony fish, on the other hand, have a skeleton composed of bone and most also have a swim bladder. These fish can regulate the amount of gas in the bladder to control their density. Thus, the fish can remain at a given level in the water without expending large amounts of energy.

Scientists plan to study nuclear fusion involving extremely dense plasmas, up to six times denser than the center of the sun. The proposed National Ignition Facility (NIF) is expected to be funded and completed by 2010. This research facility will use high-power lasers to achieve self-sustaining fusion reactions, with densities reaching up to 1,000 g/cm^3. Existing research facilities today are capable of creating plasma densities up to about 50 g/cm^3. Compare these densities with what you might find at the center of the sun, about 160 g/cm^3.

Finally, have you ever noticed the different floating characteristics of cans of the normal version of a carbonated cola beverage and a diet version? The surprising result is that the normal version usually sinks and the diet version usually floats. This has nothing to do with the amount of carbon dioxide in the two drinks. It is a result of the increase in density from the sugar added to the normal version, while the diet version has much less of an artificial sweetener that is much sweeter than sugar.

EXAMPLE 1.1

Two blocks are on a table. Block A has a volume of 30.0 cm^3 and a mass of 81.0 g. Block B has a volume of 50.0 cm^3 and a mass of 135 g. Which block has the greater density? If the two blocks have the same density, what material are they? (See Table 1.4.)

Solution

Density is defined as the ratio of the mass or weight of a substance per unit volume. Assuming the mass is distributed equally throughout the volume, you could assume that the ratio of mass to volume is the same no matter what quantity of mass and volume are measured. If you can accept this assumption, you can use equation 1.1 to determine the mass density—choosing this equation because the mass of the substance is given, not the weight:

Block A

mass (m) = 81.0 g
volume (V) = 30.0 cm^3
density = ?

$$\rho = \frac{m}{V}$$
$$= \frac{81.0 \text{ g}}{30.0 \text{ cm}^3}$$
$$= 2.70 \frac{\text{g}}{\text{cm}^3}$$

Block B

mass (m) = 135 g
volume (V) = 50.0 cm^3
density = ?

$$\rho = \frac{m}{V}$$
$$= \frac{135 \text{ g}}{50.0 \text{ cm}^3}$$
$$= 2.70 \frac{\text{g}}{\text{cm}^3}$$

As you can see, both blocks have the same mass density. Inspecting Table 1.4, you can see that aluminum has a mass density of 2.70 g/cm^3, so both blocks must be aluminum.

EXAMPLE 1.2

A rock with a volume of 4.50 cm^3 has a mass of 15.0 g. What is the density of the rock? (Answer: 3.33 g/cm^3)

A Dense Textbook?

What is the mass density of this book? Measure the length, width, and height of this book in cm, then multiply to find the volume in cm^3. Use a balance to find the mass of this book in grams. Compute the density of the book by dividing the mass by the volume. Compare the density in g/cm^3 with other substances listed in Table 1.4.

684 g

Symbols and Equations

In the previous section, the relationship of density, mass, and volume was written with symbols. Mass density was represented by ρ, the lowercase letter rho in the Greek alphabet, mass was represented by m, and volume by V. The use of such symbols is established and accepted by convention, and these symbols are like the vocabulary of a foreign language. You learn what the symbols mean by use and practice, with the understanding that each symbol stands for a very specific property or concept. The symbols actually represent **quantities,** or *measured properties.* The symbol m thus represents a quantity of mass that is specified by a number and a unit, for example, 16 g. The symbol V represents a quantity of volume that is specified by a number and a unit, such as 17 cm^3.

Symbols

Symbols usually provide a clue about which quantity they represent, such as m for mass and V for volume. However, in some cases two quantities start with the same letter, such as volume and velocity, so the uppercase letter is used for one (V for volume) and the lowercase letter is used for the other (v for velocity). There are more quantities than upper- and lowercase letters, however, so letters from the Greek alphabet are also used, for example, ρ for mass density. Sometimes a subscript is used to identify a quantity in a particular situation, such as v_i for initial, or beginning, velocity and v_f for final velocity. Some symbols are also used to carry messages; for example, the Greek letter delta (Δ) is a message that means "the change in" a value. Other message symbols are the symbol $\therefore$, which means "therefore," and the symbol $\propto$, which means "is proportional to."

Equations

Symbols are used in an **equation,** a statement that describes a relationship where *the quantities on one side of the equal sign are identical to the quantities on the other side.* Identical refers to both the numbers and the units. Thus, in the equation describing the property of density, $\rho = m/V$, the numbers on both sides of the equal sign are identical (e.g., 5 = 10/2). The units on both sides of the equal sign are also identical (e.g., g/cm^3 = g/cm^3).

Equations are used to (1) *describe a property,* (2) *define a concept,* or (3) *describe how quantities change relative to each other.* Understanding how equations are used in these three classes is basic to successful problem solving and comprehension of physical science. Each class of uses is considered separately in the following discussion.

Describing a property. You have already learned that the compactness of matter is described by the property called density. Density is a ratio of mass to a unit volume, or $\rho = m/V$. The key to understanding this property is to understand the meaning of a ratio and what "per" or "for each" means. Other examples of properties that can be defined by ratios are how fast something is moving (speed) and how rapidly a speed is changing (acceleration).

Defining a concept. A physical science concept is sometimes defined by specifying a measurement procedure. This is called an *operational definition* because a procedure is established that defines a concept as well as telling you how to measure it. Concepts of what is meant by force, mechanical work, and mechanical power and concepts involved in electrical and magnetic interactions can be defined by measurement procedures.

Describing how quantities change relative to each other. The term **variable** refers to a specific quantity of an object or event that can have different values. Your weight, for example, is a variable because it can have a different value on different days. The rate of your heartbeat, the number of times you breathe each minute, and your blood pressure are also variables. Any quantity describing an object or event can be considered a variable, including the conditions that result in such things as your current weight, pulse, breathing rate, or blood pressure.

As an example of relationships between variables, consider that your weight changes in size in response to changes in other variables, such as the amount of food you eat. With all other factors being equal, a change in the amount of food you eat results in a change in your weight, so the variables of amount of food eaten and weight change together in the same ratio.

When two variables increase (or decrease) together in the same ratio, they are said to be in **direct proportion.** When two variables are in direct proportion, *an increase or decrease in one variable results in the same relative increase or decrease in a second variable.* Recall that the symbol $\propto$ means "is proportional to," so the relationship is

$$\text{amount of food consumed} \propto \text{weight gain}$$

Variables do not always increase or decrease together in direct proportion. Sometimes one variable *increases* while a second variable *decreases* in the same ratio. This is an **inverse proportion** relationship. Other common relationships include one variable increasing in proportion to the *square* or to the *inverse square* of a second variable. Here are the forms of these four different types of proportional relationships:

Direct	$a \propto b$
Inverse	$a \propto 1/b$
Square	$a \propto b^2$
Inverse square	$a \propto 1/b^2$

Proportionality Statements

Proportionality statements describe in general how two variables change relative to each other, but a proportionality statement is *not* an equation. For example, consider the last time you filled your fuel tank at a service station (Figure 1.12). You could say that the volume of gasoline in an empty tank you are filling is directly proportional to the amount of time that the fuel pump was running, or

$$\text{volume} \propto \text{time}$$

This is not an equation because the numbers and units are not identical on both sides. Considering the units, for example, it should be clear that minutes do not equal liters; they are two different quantities. To make a statement of proportionality into an equation, you need to apply a **proportionality constant,** which is sometimes given the symbol k. For the fuel pump example the equation is

$$\text{volume} = (\text{time})(\text{constant})$$

or

$$V = tk$$

Inverse Square Relationship

An inverse square relationship between energy and distance is found in light, sound, gravitational force, electric fields, nuclear radiation, and any other phenomena that spread equally in all directions from a source.

Box Figure 1.2 could represent any of the phenomena that have an inverse square relationship, but let us assume it is showing a light source and how the light spreads at a certain distance (d), at twice that distance (2d), and at three times that distance (3d). As you can see, light twice as far from the source is spread over four times the area, and will therefore have one-fourth the intensity. This is the same as $\frac{1}{2^2}$, or $\frac{1}{4}$.

Light three times as far from the source is spread over nine times the area and will therefore have one-ninth the intensity. This is the same as $\frac{1}{3^2}$, or $\frac{1}{9}$, again showing an inverse square relationship.

You can measure the inverse square relationship by moving an overhead projector so its light is shining on a wall (see distance d in Box Figure 1.2). Use a light meter or some other way of measuring the intensity of light. Now move the projector to double the distance from the wall. Measure the increased area of the projected light on the wall, and again measure the intensity of the light. What relationship did you find between the light intensity and distance?

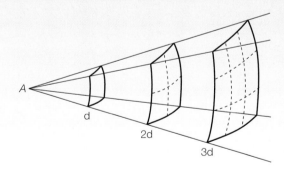

BOX FIGURE 1.2

How much would light moving from point *A* spread out at twice the distance (2d) and three times the distance (3d)? What would this do to the brightness of the light?

FIGURE 1.12

The volume of fuel you have added to the fuel tank is directly proportional to the amount of time that the fuel pump has been running. This relationship can be described with an equation by using a proportionality constant.

In the example, the constant is the flow of gasoline from the pump in L/min (a ratio). Assume the rate of flow is 40 L/min. In units, you can see why the statement is now an equality.

$$L = (\min)\left(\frac{L}{\min}\right)$$

$$L = \frac{\min \times L}{\min}$$

$$L = L$$

A proportionality constant in an equation might be a **numerical constant,** a constant that is without units. Such numerical constants are said to be dimensionless, such as 2 or 3. Some of the more important numerical constants have their own symbols, for example, the ratio of the circumference of a circle to its diameter is known as π (pi). The numerical constant of π does not have units because the units cancel when the ratio is simplified by division (Figure 1.13). The value of π is usually rounded to 3.14, and an example of using this numerical constant in an equation is that the area of a circle equals π times the radius squared ($A = \pi r^2$).

The flow of gasoline from a pump is an example of a constant that has dimensions (40 L/min). Of course the value of this constant will vary with other conditions, such as the particular fuel pump used and how far the handle on the pump hose is depressed, but it can be considered to be a constant under the same conditions for any experiment.

Problem Solving Made Easy

The activity of problem solving is made easier by using certain techniques that help organize your thinking. One such technique is to follow a format, such as the following procedure:

Step 1: Read through the problem and *make a list* of the variables with their symbols on the left side of the page, including the unknown with a question mark.

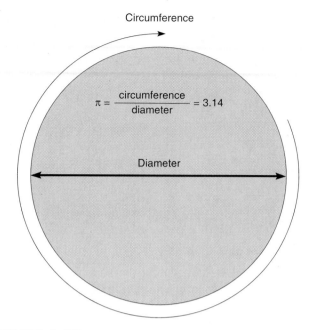

Circumference

$$\pi = \frac{\text{circumference}}{\text{diameter}} = 3.14$$

Diameter

FIGURE 1.13

The ratio of the circumference of *any* circle to the diameter of that circle is always π, a numerical constant that is usually rounded to 3.14. Pi does not have units because they cancel in the ratio.

Step 2: Inspect the list of variables and the unknown and identify the equation that expresses a relationship between these variables. A list of equations discussed in each chapter is found at the end of that chapter. *Write the equation* on the right side of your paper, opposite the list of symbols and quantities.

Step 3: If necessary, *solve the equation* for the variable in question. This step must be done before substituting any numbers or units in the equation. This simplifies things and keeps down confusion that might otherwise result. If you need help solving an equation, see the section on this topic in appendix A.

Step 4: If necessary, *convert unlike units* so they are all the same. For example, if a time is given in seconds and a speed is given in kilometers per hour, you should convert the km/h to m/s. Again, this step should be done at this point in the procedure to avoid confusion or incorrect operations in a later step. If you need help converting units, see the section on this topic in appendix A.

Step 5: Now you are ready to *substitute the number value and unit* for each symbol in the equation (except the unknown). Note that it might sometimes be necessary to perform a "subroutine" to find a missing value and unit for a needed variable.

Step 6: Do the indicated *mathematical operations* on the numbers and on the units. This is easier to follow if you first separate the numbers and units, as shown in the example that follows and in the examples throughout this text. Then perform the indicated operations on the numbers and units as separate steps, showing all work. If you are not sure how to read the indicated operations, see the section on "Symbols and Operations" in appendix A.

Step 7: *Draw a box* around your answer (numbers and units) to communicate that you have found what you were looking for. The box is a signal that you have finished your work on this problem.

For an example problem, use the equation from the previous section describing the variables of a fuel pump, $V = tk$, to predict how long it will take to fill an empty 80-liter tank. Assume $k = 40$ L/min.

Step 1
$V = 80$ L

$k = 80$ L min

$t = ?$

(no conversion needed for this problem)   **Step 4**

$V = tk$ **Step 2**

$\dfrac{V}{k} = \dfrac{tk}{k}$ **Step 3**

$t = \dfrac{V}{k}$

$t = \dfrac{80 \text{ L}}{40 \dfrac{\text{L}}{\text{min}}}$ **Step 5**

$= \dfrac{80}{40}\dfrac{\text{L}}{1} \times \dfrac{\text{min}}{\text{L}}$ **Step 6**

$= \boxed{2 \text{ min}}$ **Step 7**

Note that procedure step 4 was not required in this solution.

This formatting procedure will be demonstrated throughout this text in example problems and in the solutions to problems found in appendix D. Note that each of the chapters with problems has parallel exercises. The exercises in groups A and B cover the same concepts. If you cannot work a problem in group B, look for the parallel problem in group A. You will find a solution to this problem, in the previously described format, in appendix D. Use this parallel problem solution as a model to help you solve the problem in group B. If you follow the suggested formatting procedures and seek help from the appendix as needed, you will find that problem solving is a simple, fun activity that helps you to learn to think in a new way. Here are some more considerations that will prove helpful.

1. Read the problem carefully, perhaps several times, to understand the problem situation. If possible, make a sketch to help you visualize and understand the problem in terms of the real world.
2. Be alert for information that is not stated directly. For example, if a moving object "comes to a stop," you know that the final velocity is zero, even though this was not stated outright. Likewise, questions about "how far?" are usually asking a question about distance, and questions about "how long?" are usually asking a question about time. Such information can be very important in procedure step 1, the listing of quantities and their symbols. Overlooked or missing quantities and symbols can make it difficult to identify the appropriate equation.
3. Understand the meaning and concepts that an equation represents. An equation represents a *relationship* that exists between variables. Understanding the relationship helps you to identify the appropriate equation or equations by inspection of the list of known and unknown quantities (procedure step 2). You will find a list of the equations being considered at the end of each chapter. Information about the meaning and the concepts that an equation represents is found within each chapter.

4. Solve the equation before substituting numbers and units for symbols (procedure step 3). A helpful discussion of the mathematical procedures required, with examples, is in appendix A.

5. Note whether the quantities are in the same units. A mathematical operation requires the units to be the same; for example, you cannot add nickels, dimes, and quarters until you first convert them all to the same unit of money. Likewise, you cannot correctly solve a problem if one time quantity is in seconds and another time quantity is in hours. The quantities must be converted to the same units before anything else is done (procedure step 4). There is a helpful section on how to use conversion ratios in appendix A.

6. Perform the required mathematical operations on the numbers and the units as if they were two separate problems (procedure step 6). You will find that following this step will facilitate problem-solving activities because the units you obtain will tell you if you have worked the problem correctly. If you just write the units that you think should appear in the answer, you have missed this valuable self-check.

7. Be aware that not all learning takes place in a given time frame and that solutions to problems are not necessarily arrived at "by the clock." If you have spent a half an hour or so unsuccessfully trying to solve a particular problem, move on to another problem or do something entirely different for a while. Problem solving often requires time for something to happen in your brain. If you move on to some other activity, you might find that the answer to a problem that you have been stuck on will come to you "out of the blue" when you are not even thinking about the problem. This unexpected revelation of solutions is common to many real-world professions and activities that involve thinking.

Example Problem

Mercury is a liquid metal with a mass density of 13.6 g/cm³. What is the mass of 10.0 cm³ of mercury?

Solution

The problem gives two known quantities, the mass density (ρ) of mercury and a known volume (V), and identifies an unknown quantity, the mass (m) of that volume. Make a list of these quantities:

$\rho = 13.6 \text{ g/cm}^3$
$V = 10.0 \text{ cm}^3$
$m = ?$

The appropriate equation for this problem is the relationship between mass density (ρ), mass (m), and volume (V):

$$\rho = \frac{m}{V}$$

The unknown in this case is the mass, m. Solving the equation for m, by multiplying both sides by V, gives:

$$V\rho = \frac{mV}{V}$$
$$V\rho = m, \text{ or}$$
$$m = V\rho$$

Now you are ready to substitute the known quantities in the equation:

$$m = \left(13.6 \frac{\text{g}}{\text{cm}^3}\right)(10.0 \text{ cm}^3)$$

And perform the mathematical operations on the numbers and on the units:

$$m = (13.6)(10.0) \left(\frac{\text{g}}{\text{cm}^3}\right)(\text{cm}^3)$$
$$= 136 \frac{\text{g}\cdot\text{cm}^3}{\text{cm}^3}$$
$$= \boxed{136 \text{ g}}$$

The Nature of Science

Most humans are curious, at least when they are young, and are motivated to understand their surroundings. These traits have existed since antiquity and have proven to be a powerful motivation. In recent times the need to find out has motivated the launching of space probes to learn what is "out there," and humans have visited the moon to satisfy their curiosity. Curiosity and the motivation to understand nature were no less powerful in the past than today. Over two thousand years ago, the Greeks lacked the tools and technology of today and could only make conjectures about the workings of nature. These early seekers of understanding are known as *natural philosophers,* and they observed, thought about, and wrote about the workings of all of nature. They are called philosophers because their understandings came from reasoning only, without experimental evidence. Nonetheless, some of their ideas were essentially correct and are still in use today. For example, the idea of matter being composed of *atoms* was first reasoned by certain Greeks in the fifth century B.C. The idea of *elements,* basic components that make up matter, was developed much earlier but refined by the ancient Greeks in the fourth century B.C. The concept of what the elements are and the concept of the nature of atoms have changed over time, but the idea first came from ancient natural philosophers.

The Scientific Method

Some historians identify the time of Galileo and Newton, approximately three hundred years ago, as the beginning of modern science. Like the ancient Greeks, Galileo and Newton were interested in studying all of nature. Since the time of Galileo and Newton, the content of physical science has increased in scope and specialization, but the basic means of acquiring understanding, the scientific investigation, has changed little. A *scientific investigation* provides understanding through *experimental evidence* as opposed to the conjectures based on the "thinking only" approach of the ancient natural philosophers. In the next chapter, for example, you will learn how certain ancient Greeks described how objects fall toward the earth with a thought-out, or reasoned, explanation. Galileo, on the other hand, changed how people thought of falling objects by developing explanations from both creative thinking and precise measurement of physical quantities, providing experimental evidence for his explanations. Experimental evidence provides explanations today, much as it did for Galileo, as relationships are found from precise measurements of physical quantities. Thus, scientific knowledge about nature has grown as measurements and investigations have led to understandings that lead to further measurements and investigations.

What is a scientific investigation, and what methods are used to conduct one? Attempts have been made to describe scientific methods in a series of steps (define problem, gather data, make hypothesis, test, make conclusion), but no single description has ever been satisfactory to all concerned. Scientists do similar things in investigations, but there are different approaches and different ways to evaluate what is found. Overall, the similar things might look like this:

1. Observe some aspect of nature.
2. Propose an explanation for something observed.
3. Use the explanation to make predictions.
4. Test predictions by doing an experiment or by making more observations.
5. Modify explanation as needed.
6. Return to step 3.

The exact approach used depends on the individual doing the investigation and on the field of science being studied.

Another way to describe what goes on during a scientific investigation is to consider what can be generalized. There are at least three separate activities that seem to be common to scientists in different fields as they conduct scientific investigations, and these generalizations look like this:

• Collecting observations
• Developing explanations
• Testing explanations

No particular order or routine can be generalized about these common elements. In fact, individual scientists might not even be involved in all three activities. Some, for example, might spend all of their time out in nature, "in the field" collecting data and generalizing about their findings. This is an acceptable means of investigation in some fields of science. Other scientists might spend all of their time indoors at computer terminals developing theoretical equations to explain the generalizations made by others. Again, the work at a computer terminal is an acceptable means of scientific investigation. Thus, many of today's specialized scientists never engage in a five-step process. This is one reason why many philosophers of science argue that there is no such thing as *the* scientific method. There are common activities of observing, explaining, and testing in scientific investigations in different fields, and these activities will be discussed next.

Explanations and Investigations

Explanations in the natural sciences are concerned with things or events observed, and there can be several different ways to develop or create explanations. In general, explanations can come from the results of experiments, from an educated guess, or just from imaginative thinking. In fact, there are even several examples in the history of science of valid explanations being developed from dreams. Explanations go by various names, each depending on intended use or stage of development. For example, an explanation in an early stage of development is sometimes called a *hypothesis*. A **hypothesis** is a tentative thought- or experiment-derived explanation. It must be compatible with all observations and provide understanding of some aspect of nature, but the key word here is *tentative*. A hypothesis is tested by experiment and is rejected, or modified, if a single observation or test does not fit. The successful testing of a hypothesis may lead to the design of experiments, or it could lead to the development of another hypothesis, which could, in turn, lead to the design of yet more experiments, which could lead to . . . As you can see, this is a branching, ongoing process that is very difficult to describe in specific terms. In addition, it can be difficult to identify an endpoint in the process that you could call a conclusion. The search for new concepts to explain experimental evidence may lead from hypothesis to a new theory, which results in more new hypotheses. This is why one of the best ways to understand scientific methods is to study the history of science. Or do the activity of science yourself by planning, then conducting experiments.

Testing a Hypothesis

In some cases a hypothesis may be tested by simply making some simple observations. For example, suppose you hypothesized that the height of a bounced ball depends only on the height from which the ball is dropped. You could test this by observing different balls being dropped from several different heights and recording how high each bounced.

Another common method for testing a hypothesis involves devising an experiment. An **experiment** is a re-creation of an event or occurrence in a way that enables a scientist to support or disprove a hypothesis. This can be difficult, since an event can be influenced by a great many different things. For example, suppose someone tells you that soup heats to the boiling point faster than water. Is this true? How can you find the answer to this question? The time required to boil a can of soup might depend on a number of things: the composition of the soup, how much soup is in the pan, what kind of pan is used, the nature of the stove, the size of the burner, how high the temperature is set, environmental factors such as the humidity and temperature, and more factors. It might seem that answering a simple question about the time involved in boiling soup is an impossible task. To help unscramble such situations, scientists use what is known as a *controlled experiment*. A **controlled experiment** compares two situations in which all the influencing factors are identical except one. The situation used as the basis of comparison is called the *control group* and the other is called the *experimental group*. The single influencing factor that is allowed to be different in the experimental group is called the *experimental variable*.

The situation involving the time required to boil soup and water would have to be broken down into a number of simple questions. Each question would provide the basis on which experimentation would occur. Each experiment would provide information about a small part of the total process of heating liquids. For example, in order to test the hypothesis that soup boils faster than water, an experiment could be performed in which soup is brought to a boil (the experimental group), while water is brought to a boil in the control group. Every factor in the control group is *identical* to the factors in the experimental group except the experimental variable—the soup factor. After the experiment, the new data (facts) are gathered and analyzed. If there were no differences between the two groups, you could conclude that the soup variable evidently did not have a cause-and-effect relationship with the time needed to come to a boil (i.e., soup was not responsible for the time to boil). However, if there were a difference, it would be likely that this variable was responsible for the difference between the control and experimental groups. In the case of the time to come to a boil, you would find that soup indeed does boil faster than water alone. If you doubt this, why not do the experiment yourself?

Accept Results?

Scientists are not likely to accept the results of a single experiment, since it is possible that a random event that had nothing to do with the experiment could have affected the results and caused people to think there was a cause-and-effect relationship when none existed. For example, the density of soup is greater than the density of water, and this might be the important factor. A way to overcome this difficulty would be to test a number of different kinds of soup with different densities. When there is only one variable, many replicates (copies) of the same experiment are conducted, and the consistency of the results determines how convincing the experiment is.

Furthermore, scientists often apply statistical tests to the results to help decide in an impartial manner if the results obtained are *valid* (meaningful; fit with other knowledge), *reliable* (give the same results repeatedly), and show cause-and-effect or if they are just the result of random events.

Other Considerations

As you can see from the discussion of the nature of science, a scientific approach to the world requires a certain way of thinking. There is an insistence on ample supporting evidence by numerous studies rather than easy acceptance of strongly stated opinions. Scientists must separate opinions from statements of fact. A scientist is a healthy skeptic.

Careful attention to detail is also important. Since scientists publish their findings and their colleagues examine their work, there is a strong desire to produce careful work that can be easily defended. This does not mean that scientists do not speculate and state opinions. When they do, however, they take great care to clearly distinguish fact from opinion.

There is also a strong ethic of honesty. Scientists are not saints, but the fact that science is conducted out in the open in front of one's peers tends to reduce the incidence of dishonesty. In addition, the scientific community strongly condemns and severely penalizes those who steal the ideas of others, perform shoddy science, or falsify data. Any of these infractions could lead to the loss of one's job and reputation.

Science is also limited by the ability of people to pry understanding from the natural world. People are fallible and do not always come to the right conclusions, because information is lacking or misinterpreted, but science is self-correcting. As new information is gathered, old, incorrect ways of thinking must be changed or discarded. For example, at one time people were sure that the sun went around the earth. They observed that the sun rose in the east and traveled across the sky to set in the west. Since they could not feel the earth moving, it seemed perfectly logical that the sun traveled around the earth. Once they understood that the earth rotated on its axis, people began to understand that the rising and setting of the sun could be explained in other ways. A completely new concept of the relationship between the sun and the earth developed.

Although this kind of study seems rather primitive to us today, this change in thinking about the sun and the earth was a very important step in understanding the universe and how the various parts are related to one another. This background information was built upon by many generations of astronomers and space scientists, and it finally led to space exploration.

People also need to understand that science cannot answer all the problems of our time. Although science is a powerful tool, there are many questions it cannot answer and many problems it cannot solve. The behavior and desires of people generate most of the problems societies face. Famine, drug abuse, and pollution are human-caused and must be resolved by humans. Science may provide some tools for social planners, politicians, and ethical thinkers, but science does not have, nor does it attempt to provide, answers for the problems of the human race. Science is merely one of the tools at our disposal.

Pseudoscience

Pseudoscience (*pseudo-* means false) is not science, but it uses the appearance or language of science to mislead people into thinking that something has scientific validity. Absurd claims that are clearly pseudoscience sometimes appear to gain public acceptance because of promotion in the press. Thus some people continue to believe stories that psychics can really help solve puzzling crimes, that perpetual energy machines exist, or that sources of water can be found by a person with a forked stick. Such claims could be directly tested and disposed of if they fail the test, but this process is generally ignored. In addition to experimentally testing a claim that appears to be pseudoscience, here are some questions that should be considered:

1. What is the academic background and scientific experience of the person promoting the claim?
2. How many articles have been published by the person in peer-reviewed scientific journals?
3. Has the person given invited scientific talks at universities and national professional organization meetings?
4. Has the claim been researched and published by the person in a peer-reviewed scientific journal *and* have other scientists independently validated the claim?

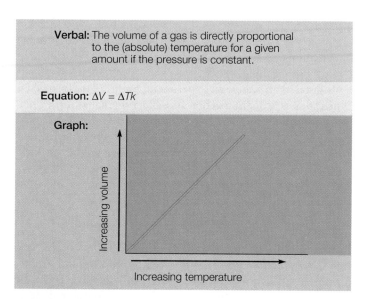

Verbal: The volume of a gas is directly proportional to the (absolute) temperature for a given amount if the pressure is constant.

Equation: $\Delta V = \Delta Tk$

Graph:

FIGURE 1.14

A relationship between variables can be described in at least three different ways: (1) verbally, (2) with an equation, and (3) with a graph. This figure illustrates the three ways of describing the relationship known as Charles' law.

Laws and Principles

Sometimes you can observe a series of relationships that seem to happen over and over again. There is a popular saying, for example, that "if anything can go wrong, it will." This is called Murphy's law. It is called a *law* because it describes a relationship between events that seems to happen time after time. If you drop a slice of buttered bread, for example, it can land two ways, butter side up or butter side down. According to Murphy's law, it will land butter side down. With this example, you know at least one way of testing the validity of Murphy's law.

Another "popular saying" type of relationship seems to exist between the cost of a houseplant and how long it lives. You could call it the "law of houseplant longevity" that the life span of a houseplant is inversely proportional to its purchase price. This "law" predicts that a ten-dollar houseplant will wilt and die within a month, but a fifty-cent houseplant will live for years. The inverse relationship is between the variables of (1) cost and (2) life span, meaning the more you pay for a plant, the shorter the time it will live. This would also mean that inexpensive plants will live for a long time. Since the relationship seems to occur time after time, it is called a "law."

A **scientific law** describes an important relationship that is observed in nature to occur consistently time after time. Basically, scientific laws describe *what* happens in nature. The law is often identified with the name of a person associated with the formulation of the law. For example, with all other factors being equal, an increase in the temperature of the air in a balloon results in an increase in its volume. Likewise, a decrease in the temperature results in a decrease in the total volume of the balloon. The volume of the balloon varies directly with the temperature of the air in the balloon, and this can be observed to occur consistently time after time. This relationship was first discovered in the latter part of the eighteenth century by two French scientists, A.C. Charles and Joseph Gay-Lussac. Today, the relationship is sometimes called *Charles' law* (Figure 1.14). When you read about a scientific *law*, you should remember that a law is a statement that means something about a relationship that you can observe time after time in nature.

Have you ever heard someone state that something behaved a certain way *because* of a scientific principle or law? For example, a big truck accelerated slowly *because* of Newton's laws of motion. Perhaps this person misunderstands the nature of scientific principles and laws. Scientific principles and laws do not dictate the behavior of objects; they simply describe it. They do not say how things ought to act but rather how things *do* act. A scientific principle or law is *descriptive;* it describes how things act.

A **scientific principle** describes a more specific set of relationships than is usually identified in a law. The difference between a scientific principle and a scientific law is usually one of the extent of the phenomena covered by the explanation, but there is not always a clear distinction between the two. As an example of a scientific principle, consider Archimedes' principle. This principle is concerned with the relationship between an object, a fluid, and buoyancy, which is a specific phenomenon.

Models and Theories

Often the part of nature being considered is too small or too large to be visible to the human eye, and the use of a *model* is needed. A **model** (Figure 1.15) is a description of a theory or idea that accounts for all known properties. The description can come in many different forms, such as a physical model, a computer model, a sketch, an analogy, or an equation. No one has ever seen the whole solar system, for example, and all you can see in the real world is the movement of the sun, moon, and planets against a background of stars. A physical model or sketch of the solar system, however, will give you a pretty good idea of what the solar system might look like. The physical model and the sketch are both models, since they both give you a mental picture of the solar system.

A

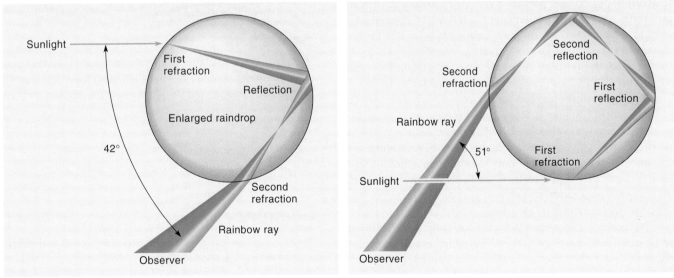

B

FIGURE 1.15

A model helps you visualize something that cannot be observed. You cannot observe what is making a double rainbow, for example, but models of light entering the upper and lower surfaces of a raindrop help you visualize what is happening. The drawings in B serve as a model that explains how a double rainbow is produced (also see "The Rainbow" in chapter 7).

At the other end of the size scale, models of atoms and molecules are often used to help us understand what is happening in this otherwise invisible world. A container of small, bouncing rubber balls can be used as a model to explain the relationships of Charles' law. This model helps you see what happens to invisible particles of air as the temperature, volume, or pressure of the gas changes. Some models are better than others are, and models constantly change as our understanding evolves. Early twentieth-century models of atoms, for example, were based on a "planetary model," in which electrons moved around the nucleus like planets around the sun. Today, the model has changed as our understanding of the nature of atoms has changed. Electrons are now pictured as vibrating with certain wavelengths, which can make standing waves only at certain distances from the nucleus. Thus the model of the atom changed from one that views electrons as solid particles to one that views them like vibrations on a string.

The most recently developed scientific theory was refined and expanded during the 1970s. This theory concerns the surface of the earth, and it has changed our model of what the earth is like. At first, the basic idea of today's accepted theory was pure and simple conjecture. The term *conjecture* usually means an explanation or idea based on speculation, or one based on trivial grounds without any real evidence. Scientists would look at a map of Africa and South America, for example, and mull over how the two continents look like pieces of a picture puzzle that had moved apart (Figure 1.16). Any talk of moving continents was considered conjecture, because it was not based on anything acceptable as real evidence.

Many years after the early musings about moving continents, evidence was collected from deep-sea drilling rigs that the ocean floor becomes progressively older toward the African and South American continents. This was good enough evidence to establish the "seafloor spreading hypothesis" that described the two continents moving apart.

If a hypothesis survives much experimental testing and leads, in turn, to the design of new experiments with the generation of new hypotheses that can be tested, you now have a working *theory.* A **theory** is defined as a broad working hypothesis that is based on extensive experimental evidence. A scientific theory tells you *why* something happens. For example, the plate tectonic theory describes how the continents have moved apart, just like pieces of a picture puzzle. Is this the same idea that was once considered conjecture? Sort of, but this time it is supported by experimental evidence.

The term *scientific theory* is reserved for historic schemes of thought that have survived the test of detailed examination for long periods of time. The *atomic theory,* for example, was developed in the late 1800s and has been the subject of extensive investigation and experimentation over the last century. The atomic theory and other scientific theories form the framework of scientific thought and experimentation today. Scientific theories point to new ideas about the behavior of nature, and these ideas result in more experiments, more data to collect, and more explanations to develop. All of this may lead to a slight modification of an existing theory, a major modification, or perhaps the

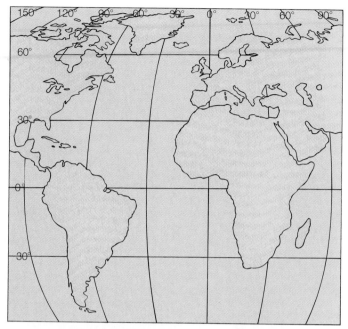

A

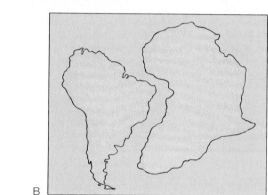

B

FIGURE 1.16

(*A*) Normal position of the continents on a world map. (*B*) A sketch of South America and Africa, suggesting that they once might have been joined together and subsequently separated by continental drift.

creation of an entirely new theory. These activities are all part of the continuing attempt to satisfy our curiosity about nature.

Summary

Physical science is a search for order in our physical surroundings. People have *concepts,* or mental images, about material *objects* and intangible *events* in their surroundings. Concepts are used for thinking and communicating. Concepts are based on *properties,* or attributes that describe a thing or event. Every property implies a *referent* that describes the property. Referents are not always explicit, and most communications require assumptions. Measurement brings precision to descriptions by using numbers and standard units for referents to communicate "exactly how much of exactly what."

Measurement is a process that uses a well-defined and agreed-upon *referent* to describe a *standard unit.* The unit is compared to the property being defined by an *operation* that determines the *value* of the unit by *counting*. Measurements are always reported with a *number*, or value, and a *name* for the unit.

The two major *systems* of standard units are the *English system* and the *metric system*. The English system uses standard units that were originally based on human body parts, and the metric system uses standard units based on referents found in nature. The metric system also uses a system of prefixes to express larger or smaller amounts of units. The metric standard units for length, mass, and time are the *meter, kilogram,* and *second.*

Measurement information used to describe something is called *data.* One way to extract meanings and generalizations from data is to use a *ratio,* a simplified relationship between two numbers. Density is a ratio of mass to volume, or $\rho = m/V$.

Symbols are used to represent *quantities,* or measured properties. Symbols are used in *equations,* which are shorthand statements that describe a relationship where the quantities (both number values and units) are identical on both sides of the equal sign. Equations are used to (1) *describe* a property, (2) *define* a concept, or (3) *describe* how *quantities change* together.

Quantities that can have different values at different times are called *variables.* Variables that increase or decrease together in the same ratio are said to be in *direct proportion.* If one variable increases while the other decreases in the same ratio, the variables are in *inverse proportion.* Proportionality statements are not necessarily equations. A *proportionality constant* can be used to make such a statement into an equation. Proportionality constants might have numerical value only, without units, or they might have both value and units.

Modern science began about three hundred years ago during the time of Galileo and Newton. Since that time, *scientific investigation* has been used to provide *experimental evidence* about nature. *Methods* used to conduct scientific investigations can be generalized as *collecting observations, developing explanations,* and *testing explanations.*

A *hypothesis* is a tentative explanation that is accepted or rejected based on experimental data. Experimental data can come from *observations* or from a *controlled experiment.* The controlled experiment compares two situations that have all the influencing factors identical except one. The single influencing variable being tested is called the *experimental variable,* and the group of variables that form the basis of comparison is called the *control group.*

An accepted hypothesis may result in a *principle,* an explanation concerned with a specific range of phenomena, or a *scientific law,* an explanation concerned with important, wider-ranging phenomena. Laws are sometimes identified with the name of a scientist and can be expressed verbally, with an equation, or with a graph.

A *model* is used to help understand something that cannot be observed directly, explaining the unknown in terms of things already understood. Physical models, mental models, and equations are all examples of models that explain how nature behaves. A *theory* is a broad, detailed explanation that guides development and interpretations of experiments in a field of study.

Summary of Equations

1.1

$$\text{mass density} = \frac{\text{mass}}{\text{volume}}$$

$$\rho = \frac{m}{V}$$

KEY TERMS

area (p. **9**)
controlled experiment (p. **16**)
data (p. **9**)
density (p. **10**)
direct proportion (p. **12**)
English system (p. **5**)
equation (p. **12**)
experiment (p. **16**)
fundamental properties (p. **6**)
hypothesis (p. **16**)
inverse proportion (p. **12**)
kilogram (p. **7**)
liter (p. **8**)
measurement (p. **4**)
meter (p. **6**)

metric system (p. **6**)
model (p. **18**)
numerical constant (p. **13**)
properties (p. **3**)
proportionality constant (p. **13**)
quantities (p. **12**)
ratio (p. **9**)
referent (p. **3**)
scientific law (p. **18**)
scientific principle (p. **18**)
second (p. **7**)
standard unit (p. **5**)
theory (p. **20**)
unit (p. **4**)
variable (p. **12**)
volume (p. **9**)

APPLYING THE CONCEPTS

1. The process of comparing a property of an object to a well-defined and agreed-upon referent is called
   a. generalizing.
   b. measurement.
   c. graphing.
   d. scientific investigation.

2. The height of an average person is closest to
   a. 1.0 m.
   b. 1.5 m.
   c. 2.5 m.
   d. 3.5 m.

3. Which of the following standard units is defined in terms of an object as opposed to an event?
   a. kilogram
   b. meter
   c. second
   d. None of the above is correct.

4. One-half liter of water has a mass of
   a. 0.5 g.
   b. 5 g.
   c. 50 g.
   d. 500 g.

5. A cubic centimeter (cm^3) of water has a mass of about 1
   a. mL.
   b. kg.
   c. g.
   d. dm.

6. Measurement information that is used to describe something is called
   a. referents.
   b. properties.
   c. data.
   d. a scientific investigation.

7. The property of volume is a measure of
   a. how much matter an object contains.
   b. how much space an object occupies.
   c. the compactness of matter in a certain size.
   d. the area on the outside surface.

8. As the volume of a cube becomes larger and larger, the surface-area-to-volume ratio
   a. increases.
   b. decreases.
   c. remains the same.
   d. sometimes increases and sometimes decreases.

9. If you consider a very small portion of a material that is the same throughout, the density of the small sample will be
   a. much less.
   b. slightly less.
   c. the same.
   d. greater.

10. Symbols that are used in equations represent
    a. a message.
    b. specific properties.
    c. quantities, or measured properties.
    d. all of the above.

11. An equation is composed of symbols in such a way that
    a. the numbers and units on both sides are always equal.
    b. the units are equal, but the numbers are not because one is unknown.
    c. the numbers are equal, but the units are not equal.
    d. neither the numbers nor units are equal because of the unknown.

12. The symbol $\therefore$ has a meaning of
    a. is proportional to.
    b. the change in.
    c. therefore.
    d. however.

13. Quantities, or measured properties, that are capable of changing values are called
    a. data.
    b. variables.
    c. proportionality constants.
    d. dimensionless constants.

14. A proportional relationship that is represented by the symbols $a \propto 1/b$ represents which of the following relationships?
    a. direct proportion
    b. inverse proportion
    c. direct square proportion
    d. inverse square proportion

15. A hypothesis concerned with a specific phenomenon is found to be acceptable through many experiments over a long period of time. This hypothesis usually becomes known as a
    a. scientific law.
    b. scientific principle.
    c. theory.
    d. model.

16. A scientific law can be expressed as
    a. a written concept.
    b. an equation.
    c. a graph.
    d. all of the above.

17. The symbol $\propto$ has a meaning of
    a. almost infinity.
    b. the change in.
    c. is proportional to.
    d. therefore.

18. Which of the following symbols represents a measured property of the compactness of matter?
    a. m
    b. ρ
    c. V
    d. Δ

Answers

1. b 2. b 3. a 4. d 5. c 6. c 7. b 8. b 9. c 10. d 11. a 12. c 13. b 14. b 15. b 16. d 17. c 18. b

QUESTIONS FOR THOUGHT

1. What is a concept?
2. What two things does a measurement statement always contain? What do the two things tell you?
3. Other than familiarity, what are the advantages of the English system of measurement?
4. Describe the metric standard units for length, mass, and time.
5. Does the density of a liquid change with the shape of a container? Explain.
6. Does a flattened pancake of clay have the same density as the same clay rolled into a ball? Explain.
7. What is an equation? How are equations used in the physical sciences?
8. Compare and contrast a scientific principle and a scientific law.
9. What is a model? How are models used?
10. Are all theories always completely accepted or completely rejected? Explain.

The exercises in groups A and B cover the same concepts. Solutions to group A exercises are located in appendix D.

Note: *You will need to refer to Table 1.4 to complete some of the following exercises.*

Group A

1. What is your height in meters? In centimeters?

2. What is the mass density of mercury if 20.0 cm^3 has a mass of 272 g?

3. What is the mass of a 10.0 cm^3 cube of lead?

4. What is the volume of a rock with a mass density of 3.00 g/cm^3 and a mass of 600 g?

5. If you have 34.0 g of a 50.0 cm^3 volume of one of the substances listed in Table 1.4, which one is it?

6. What is the mass of water in a 40 L aquarium?

7. A 2.1 kg pile of aluminum cans is melted, then cooled into a solid cube. What is the volume of the cube?

8. A cubic box contains 1,000 g of water. What is the length of one side of the box in meters? Explain your reasoning.

9. A loaf of bread (volume 3,000 cm^3) with a density of 0.2 g/cm^3 is crushed in the bottom of a grocery bag into a volume of 1,500 cm^3. What is the density of the mashed bread?

10. According to Table 1.4, what volume of copper would be needed to balance a 1.00 cm^3 sample of lead on a two-pan laboratory balance?

Group B

1. What is your mass in kilograms? In grams?

2. What is the mass density of iron if 5.0 cm^3 has a mass of 39.5 g?

3. What is the mass of a 10.0 cm^3 cube of copper?

4. If ice has a mass density of 0.92 g/cm^3, what is the volume of 5,000 g of ice?

5. If you have 51.5 g of a 50.0 cm^3 volume of one of the substances listed in Table 1.4, which one is it?

6. What is the mass of gasoline ($\rho = 0.680$ g/cm^3) in a 94.6 L gasoline tank?

7. What is the volume of a 2.00 kg pile of iron cans that are melted, then cooled into a solid cube?

8. A cubic tank holds 1,000.0 kg of water. What are the dimensions of the tank in meters? Explain your reasoning.

9. A hot dog bun (volume 240 cm^3) with a density of 0.15 g/cm^3 is crushed in a picnic cooler into a volume of 195 cm^3. What is the new density of the bun?

10. According to Table 1.4, what volume of iron would be needed to balance a 1.00 cm^3 sample of lead on a two-pan laboratory balance?

Information about the mass of a hot air balloon and forces on the balloon will enable you to predict if it is going to move up, down, or drift across the river. This chapter is about such relationships among force, mass, and changes in motion.

CHAPTER

2

Motion

In chapter 1, you learned some "tools and rules" and some techniques for finding order in your physical surroundings. Order is often found in the form of patterns, or relationships between quantities that are expressed as equations. Recall that equations can be used to (1) describe properties, (2) define concepts, and (3) describe how quantities change relative to each other. In all three uses, patterns are quantified, conceptualized, and used to gain a general understanding about what is happening in nature.

In the study of physical science, certain parts of nature are often considered and studied together for convenience. One of the more obvious groupings involves *movement.* Most objects around you spend a great deal of time sitting quietly without motion. Buildings, rocks, utility poles, and trees rarely, if ever, move from one place to another. Even things that do move from time to time sit still for a great deal of time. This includes you, automobiles, and bicycles (Figure 2.1). On the other hand, the sun, the moon, and starry heavens seem to always move, never standing still. Why do things stand still? Why do things move?

Questions about motion have captured the attention of people for thousands of years. But the ancient people answered questions about motion with stories of mysticism and spirits that lived in objects. It was during the classic Greek culture, between 600 B.C. and 300 B.C., that people began to look beyond magic and spirits. One particular Greek philosopher, Aristotle, wrote a theory about the universe that offered not only explanations about things such as motion but also offered a sense of beauty, order, and perfection. The theory seemed to fit with other ideas that people had and was held to be correct for nearly two thousand years after it was written. It was not until the work of Galileo and Newton during the 1600s that a new, correct understanding about motion was developed. The development of ideas about motion is an amazing and absorbing story. You will learn in this chapter how to describe and use some properties of motion. This will provide some basic understandings about motion and will be very helpful in understanding some important aspects of astronomy and the earth sciences, as well as the movement of living things.

Describing Motion

Motion is one of the more common events in your surroundings. You can see motion in natural events such as clouds moving, rain and snow falling, and streams of water moving, all in a never-ending cycle. Motion can also be seen in the activities of people who walk, jog, or drive various machines from place to place. Motion is so common that you would think everyone would intuitively understand the concepts of motion, but history indicates that it was only during the past three hundred years or so that people began to understand motion correctly. Perhaps the correct concepts are subtle and contrary to common sense, requiring a search for simple, clear concepts in an otherwise complex situation. The process of finding such order in a multitude of sensory impressions by taking measurable data, and then inventing a concept to describe what is happening, is the activity called *science.* We will now apply this process to motion.

What is motion? Consider a ball that you notice one morning in the middle of a lawn. Later in the afternoon, you notice that the ball is at the edge of the lawn, against a fence, and you wonder if the wind or some person moved the ball. You do not know if the wind blew it at a steady rate, if many gusts of wind moved it, or even if some children kicked it all over the yard. All you know for sure is that the ball has been moved because it is in a different position after some time passed. These are the two important aspects of motion: (1) a change of position and (2) the passage of time.

If you did happen to see the ball rolling across the lawn in the wind, you would see more than the ball at just two locations. You would see the ball moving continuously. You could consider, however, the ball in continuous motion to be a series of individual locations with very small time intervals. Moving involves a change of position during some time period. Motion is the act or process of something changing position.

The motion of an object is usually described with respect to something else that is considered to be not moving. (Such a stationary object is said to be "at rest.") Imagine that you are traveling in an automobile with another person. You know that you are moving across the land outside the car since your location on the highway changes from one moment to another. Observing your fellow passenger, however, reveals no change of position. You are in motion relative to the highway outside the car. You are not in motion relative to your fellow passenger. Your motion, and the motion of any other object or body, is the process of a change in position *relative* to some reference object or location. Thus *motion* can be defined as the act or process of changing position relative to some reference during a period of time.

Measuring Motion

You have learned that objects can be described by measuring certain fundamental properties such as mass and length. Since motion involves (1) a change of *position,* and (2) the passage of

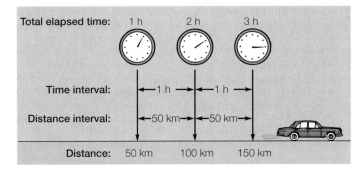

Time interval: ◄─1 h─►◄─1 h─►

Distance interval: ◄─50 km─►◄─50 km─►

Distance: 50 km   100 km   150 km

FIGURE 2.2

If you know the value of any two of the three variables of distance, time, and speed, you can find the third. What is the average speed of this car? Two ways of finding the answer are in Figure 2.3.

The units used to describe speed are usually miles/hour (mi/h), kilometers/hour (km/h), or meters/second (m/s).

Let's go back to your car that is moving over a straight highway and imagine you are driving to cover equal distances in equal periods of time. If you use a stopwatch to measure the time required to cover the distance between highway mile markers (those little signs with numbers along major highways), the time intervals will all be equal. You might find, for example, that one minute lapses between each mile marker. Such a uniform straight-line motion that covers equal distances in equal periods of time is the simplest kind of motion.

If your car were moving over equal distances in equal periods of time, it would have a *constant speed* (Figure 2.2). This means that the car is neither speeding up nor slowing down. It is usually difficult to maintain a constant speed. Other cars and distractions such as interesting scenery cause you to reduce your speed. At other times you increase your speed. If you calculate your speed over an entire trip, you are considering a large distance between two places and the total time that elapsed. The increases and decreases in speed would be averaged. Therefore, most speed calculations are for an *average speed*. The speed at any specific instant is called the *instantaneous speed*. To calculate the instantaneous speed, you would need to consider a very short time interval—one that approaches zero. An easier way would be to use the speedometer, which shows the speed at any instant.

Constant, instantaneous, or average speeds can be measured with any distance and time units. Common units in the English system are miles/hour and feet/second. Metric units for speed are commonly kilometers/hour and meters/second. The ratio of any distance to time is usually read as distance per time, such as miles per hour. The "per" means "for each."

It is easier to study the relationships between quantities if you use symbols instead of writing out the whole word. The letter *v* can be used to stand for speed, the letter *d* can be used to stand for distance, and the letter *t*, to stand for time. A bar over the *v* ($\bar{v}$) is a

FIGURE 2.1

The motion of this windsurfer, and of other moving objects, can be described in terms of the distance covered during a certain time period.

time, the motion of objects can be described by using combinations of the fundamental properties of length and time. These combinations of measurement describe three properties of motion: *speed, velocity,* and *acceleration.*

Speed

Suppose you are in a car that is moving over a straight road. How could you describe your motion? You need at least two measurements, (1) the distance you have traveled, and (2) the time that has elapsed while you covered this distance. Such a distance and time can be expressed as a ratio that describes your motion. This ratio is a property of motion called **speed,** which is a measure of how fast you are moving. Speed is defined as distance per unit of time, or

$$\text{speed} = \frac{\text{distance}}{\text{time}}$$

symbol that means average (it is read "v-bar" or "v-average"). The relationship between average speed, distance, and time is therefore

$$\bar{v} = \frac{d}{t}$$

equation 2.1

This is one of the three types of equations that were discussed earlier, and in this case the equation defines a motion property. You can use this relationship to find average speed. For example, suppose a car travels 150 mi in 3 h. What was the average speed? Since $d = 150$ mi, and $t = 3$ h, then

$$\bar{v} = \frac{150 \text{ mi}}{3 \text{ h}}$$

$$= 50 \frac{\text{mi}}{\text{h}}$$

As with other equations, you can mathematically solve the equation for any term as long as two variables are known (Figure 2.3). For example, suppose you know the speed and the time but want to find the distance traveled. You can solve this by first writing the relationship

$$\bar{v} = \frac{d}{t}$$

and then multiplying both sides of the equation by t (to get d on one side by itself),

$$(\bar{v})(t) = \frac{(d)(\cancel{t})}{\cancel{t}}$$

and the t's on the right cancel, leaving

$$\bar{v}t = d \quad \text{or} \quad d = \bar{v}t$$

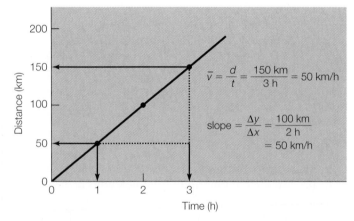

FIGURE 2.3

Speed is distance per unit of time, which can be calculated from the equation or by finding the slope of a distance-versus-time graph. This shows both ways of finding the speed of the car shown in Figure 2.2.

If the $\bar{v}$ is 50 mi/h and the time traveled is 2 h, then

$$d = \left(50 \frac{\text{mi}}{\text{h}} \right)(2 \text{ h})$$

$$= (50)(2) \left(\frac{\text{mi}}{\text{h}} \right)(\text{h})$$

$$= 100 \frac{(\text{mi})(\cancel{\text{h}})}{\cancel{\text{h}}}$$

$$= 100 \text{ mi}$$

Notice how both the numerical values and the units were treated mathematically. See "Problem Solving" in chapter 1 for more information.

EXAMPLE 2.1

The driver of a car moving at 72.0 km/h drops a road map on the floor. It takes him 3.00 seconds to locate and pick up the map. How far did he travel during this time?

Solution

The car has a speed of 72.0 km/h and the time factor is 3.00 s, so km/h must be converted to m/s. From inside the front cover of this book, the conversion factor is 1 km/h = 0.2778 m/s, so

$$\bar{v} = \frac{0.2778 \frac{\text{m}}{\text{s}}}{\frac{\text{km}}{\text{h}}} \times 72.0 \frac{\text{km}}{\text{h}}$$

$$= (0.2778)(72.0) \frac{\text{m}}{\text{s}} \times \frac{\text{h}}{\text{km}} \times \frac{\text{km}}{\text{h}}$$

$$= 20.0 \frac{\text{m}}{\text{s}}$$

The relationship between the three variables, $\bar{v}$, t, and d, is found in equation 2.1: $\bar{v} = d/t$.

$$\bar{v} = 20.0 \frac{\text{m}}{\text{s}} \qquad \bar{v} = \frac{d}{t}$$

$$t = 3.00 \text{ s} \qquad \bar{v}t = \frac{d\cancel{t}}{\cancel{t}}$$

$$d = ? \qquad d = \bar{v}t$$

$$= \left(20.0 \frac{\text{m}}{\text{s}} \right)(3.00 \text{ s})$$

$$= (20.0)(3.00) \frac{\text{m}}{\text{s}} \times \frac{\text{s}}{1}$$

$$= \boxed{60.0 \text{ m}}$$

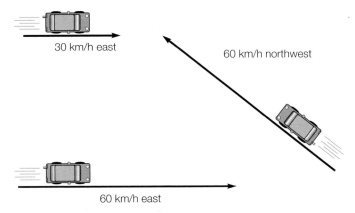

A bicycle has an average speed of 8.00 km/h. How far will it travel in 10.0 seconds? (Answer: 22.2 m)

Concepts Applied

Style Speeds

Observe how many different styles of walking you can identify in students walking across the campus. Identify each style with a descriptive word or phrase.

Is there any relationship between any particular style of walking and the speed of walking? You could find the speed of walking by measuring a distance, such as the distance between two trees, then measuring the time required for a student to walk the distance. Find the average speed for each identified style of walking by averaging the walking speeds of ten people.

Report any relationships you find between styles of walking and the average speed of people with each style. Include any problems you found in measuring, collecting data, and reaching conclusions.

Concepts Applied

How Fast Is a Stream?

A stream is a moving body of water. How could you measure the speed of a stream? Would timing how long it takes a floating leaf to move a measured distance help?

What kind of relationship, if any, would you predict for the speed of a stream and a recent rainfall? Would you predict a direct relationship? Make some measurements of stream speeds and compare your findings to recent rainfall amounts.

Velocity

The word "velocity" is sometimes used interchangeably with the word "speed," but there is a difference. **Velocity** describes the *speed and direction* of a moving object. For example, a speed might be described as 60 km/h. A velocity might be described as

30 km/h east

60 km/h northwest

60 km/h east

FIGURE 2.4
Here are three different velocities represented by three different arrows. The length of each arrow is proportional to the speed, and the arrowhead shows the direction of travel.

60 km/h to the west. To produce a change in velocity, either the speed or the direction is changed (or both are changed). A satellite moving with a constant speed in a circular orbit around the earth does not have a constant velocity since its direction of movement is constantly changing. Velocity can be represented graphically with arrows. The lengths of the arrows are proportional to the magnitude, and the arrowheads indicate the direction (Figure 2.4).

Acceleration

Motion can be changed in three different ways: (1) by changing the speed, (2) by changing the direction of travel, or (3) combining both of these by changing both the speed and direction of travel at the same time. Since velocity describes both the speed and the direction of travel, any of these three changes will result in a change of velocity. You need at least one additional measurement to describe a change of motion, which is how much time elapsed while the change was taking place. The change of velocity and time can be combined to define the *rate* at which the motion was changed. This rate is called **acceleration.** Acceleration is defined as a change of velocity per unit time, or

$$\text{acceleration} = \frac{\text{change of velocity}}{\text{time elapsed}}$$

Another way of saying "change in velocity" is the final velocity minus the initial velocity, so the relationship can also be written as

$$\text{acceleration} = \frac{\text{final velocity} - \text{initial velocity}}{\text{time elapsed}}$$

Acceleration due to a change in speed only can be calculated as follows. Consider a car that is moving with a constant, straight-line velocity of 60 km/h when the driver accelerates to 80 km/h. Suppose it takes 4 s to increase the velocity of 60 km/h to 80 km/h.

The change in velocity is therefore 80 km/h minus 60 km/h, or 20 km/h. The acceleration was

$$\text{acceleration} = \frac{80\,\dfrac{km}{h} - 60\,\dfrac{km}{h}}{4s}$$

$$= \frac{20\,\dfrac{km}{h}}{4\,s}$$

$$= 5\,\frac{km/h}{s}\ \text{or}$$

$$= 5\ km/h/s$$

The average acceleration of the car was 5 km/h for each ("per") second. This is another way of saying that the velocity increases an average of 5 km/h in each second. The velocity of the car was 60 km/h when the acceleration began (initial velocity). At the end of 1 s, the velocity was 65 km/h. At the end of 2 s, it was 70 km/h; at the end of 3 s, 75 km/h; and at the end of 4 s (total time elapsed), the velocity was 80 km/h (final velocity). Note how fast the velocity is changing with time. In summary,

start (initial velocity)	60 km/h
first second	65 km/h
second second	70 km/h
third second	75 km/h
fourth second (final velocity)	80 km/h

As you can see, acceleration is really a description of how fast the speed is changing (Figure 2.5); in this case, it is increasing 5 km/h each second.

Usually, you would want all the units to be the same, so you would convert km/h to m/s. A change in velocity of 5.0 km/h converts to 1.4 m/s and the acceleration would be 1.4 m/s/s. The units m/s per s mean what change of velocity (1.4 m/s) is occurring every second. The combination m/s/s is rather cumbersome, so it is typically treated mathematically to simplify the expression (to simplify a fraction, invert the divisor and multiply, or m/s × 1/s = m/s²). Remember that the expression 1.4 m/s² means the same as 1.4 m/s per s, a change of velocity in a given time period.

The relationship among the quantities involved in acceleration can be represented with the symbols a for average acceleration, v_f for final velocity, v_i for initial velocity, and t for time. The relationship is

$$a = \frac{v_f - v_i}{t}$$

equation 2.2

As in other equations, any one of these quantities can be found if the others are known. For example, solving the equation for the final velocity, v_f, yields:

$$v_f = at + v_i$$

In problems where the initial velocity is equal to zero (starting from rest), the equation simplifies to

$$v_f = at$$

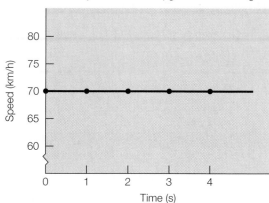

$$a = \frac{v_f - v_i}{t} = \frac{70\ km/h - 70\ km/h}{4\ s} = 0\ \frac{km/h}{s}$$

A

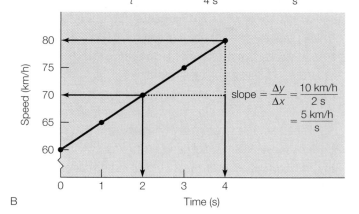

$$a = \frac{v_f - v_i}{t} = \frac{80\ km/h - 60\ km/h}{4\ s} = 5\ \frac{km/h}{s}$$

$$\text{slope} = \frac{\Delta y}{\Delta x} = \frac{10\ km/h}{2\ s}$$

$$= \frac{5\ km/h}{s}$$

B

FIGURE 2.5

(A) This graph shows how the speed changes per unit of time while driving at a constant 70 km/h in a straight line. As you can see, the speed is constant, and for straight-line motion the acceleration is 0. (B) This graph shows the speed increasing from 60 km/h to 80 km/h for 5 s. The acceleration, or change of velocity per unit of time, can be calculated either from the equation for acceleration or by calculating the slope of the straight-line graph. Both will tell you how fast the motion is changing with time.

Recall from chapter 1 that the symbol Δ means "the change in" a value. Therefore, equation 2.1 for speed could be written

$$\bar{v} = \frac{\Delta d}{t}$$

and equation 2.2 for acceleration could be written

$$a = \frac{\Delta v}{t}$$

This shows that both equations are a time rate of change. Speed is a time rate change of *distance*. Acceleration is a time rate change of *velocity*. The time rate of change of something is an important concept that you will meet again in the next chapter.

The super-speed magnetic levitation (maglev) train is a completely new technology based on magnetically suspending a train 3 to 10 cm (about 1 to 4 in) above a monorail, then moving it along with a magnetic field that travels along the monorail guides. The maglev train does not have friction between wheels and the rails since it does not have wheels. This lack of resistance at the easily manipulated magnetic fields makes very short acceleration distances possible. For example, a German maglev train can accelerate from 0 to 300 km/h (about 185 mi/h) over a distance of just 5 km (about 3 mi). A conventional train with wheels requires about 30 km (about 19 mi) to reach the same speed from a standing start. The maglev is attractive for short runs because of its superior acceleration and braking abilities. It is also attractive for longer runs because of its high top speed— up to about 500 km/h (about 310 mi/h). Today, only an aircraft can match such a speed.

EXAMPLE 2.3

A bicycle moves from rest to 5 m/s in 5 s. What was the acceleration?

Solution

$$v_i = 0 \text{ m/s}$$
$$v_f = 5 \text{ m/s}$$
$$t = 5 \text{ s}$$
$$a = ?$$

$$a = \frac{v_f - v_i}{t}$$

$$= \frac{5 \text{ m/s} - 0 \text{ m/s}}{5 \text{ s}}$$

$$= \frac{5}{5} \frac{\text{m/s}}{\text{s}}$$

$$= 1 \left(\frac{\text{m}}{\text{s}}\right)\left(\frac{1}{\text{s}}\right)$$

$$= \boxed{1 \frac{\text{m}}{\text{s}^2}}$$

EXAMPLE 2.4

An automobile uniformly accelerates from rest at 15 ft/s^2 for 6 s. What is the final velocity in ft/s? (Answer: 90 ft/s)

So far, you have learned only about straight-line, uniform acceleration that results in an increased velocity. There are also other changes in the motion of an object that are associated with acceleration. One of the more obvious is a change that results in a decreased velocity. Your car's brakes, for example, can slow your car or bring it to a complete stop. This is *negative acceleration,* which is sometimes called *deceleration.* Another change in the motion of an object is a change of direction. Velocity encompasses both the rate of motion and direction, so

a change of direction is an acceleration. The satellite moving with a constant speed in a circular orbit around the earth is constantly changing its direction of movement. It is therefore constantly accelerating because of this constant change in its motion. Your automobile has three devices that could change the state of its motion. Your automobile therefore has three accelerators—the gas pedal (which can increase magnitude of velocity), the brakes (which can decrease magnitude of velocity), and the steering wheel (which can change direction of velocity). (See Figure 2.6.) The important thing to remember is that acceleration results from any *change* in the motion of an object.

The final velocity (v_f) and the initial velocity (v_i) are different variables than the average velocity ($\bar{v}$). You cannot use an initial or final velocity for an average velocity. You may, however, calculate an average velocity ($\bar{v}$) from the other two variables as long as the acceleration taking place between the initial and final velocities is uniform. An example of such a uniform change would be an automobile during a constant, straight-line acceleration. To find an average velocity *during* a uniform

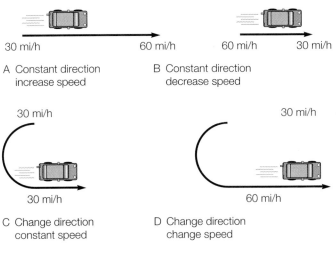

FIGURE 2.6

Four different ways (A–D) to accelerate a car.

acceleration, you add the initial velocity and the final velocity and divide by 2. This averaging can be done for a uniform acceleration that is increasing the velocity or for one that is decreasing the velocity. In symbols,

$$\bar{v} = \frac{v_f + v_i}{2}$$

equation 2.3

EXAMPLE 2.5

An automobile moving at 88 ft/s comes to a stop in 10.0 s when the driver slams on the brakes. How far did the car travel while stopping?

Solution

The car has an initial velocity of 88 ft/s (v_i) and the final velocity of 0 ft/s (v_f) is implied. The time of 10.0 s (t) is given. The problem asked for the distance (d). The relationship given between $\bar{v}$, t, and d is given in equation 2.1, $\bar{v} = d/t$, which can be solved for d. The average velocity ($\bar{v}$), however, is not given but can be found from equation 2.3,

$$\bar{v} = \frac{v_f + v_i}{2}$$

$$\bar{v} = \frac{d}{t} \quad \therefore \quad d = \bar{v} \cdot t$$

$v_i = 88$ ft/s

$v_f = 0$ ft/s

$t = 10.0$ s

$\bar{v} = ?$

$d = ?$

Since $\bar{v} = \frac{v_f + v_i}{2}$,

you can substitute $\left(\frac{v_f + v_i}{2}\right)$ for $\bar{v}$, and

$$d = \left(\frac{v_f + v_i}{2}\right)(t)$$

$$= \left(\frac{0\,\frac{ft}{s} + 88\,\frac{ft}{s}}{2}\right)(10.0 \text{ s})$$

$$= 44 \times 10.0 \frac{ft}{s} \times s$$

$$= 440 \frac{ft \cdot s}{s}$$

$$= \boxed{440 \text{ ft}}$$

EXAMPLE 2.6

What was the deceleration of the automobile in example 2.5?
(Answer: -8.8 ft/s^2)

Acceleration Patterns

Suppose the radiator in your car has a leak and drops fall constantly, one every second. What pattern would the drops make on the pavement when you accelerate the car from a stoplight? What pattern would they make when you drive at a constant speed? What pattern would you observe as the car comes to a stop? Use a marker to make dots on a sheet of paper that illustrate (1) acceleration, (2) constant speed, and (3) negative acceleration. Use words to describe the acceleration in each situation.

Forces

The Greek philosopher Aristotle considered some of the first ideas about the causes of motion back in the fourth century B.C. However, he had it all wrong when he reportedly stated that a dropped object falls at a constant speed that is determined by its weight. He also incorrectly thought that an object moving across the earth's surface requires a continuously applied force in order to continue moving. These ideas were based on observing and thinking, not measurement, and no one checked to see if they were correct. It would take about two thousand years before people began to correctly understand motion.

Aristotle did recognize an association between force and motion, and this much was acceptable. It is partly correct because a force is closely associated with *any* change of motion, as you will see. This section introduces the concept of a force, which will be developed more fully when the relationship between forces and motion is considered.

A **force** is a push or a pull that is capable of changing the state of motion of an object. Consider, for example, the movement of a ship from the pushing of two tugboats (Figure 2.7). Tugboats can vary the strength of the force exerted on a ship, but they can also push in different directions. What effect does direction have on two forces acting on an object? If the tugboats were side by side, pushing in the same direction, the overall force is the sum of the two forces. If they act in exactly opposite directions, one pushing on each side of the ship, the overall force is the difference between the strength of the two forces. If they have the same strength, the overall effect is to cancel each other without producing any motion. The **net force** is the sum of all the forces acting on an object. Net force means "final," after the forces are added (Figure 2.8).

When two parallel forces act in the same direction, they can be simply added. In this case, there is a net force that is equivalent to the sum of the two forces. When two parallel forces act in opposite directions, the net force is the difference in the direction of the larger force. When two forces act neither in a way that is exactly together nor exactly opposite each other, the result will be like a new, different net force having a new direction and strength.

FIGURE 2.7

The rate of movement and the direction of movement of this ship are determined by a combination of direction and size of force from each of the tugboats. Which direction are the two tugboats pushing? What evidence would indicate that one tugboat is pushing with a greater force? If the tugboat by the numbers is pushing with a greater force and the back tugboat is keeping the ship from moving, what will happen?

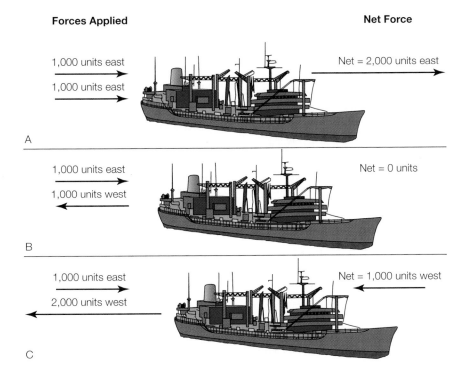

Forces Applied   **Net Force**

1,000 units east
1,000 units east   Net = 2,000 units east

A

1,000 units east
1,000 units west   Net = 0 units

B

1,000 units east
2,000 units west   Net = 1,000 units west

C

FIGURE 2.8

(A) When two parallel forces are acting on the ship in the same direction, the net force is the two forces added together. (B) When two forces are opposite and of equal size, the net force is zero. (C) When two parallel forces are not of equal size, the net force is the difference in the direction of the larger force.

Forces have a strength and direction that can be represented by force arrows. The tail of the arrow is placed on the object that feels the force, and the arrowhead points in the direction in which the force is exerted. The length of the arrow is proportional to the strength of the force. The use of force arrows helps you visualize and understand all the forces and how they contribute to the net force.

Horizontal Motion on Land

Everyday experience seems to indicate that Aristotle's idea about horizontal motion on the earth's surface is correct. After all, moving objects that are not pushed or pulled do come to rest in a short period of time. It would seem that an object keeps moving only if a force continues to push it. A moving automobile will slow and come to rest if you turn off the ignition. Likewise, a ball that you roll along the floor will slow until it comes to rest. Is the natural state of an object to be at rest, and is a force necessary to keep an object in motion? This is exactly what people thought until Galileo published his book *Two New Sciences* in 1638, which described his findings about motion. The book had three parts that dealt with uniform motion, accelerated motion, and projectile motion. Galileo described details of simple experiments, measurements, calculations, and thought experiments as he developed definitions and concepts of motion. In one of his thought experiments, Galileo presented an argument against Aristotle's view that a force is needed to keep an object in motion. Galileo imagined an object (such as a ball) moving over a horizontal surface without the force of friction. He concluded that the object would move forever with a constant velocity as long as there was no unbalanced force acting to change the motion.

Why does a rolling ball slow to a stop? You know that a ball will roll farther across a smooth, waxed floor such as a bowling lane than it will across a floor covered with carpet. The rough carpet offers more resistance to the rolling ball. The resistance of the floor friction is shown by a force arrow, F_{floor}, in Figure 2.9. This force, along with the force arrow for air resistance, F_{air}, opposes the forward movement of the ball. Notice the dashed line arrow in part A of Figure 2.9. There is no other force applied to the ball, so the rolling speed decreases until the ball finally comes to a complete stop. Now imagine what force you would need to exert by pushing with your hand, moving along with the ball to keep it rolling at a uniform rate. An examination of the forces in part B of Figure 2.9 can help you determine the amount of force. The force you apply, $F_{applied}$, must counteract the resistance forces. It opposes the forces that are slowing down the ball as illustrated by the direction of the arrows. To determine how much force you should apply, look at the arrow equation. $F_{applied}$ has the same length as the sum of the two resistance forces, but it is in the opposite direction of the resistance forces. Therefore, the overall force, F_{net}, is zero. The ball continues to roll at a uniform rate when you *balance* the force opposing its motion. It is reasonable, then, that if there were no opposing forces, you would

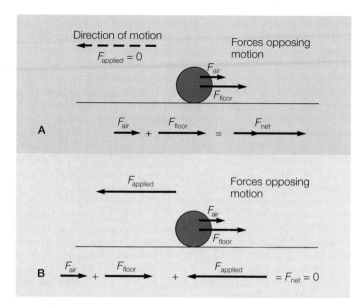

FIGURE 2.9

(A) This ball is rolling to your left with no forces in the direction of motion. The sum of the force of floor friction (F_{floor}) and the force of air friction (F_{air}) results in a net force opposing the motion, so the ball slows to a stop. (B) A force is applied to the moving ball, perhaps by a hand that moves along with the ball. The force applied ($F_{applied}$) equals the sum of the forces opposing the motion, so the ball continues to move with a constant velocity.

not need to apply a force to keep it rolling. This was the kind of reasoning that Galileo did when he discredited the Aristotelian view that a force was necessary to keep an object moving. Galileo concluded that a moving object would continue moving with a constant velocity if no unbalanced forces were applied, that is, if the net force were zero.

It could be argued that the difference in Aristotle's and Galileo's views of forced motion is really a degree of analysis. After all, moving objects on the earth do come to rest unless continuously pushed or pulled. But Galileo's conclusion describes *why* they must be pushed or pulled and reveals the true nature of the motion of objects. Aristotle argued that the natural state of objects is to be at rest, and he tried to explain why objects move. Galileo, on the other hand, argued that it is just as natural for objects to be moving, and he tried to explain why they come to rest. Galileo called the behavior of matter that causes it to persist in its state of motion **inertia**. Inertia is the *tendency of an object to remain in unchanging motion or at rest in the absence of an unbalanced force* (friction, gravity, or whatever). The development of this concept changed the way people viewed the natural state of an object and opened the way for further understandings about motion. Today, it is understood that a satellite moving through free space will continue to do so with no unbalanced forces acting on it (Figure 2.10A). An unbalanced force is needed to slow the satellite (Figure 2.10B), increase its speed (Figure 2.10C), or change its direction of travel (Figure 2.10D).

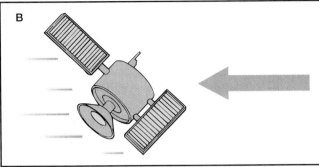

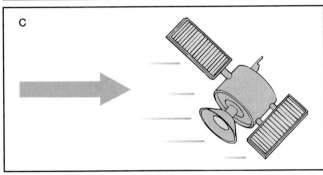

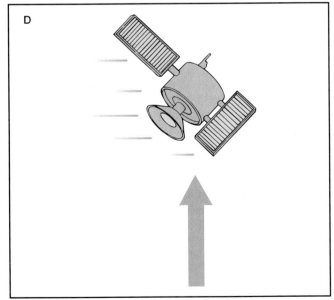

FIGURE 2.10

Examine the four illustrations and explain how together they illustrate inertia.

Falling Objects

Did you ever wonder what happens to a falling rock during its fall? Aristotle reportedly thought that a rock falls at a uniform speed that is proportional to its weight. Thus, a heavy rock would fall at a faster uniform speed than a lighter rock. As stated in a popular story, Galileo discredited Aristotle's conclusion by dropping a solid iron ball and a solid wooden ball simultaneously from the top of the Leaning Tower of Pisa (Figure 2.11). Both balls, according to the story, hit the ground nearly at the same time. To do this, they would have to fall with the same velocity. In other words, the velocity of a falling object does not depend on its weight. Any difference in freely falling bodies is explainable by air resistance. Soon after the time of Galileo the air pump was invented. The air pump could be used to remove the air from a glass tube. The effect of air resistance on falling objects could then be demonstrated by comparing how objects fall in the air with how they fall in an evacuated glass tube. You know that a coin falls faster than a feather when both are dropped in the air. A feather and heavy coin will fall together in the near vacuum of an evacuated glass tube because the effect of air resistance on the feather has been removed. When objects fall toward the earth without considering air resistance, they are said to be in **free fall.** Free fall considers only gravity and neglects air resistance.

Concepts Applied

Falling Bodies

Galileo concluded that all objects fall together, with the same acceleration, when the upward force of air resistance is removed. It would be most difficult to remove air from the room, but it is possible to do some experiments that provide some evidence of how air influences falling objects.

1. Take a sheet of paper and your textbook and drop them side by side from the same height. Note the result.
2. Place the sheet of paper on top of the book and drop them at the same time. Do they fall together?
3. Crumple the sheet of paper into a loose ball and drop the ball and book side by side from the same height.
4. Crumple a sheet of paper into a very tight ball and again drop the ball and book side by side from the same height.

Explain any evidence you found concerning how objects fall.

Galileo concluded that light and heavy objects fall together in free fall, but he also wanted to know the details of what was going on while they fell. He now knew that the velocity of an object in free fall was *not* proportional to the weight of the object. He observed that the velocity of an object in free fall *increased* as the object fell and reasoned from this that the velocity of the falling

Galileo was one of the first to recognize the role of friction in opposing motion. As shown in Figure 2.9, friction with the surface and air friction combine to produce a net force that works against anything that is moving on the surface. This article is about air friction and some techniques that bike riders use to reduce that opposing force—perhaps giving them an edge in a close race.

The bike riders in Box Figure 2.1 are forming a single-file line, called a *paceline*, because the slipstream reduces the air resistance for a closely trailing rider. Cyclists say that riding in the slipstream of another cyclist will save much of their energy. They can move up to 5 mi/h faster than they would expending the same energy riding alone.

In a sense, riding in a slipstream means that you do not have to push as much air out of your way. It has been estimated that at 20 mi/h a cyclist must move a little less than half a ton of air out of the way every minute. One of the earliest demonstrations of how a slipstream can help a cyclist was done back about the turn of the century. Charles Murphy had a special bicycle trail built down the middle of a railroad track. Riding very close behind a special train caboose, Murphy was

BOX FIGURE 2.1

The object of the race is to be in the front, to finish first. If this is true, why are these racers forming a single file line?

able to reach a speed of over 60 mi/h for a one-mile course. More recently, cyclists have reached over 125 mi/h by following close, in the slipstream of a race car.

Along with the problem of moving about a half-ton of air out of the way every minute, there are two basic factors related to air resistance. These are (1) a turbulent versus a smooth flow of air, and (2) the problem of frictional drag. A turbulent flow of air contributes to air resistance because it causes the air to separate slightly on the back side, which increases the pressure on the

front of the moving object. This is why racing cars, airplanes, boats, and other racing vehicles are streamlined to a teardrop-like shape. This shape is not as likely to have the lower-pressure-producing air turbulence behind (and resulting greater pressure in front) because it smoothes, or streamlines the air flow.

The frictional drag of air is similar to the frictional drag that occurs when you push a book across a rough tabletop. You know that smoothing the rough tabletop will reduce the frictional drag on the book. Likewise, the smoothing of a surface exposed to moving air will reduce air friction. Cyclists accomplish this "smoothing" by wearing smooth lycra clothing, and by shaving hair from arm and leg surfaces that are exposed to moving air. Each hair contributes to the overall frictional drag, and removal of the arm and leg hair can thus result in seconds saved. This might provide enough of an edge to win a close race. Shaving legs and arms, together with the wearing of lycra or some other tight, smooth-fitting garments, are just a few of the things a cyclist can do to gain an edge. Perhaps you will be able to think of more ways to reduce the forces that oppose motion.

FIGURE 2.11

According to a widespread story, Galileo dropped two objects with different weights from the Leaning Tower of Pisa. They reportedly hit the ground at about the same time, discrediting Aristotle's view that the speed during the fall is proportional to weight.

object would have to be (1) somehow proportional to the *time* of fall and (2) somehow proportional to the *distance* the object fell. If the time and distance were both related to the velocity of a falling object, how were they related to one another? To answer this question, Galileo made calculations involving distance, velocity, and time and, in fact, introduced the concept of acceleration. The relationships between these variables are found in the same three equations that you have already learned. Let's see how the equations can be rearranged to incorporate acceleration, distance, and time for an object in free fall.

Step 1: Equation 2.1 gives a relationship between average velocity ($\bar{v}$), distance (d), and time (t). Solving this equation for distance gives

$$d = \bar{v}t$$

Step 2: An object in free fall should have uniformly accelerated motion, so the average velocity could be calculated from equation 2.3,

$$\bar{v} = \frac{v_f + v_i}{2}$$

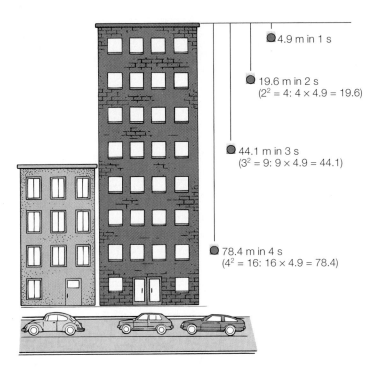

FIGURE 2.12

An object dropped from a tall building covers increasing distances with every successive second of falling. The distance covered is proportional to the square of the time of falling ($d \propto t^2$).

Substituting this equation in the rearranged equation 2.1, the distance relationship becomes

$$d = \left(\frac{v_f + v_i}{2}\right)(t)$$

Step 3: The initial velocity of a falling object is always zero just as it is dropped, so the v_i can be eliminated,

$$d = \left(\frac{v_f}{2}\right)(t)$$

Step 4: Now you want to get acceleration into the equation in place of velocity. This can be done by solving equation 2.2 for the final velocity (v_f), then substituting. The initial velocity (v_i) is again eliminated because it equals zero.

$$a = \frac{v_f - v_i}{t}$$

$$v_f = at$$

$$d = \left(\frac{at}{2}\right)(t)$$

Step 5: Simplifying, the equation becomes

$$d = \frac{1}{2}at^2$$

equation 2.4

Thus, Galileo reasoned that a freely falling object should cover a distance *proportional to the square of the time of the fall* ($d \propto t^2$).

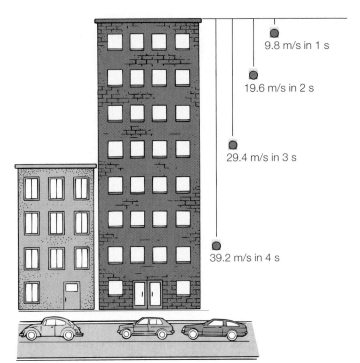

FIGURE 2.13

The velocity of a falling object increases at a constant rate, 9.8 m/s².

In other words the object should fall 4 times as far in 2 s as in 1 s ($2^2 = 4$), 9 times as far in 3 s ($3^2 = 9$), and so on. Compare this prediction with Figure 2.12.

Galileo checked this calculation by rolling balls on an inclined board with a smooth groove in it. He used the inclined board to slow the motion of descent in order to measure the distance and time relationships, a necessary requirement since he lacked the accurate timing devices that exist today. He found, as predicted, that the falling balls moved through a distance proportional to the square of the time of falling. This also means that the *velocity of the falling object increased at a constant rate*, as shown in Figure 2.13. Recall that a change of velocity during some time period is called *acceleration*. In other words, a falling object *accelerates* toward the surface of the earth.

Since the velocity of a falling object increases at a constant rate, this must mean that falling objects are *uniformly accelerated* by the force of gravity. *All objects in free fall experience a constant acceleration.* During each second of fall, the object gains 9.8 m/s (32 ft/s) in velocity. This gain is the acceleration of the falling object, 9.8 m/s² (32 ft/s²).

The acceleration of objects falling toward the earth varies slightly from place to place on the earth's surface because of the earth's shape and spin. The acceleration of falling objects decreases from the poles to the equator and also varies from place to place because the earth's mass is not distributed equally. The value of 9.8 m/s² (32 ft/s²) is an approximation that is fairly close to, but not exactly, the acceleration due to gravity in any particular location. The acceleration due to gravity is important in a number of situations, so the acceleration from this force is given a special symbol, **g**.

There are two different meanings for the term "free fall." In physics, "free fall" means the unconstrained motion of a body in a gravitational field, without considering air resistance. Without air resistance all objects are assumed to accelerate toward the surface at 9.8 m/s².

In the sport of skydiving, "free fall" means falling within the atmosphere without a drag-producing device such as a parachute. Air provides a resisting force that opposes the motion of a falling object, and the net force is the difference between the downward force (weight) and the upward force of air resistance. The weight of the falling object depends on the mass and acceleration from gravity, and this is the force downward. The resisting force is determined by at least two variables, (1) the area of the object exposed to the airstream, and (2) the speed of the falling object. Other variables such as streamlining, air temperature, and turbulence play a role, but the greatest effect seems to be from exposed area and the increased resistance as speed increases.

A skydiver's weight is constant, so the downward force is constant. Modern skydivers typically free-fall from about 3,650 m (about 12,000 ft) above the ground until about 750 m (about 2,500 ft), where they open their parachutes. After jumping from the plane, the diver at first accelerates toward the surface, reaching speeds up to about 185 to 210 km/h (about 115 to 130 mi/h). The air resistance increases with increased speed and the net force becomes less and less. Eventually, the downward weight force will be balanced by the upward air resistance force, and the net force becomes zero. The person now falls at a constant speed and we say the terminal velocity has been reached. It is possible to change your body position to vary your rate of fall up or down to 32 km/h (about 20 mi/h). However, by diving or "standing up" in free fall, experienced skydivers can learn to reach speeds of up to 290 km/h (about 180 mi/h). The record free fall speed, done without any special equipment, is 517 km/h (about 321 mi/h). Once the parachute opens, a descent rate of about 16 km/h (about 10 mi/h) is typical.

EXAMPLE 2.7

A rock that is dropped into a well hits the water in 3.0 s. Ignoring air resistance, how far is it to the water?

Solution 1

The problem concerns a rock in free fall. The time of fall (t) is given, and the problem asks for a distance (d). Since the rock is in free fall, the acceleration due to the force of gravity (g) is implied. The metric value and unit for g is 9.8 m/s², and the English value and unit is 32 ft/s². You would use the metric g to obtain an answer in meters and the English unit to obtain an answer in feet. Equation 2.4, $d = 1/2\, at^2$, gives a relationship between distance (d), time (t), and average acceleration (a). The acceleration in this case is the acceleration due to gravity (g), so

$$t = 3.0 \text{ s} \qquad d = \frac{1}{2}gt^2 (a = g = 9.8 \text{ m/s}^2)$$
$$g = 9.8 \text{ m/s}^2$$
$$d = ? \qquad d = \frac{1}{2}(9.8 \text{ m/s}^2)(3.0 \text{ s})^2$$
$$= (4.9 \text{ m/s}^2)(9.0 \text{ s}^2)$$
$$= 44 \frac{\text{m} \cdot \cancel{\text{s}}^2}{\cancel{\text{s}}^2}$$
$$= \boxed{44 \text{ m}}$$

Solution 2

You could do each step separately. Check this solution by a three-step procedure:

1. find the final velocity, v_f, of the rock from $\bar{v}_f = at$;
2. calculate the average velocity (v) from the final velocity

$$\bar{v} = \frac{v_f + v_i}{2}$$

then;

3. use the average velocity ($\bar{v}$) and the time (t) to find distance $\bar{v}(d), d = \bar{v}t$.

Note that the one-step procedure is preferred over the three-step procedure because fewer steps mean fewer possibilities for mistakes.

Compound Motion

So far we have considered two types of motion: (1) the horizontal, straight-line motion of objects moving on the surface of the earth and (2) the vertical motion of dropped objects that accelerate toward the surface of the earth. A third type of motion occurs when an object is thrown, or projected, into the air. Essentially, such a projectile (rock, football, bullet, golf ball, or

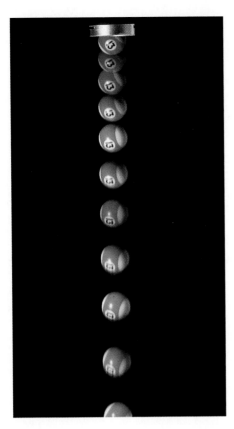

FIGURE 2.14

High-speed multiflash photograph of a freely falling billiard ball.

FIGURE 2.15

On its way up, a vertical projectile is slowed by the force of gravity until an instantaneous stop; then it accelerates back to the surface, just as another ball does when dropped from the same height. The straight up and down moving ball has been moved to the side in the sketch so we can see more clearly what is happening. Note that the falling ball has the same speed in the opposite direction that it had on the way up.

whatever) could be directed straight upward as a vertical projection, directed straight out as a horizontal projection, or directed at some angle between the vertical and the horizontal. Basic to understanding such compound motion is the observation that (1) gravity acts on objects *at all times,* no matter where they are, and (2) the acceleration due to gravity (g) is *independent of any motion* that an object may have.

Vertical Projectiles

Consider first a ball that you throw straight upward, a vertical projection. The ball has an initial velocity but then reaches a maximum height, stops for an instant, then accelerates back toward the earth. Gravity is acting on the ball throughout its climb, stop, and fall. As it is climbing, the force of gravity is accelerating it back to the earth. The overall effect during the climb is deceleration, which continues to slow the ball until the instantaneous stop. The ball then accelerates back to the surface just like a ball that has been dropped (Figure 2.14). If it were not for air resistance, the ball would return with the same speed in the opposite direction that it had initially. The velocity arrows for a ball thrown straight up are shown in Figure 2.15.

Horizontal Projectiles

Horizontal projections are easier to understand if you split the complete motion into vertical and horizontal parts. Con-

sider, for example, an arrow shot horizontally from a bow. The force of gravity accelerates the arrow downward, giving it an increasing downward velocity as it moves through the air. This increasing downward velocity is shown in Figure 2.16 as increasingly longer velocity arrows (v_v). There are no forces in the horizontal direction if you ignore air resistance, so the horizontal velocity of the arrow remains the same as shown by the v_h velocity arrows. The combination of the increasing vertical (v_v) motion and the unchanging horizontal (v_h) motion causes the arrow to follow a curved path until it hits the ground.

An interesting prediction that can be made from the shot arrow analysis is that an arrow shot horizontally from a bow will hit the ground at the same time as a second arrow that is simply dropped from the same height (Figure 2.16). Would this be true of a bullet dropped at the same time as one fired horizontally from a rifle? The answer is yes; both bullets would hit the

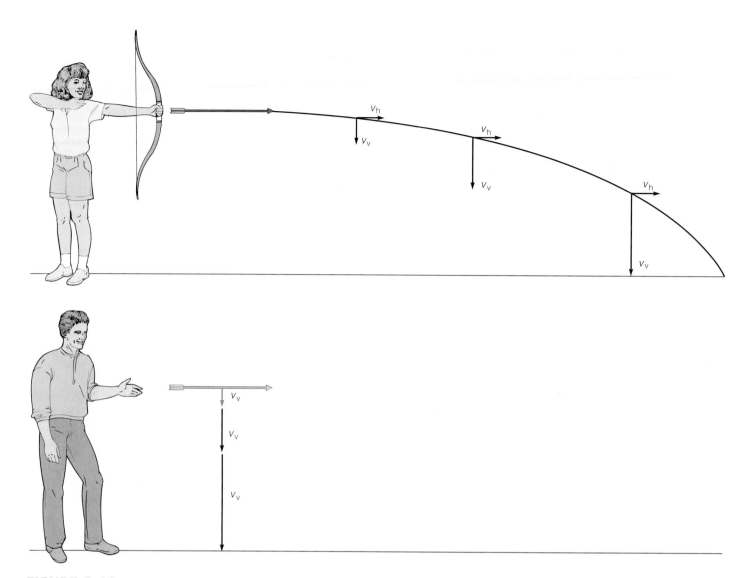

FIGURE 2.16

A horizontal projectile has the same horizontal velocity throughout the fall as it accelerates toward the surface, with the combined effect resulting in a curved path. Neglecting air resistance, an arrow shot horizontally will strike the ground at the same time as one dropped from the same height above the ground, as shown here by the increasing vertical velocity arrows.

ground at the same time. Indeed, if you ignore air resistance, all the bullets and arrows should hit the ground at the same time if dropped or shot from the same height.

Golf balls, footballs, and baseballs are usually projected upward at some angle to the horizon. The horizontal motion of these projectiles is constant as before because there are no horizontal forces involved. The vertical motion is the same as that of a ball projected directly upward. The combination of these two motions causes the projectile to follow a curved path called a *parabola,* as shown in Figure 2.17. The next time you have the opportunity, observe the path of a ball that has been projected at some angle. Note that the second half of the path is almost a reverse copy of the first half. If it were not for air resistance, the two values of the path would be exactly the same. Also note the distance that the ball travels as compared

to the angle of projection. An angle of projection of 45° results in the maximum distance of travel if air resistance is ignored and if the launch point and the landing are at the same elevation.

Three Laws of Motion

In the previous sections you learned how to describe motion in terms of distance, time, velocity, and acceleration. In addition, you learned about different kinds of motion, such as straight-line motion, the motion of falling objects, and the compound motion of objects projected up from the surface of the earth. You were also introduced, in general, to two concepts closely associated with motion: (1) that objects have inertia, a tendency

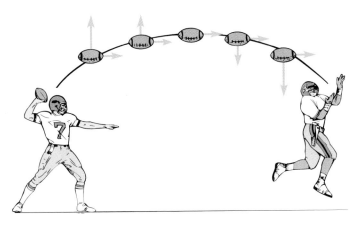

FIGURE 2.17

A football is thrown at some angle to the horizon when it is passed downfield. Neglecting air resistance, the horizontal velocity is a constant, and the vertical velocity decreases, then increases, just as in the case of a vertical projectile. The combined motion produces a parabolic path. Contrary to statements by sportscasters about the abilities of certain professional quarterbacks, it is impossible to throw a football with a "flat trajectory" because it begins to accelerate toward the surface as soon as it leaves the quarterback's hand.

to resist a change in motion, and (2) that forces are involved in a change of motion.

The relationship between forces and a change of motion is obvious in many everyday situations (Figure 2.18). When a car, bus, or plane starts moving, you feel a force on your back. Likewise, you feel a force on the bottoms of your feet when an elevator starts moving upward. On the other hand, you seem to be forced toward the dashboard if a car stops quickly, and it feels as if the floor pulls away from your feet when an elevator drops rapidly. These examples all involve patterns between forces and motion, patterns that can be quantified, conceptualized, and used to answer questions about why things move or stand still. These patterns are the subject of Newton's three laws of motion.

Newton's First Law of Motion

Newton's first law of motion is also known as the *law of inertia* and is very similar to one of Galileo's findings about motion. Recall that Galileo used the term *inertia* to describe the tendency of an object to resist changes in motion. Newton's first law describes this tendency more directly. In modern terms (not Newton's words), the **first law of motion** is as follows:

> **Every object retains its state of rest or its state of uniform straight-line motion unless acted upon by an unbalanced force.**

This means that an object at rest will remain at rest unless it is put into motion by an unbalanced force; that is, the net force must be greater than zero. Likewise, an object moving with uniform straight-line motion will retain that motion unless a net force causes it to speed up, slow down, or change its direction of travel. Thus, Newton's first law describes the tendency of an object to resist *any* change in its state of motion.

FIGURE 2.18

In a moving airplane, you feel forces in many directions when the plane changes its motion. You cannot help but notice the forces involved when there is a change of motion.

Think of Newton's first law of motion when you ride standing in the aisle of a bus. The bus begins to move, and you, being an independent mass, tend to remain at rest. You take a few steps back as you tend to maintain your position relative to the ground outside. You reach for a seat back or some part of the bus. Once you have a hold on some part of the bus it supplies the forces needed to give you the same motion as the bus and you no longer find it necessary to step backward. You now have the same motion as the bus, and no forces are involved, at least until the bus goes around a curve. You now feel a tendency to move to the side of the bus. The bus has changed its straight-line motion, but you, again being an independent mass, tend to move straight ahead. The side of the seat forces you into following the curved motion of the bus. The forces you feel when the bus starts moving or turning are a result of your tendency to remain at rest or follow a straight path until forces correct your motion so that it is the same as that of the bus (Figure 2.19).

Concepts Applied

First Law Experiment

Place a small ball on a flat part of the floor in a car, SUV, or pickup truck. First, predict what will happen to the ball in each of the following situations: (1) The vehicle moves forward from a stopped position. (2) The vehicle is moving at a constant speed. (3) The vehicle is moving at a constant speed, then turns to the right. (4) The vehicle is moving at a constant speed, then comes to a stop. Now, test your predictions, and then explain each finding in terms of Newton's first law of motion.

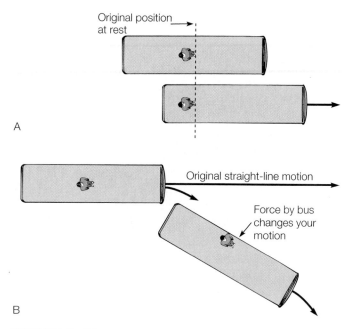

FIGURE 2.19

Top view of a person standing in the aisle of a bus. (*A*) The bus is at rest, and then starts to move forward. Inertia causes the person to remain in the original position, appearing to fall backward. (*B*) The bus turns to the right, but inertia causes the person to retain the original straight-line motion until forced in a new direction by the side of the bus.

Newton's Second Law of Motion

Newton had successfully used Galileo's ideas to describe the nature of motion. Newton's first law of motion explains that any object, once started in motion, will continue with a constant velocity in a straight line unless a force acts on the moving object. This law not only describes motion but establishes the role of a force as well. A change of motion is therefore *evidence* of the action of net force. The association of forces and a change of motion is common in your everyday experience. You have felt forces on your back in an accelerating automobile, and you have felt other forces as the automobile turns or stops. You have also learned about gravitational forces that accelerate objects toward the surface of the earth. Unbalanced forces and acceleration are involved in any change of motion. Newton's second law of motion is a relationship between *net force, acceleration,* and *mass* that describes the cause of a change of motion.

Consider the motion of you and a bicycle you are riding. Suppose you are riding your bicycle over level ground in a straight line at 10 miles per hour. Newton's first law tells you that you will continue with a constant velocity in a straight line as long as no external, unbalanced force acts on you and the bicycle. The force that you *are* exerting on the pedals seems to equal some external force that moves you and the bicycle along (more on this later). The force exerted as you move along is needed to *balance* the resisting forces of tire friction and air

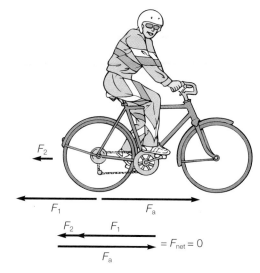

FIGURE 2.20

At a constant velocity the force of tire friction (F_1) and the force of air resistance (F_2) have a sum that equals the force applied (F_a). The net force is therefore 0.

resistance. If these resisting forces were removed you would not need to exert any force at all to continue moving at a constant velocity. The net force is thus the force you are applying minus the forces from tire friction and air resistance. The *net force* is therefore zero when you move at a constant speed in a straight line (Figure 2.20).

If you now apply a greater force on the pedals the *extra* force you apply is unbalanced by friction and air resistance. Hence there will be a net force greater than zero, and you will accelerate. You will accelerate during, and *only* during, the time that the net force is greater than zero. Likewise, you will slow down if you apply a force to the brakes, another kind of resisting friction. A third way to change your velocity is to apply a force on the handlebars, changing the direction of your velocity. Thus, *unbalanced forces* on you and your bicycle produce an *acceleration.*

Starting a bicycle from rest suggests a relationship between force and acceleration. You observe that the harder you push on the pedals, the greater your acceleration. Recall that when quantities increase or decrease together in the same ratio, they are said to be *directly proportional.* The acceleration is therefore directly proportional to the net force applied.

Suppose that your bicycle has two seats, and you have a friend who will ride with you. Suppose also that the addition of your friend on the bicycle will double the mass of the bike and riders. If you use the same net force as before, the bicycle will undergo a much smaller acceleration. In fact, with all other factors equal, doubling the mass and applying the same extra force will produce an acceleration of only half as much (Figure 2.21). An even more massive friend would reduce the acceleration even more. Recall that when a relationship between two quantities shows that one quantity increases as another decreases, in the same ratio, the quantities are said to be

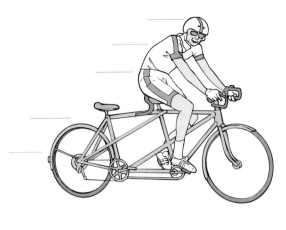

FIGURE 2.21
More mass results in less acceleration when the same force is applied. With the same force applied, the riders and bike with twice the mass will have half the acceleration, with all other factors constant. Note that the second rider is not pedaling.

inversely proportional. The acceleration of an object is therefore inversely proportional to its mass.

If we express force in appropriate units we can combine these relationships as an equation,

$$a = \frac{F}{m}$$

By solving for F we rearrange the equation into the form in which it is most often expressed,

$$F = ma$$

equation 2.5

In the metric system you can see that the units for force will be the units for mass (m) times acceleration (a). The unit for mass is kg and the unit for acceleration is m/s². The combination of these units, (kg)(m/s²), is a unit of force called the **newton** (N) in honor of Isaac Newton. So,

$$1 \text{ newton} = 1 \text{ N} = 1 \frac{\text{kg} \cdot \text{m}}{\text{s}^2}$$

Newton's second law of motion is the essential idea of his work on motion. According to this law there is always a relationship between the acceleration, a net force, and the mass of an object. Implicit in this statement are two understandings, (1) that we are talking about the net force, meaning total external force acting on an object, and (2) that the motion statement is concerned with acceleration, not velocity.

The acceleration of an object depends on *both* the *net force applied* and the *mass* of the object. The **second law of motion** is as follows:

> **The acceleration of an object is directly proportional to the net force acting on it and inversely proportional to the mass of the object.**

Until now, equations were used to *describe properties* of matter such as density, velocity, and acceleration. This is your first example of an equation that is used to *define a concept*, specifically the concept of what is meant by a force. Since the concept is defined by specifying a measurement procedure, it is also an example of an *operational definition.* You are told not only what a newton of force is but also how to go about measuring it. Notice that the newton is defined in terms of mass measured in kg and acceleration measured in m/s². Any other units must be converted to kg and m/s² before a problem can be solved for newtons of force.

EXAMPLE 2.8

A 60 kg bicycle and rider accelerate at 0.5 m/s². How much extra force was applied?

Solution

The mass (m) of 60 kg and the acceleration (a) of 0.5 m/s² are given. The problem asked for the extra force (F) needed to give the mass the acquired acceleration. The relationship is found in equation 2.5, $F = ma$.

$$m = 60 \text{ kg}$$
$$a = 0.5 \frac{\text{m}}{\text{s}^2}$$
$$F = ?$$

$$F = ma$$
$$= (60 \text{ kg})\left(0.5 \frac{\text{m}}{\text{s}^2}\right)$$
$$= (60)(0.5)(\text{kg})\left(\frac{\text{m}}{\text{s}^2}\right)$$
$$= 30 \frac{\text{kg} \cdot \text{m}}{\text{s}^2}$$
$$= \boxed{30 \text{ N}}$$

An *extra* force of 30 N beyond that required to maintain constant speed must be applied to the pedals for the bike and rider to maintain an acceleration of 0.5 m/s². (Note that the units kg·m/s² form the definition of a newton of force, so the symbol N is used.)

EXAMPLE **2.9**

What is the acceleration of a 20 kg cart if the net force on it is 40 N? (Answer: 2 m/s²)

Concepts Applied

Second Law Experiment

Tie one end of a string to a book and the other end to a large rubber band. With your index finger curled in the loop of the rubber band, pull the book across a smooth tabletop. How much the rubber band stretches will provide a rough estimate of the force you are applying. (1) Pull the book with a constant velocity across the tabletop. Compare the force required for different constant velocities. (2) Accelerate the book at different rates. Compare the force required to maintain the different accelerations. (3) Use a different book with a greater mass and again accelerate the book at different rates. How does more mass change the results?

Based on your observations, can you infer a relationship between force, acceleration, and mass?

Weight and Mass

What is the meaning of weight—is it the same concept as mass? Weight is a familiar concept to most people, and in everyday language the word is often used as having the same meaning as mass. In physics, however, there is a basic difference between weight and mass, and this difference is very important in Newton's explanation of motion and the causes of motion.

Mass is defined as the property that determines how much an object resists a change in its motion. The greater the mass the greater the *inertia,* or resistance to change in motion. Consider, for example, that it is easier to push a small car into motion than to push a large truck into motion. The truck has more mass, and therefore more inertia. Newton originally defined mass as the "quantity of matter" in an object, and this definition is intuitively appealing. However, Newton needed to measure inertia because of its obvious role in motion, and he redefined mass as a measure of inertia.

You could use Newton's second law to measure a mass by exerting a force on the mass and measuring the resulting acceleration. This is not very convenient, so masses are usually measured on a balance by comparing the force of gravity acting on a standard mass compared to the force of gravity acting on the unknown mass.

The force of gravity acting on a mass is the *weight* of an object. Weight is a force and has different units (N) than mass (kg). Since weight is a measure of the force of gravity acting on an object, the force can be calculated from Newton's second law of motion,

$$F = ma$$

or

downward force = (mass)(acceleration due to gravity)

or

$$weight = (mass)(g)$$

$$or \quad w = mg$$

equation 2.6

You learned in a previous section that g is the symbol used to represent acceleration due to gravity. Near the earth's surface, g has an approximate value of 9.8 m/s². To understand how g is applied to an object that is not moving, consider a ball you are holding in your hand. By supporting the weight of the ball you hold it stationary, so the upward force of your hand and the downward force of the ball (its weight) must add to a net force of zero. When you let go of the ball the gravitational force is the only force acting on the ball. The ball's weight is then the net force that accelerates it at g, the acceleration due to gravity. Thus, $F_{net} = w = ma = mg$. The weight of the ball never changes in a given location, so its weight is always equal to $w = mg$, even if the ball is not accelerating.

In the metric system, *mass* is measured in kilograms. The acceleration due to gravity, g, is 9.8 m/s². According to equation 2.6, weight is mass times acceleration. A kilogram multiplied by an acceleration measured in m/s² results in kg·m/s², a unit you now recognize as a force called a newton. The *unit of weight* in the metric system is therefore the *newton (N)*.

In the English system the pound is the unit of *force*. The acceleration due to gravity, g, is 32 ft/s². The force unit of a pound is defined as the force required to accelerate a unit of mass called the *slug*. Specifically, a force of 1.0 lb will give a 1.0 slug mass an acceleration of 1.0 ft/s².

The important thing to remember is that *pounds* and *newtons* are units of *force* (Table 2.1). A *kilogram,* on the other hand, is a measure of *mass*. Thus the English unit of 1.0 lb is comparable to

TABLE 2.1

Units of mass and weight in the metric and English systems of measurement

	Mass	×	Acceleration	=	Force
Metric System	kg	×	$\frac{m}{s^2}$	=	$\frac{kg \cdot m}{s^2}$ (newton)
English System	$\left(\frac{lb}{ft/s^2}\right)$	×	$\frac{ft}{s^2}$	=	lb (pound)

the metric unit of 4.5 N (or 0.22 lb is equivalent to 1.0 N). Conversion tables sometimes show how to convert from pounds (a unit of weight) to kilograms (a unit of mass). This is possible because weight and mass are proportional in a given location on the surface of the earth. Using conversion factors from inside the front cover of this book, see if you can express your weight in pounds and newtons and your mass in kg.

EXAMPLE 2.10

What is the weight of a 60.0 kg person on the surface of the earth?

Solution

A mass (m) of 60.0 kg is given, and the acceleration due to gravity (g) 9.8 m/s^2 is implied. The problem asked for the weight (w). The relationship is found in equation 2.6, $w = mg$, which is a form of $F = ma$.

$$m = 60.0 \ kg$$
$$g = 9.8\frac{m}{s^2}$$
$$w = ?$$

$$w = mg$$
$$= (60.0 \ kg)\left(9.8\frac{m}{s^2}\right)$$
$$= (60.0)(9.8) \ (kg)\left(\frac{m}{s^2}\right)$$
$$= 588 \ \frac{kg \cdot m}{s^2}$$
$$= \boxed{590 \ N}$$

EXAMPLE 2.11

A 60.0 kg person weighs 100.0 N on the moon. What is the value of g on the moon? (Answer: 1.67 m/s^2)

Concepts Applied

Apparent Weightlessness

Use a sharp pencil to make a small hole in the bottom of a Styrofoam cup. The hole should be large enough for a thin stream of water to flow from the cup, but small enough for the flow to continue for 3 or 4 seconds. Test the water flow over a sink.

Hold a finger over the hole in the cup as you fill it with water. Stand on a ladder or outside stairwell as you hold the cup out at arm's length. Move your finger, allowing a stream of water to flow from the cup, and at the same time drop the cup. Observe what happens to the stream of water as the cup is falling. Explain your observations. Also predict what you would see if you were falling with the cup.

Newton's Third Law of Motion

Newton's first law of motion states that an object retains its state of motion when the net force is zero. The second law states what happens when the net force is *not* zero, describing how an object with a known mass moves when a given force is applied. The two laws give one aspect of the concept of a force; that is, if you observe that an object starts moving, speeds up, slows down, or changes its direction of travel, you can conclude that an unbalanced force is acting on the object. Thus, any change in the state of motion of an object is *evidence* that an unbalanced force has been applied.

Newton's third law of motion is also concerned with forces and considers how a force is produced. First, consider where a force comes from. A force is always produced by the interaction of two or more objects. There is always a second object pushing or pulling on the first object to produce a force. To simplify the many interactions that occur on the earth, consider a satellite in space. According to Newton's second law ($F = ma$), a force must be applied to change the state of motion of the satellite. What is a possible source of such a force? Perhaps an astronaut pushes on the satellite for 1 second. The satellite would accelerate *during* the application of the force, then move away from the original position at some constant velocity. The astronaut would also move away from the original position, but in the opposite direction (Figure 2.22). A *single* force *does not exist* by itself. There is always a matched and opposite force that occurs at the same time. Thus, the astronaut exerted a momentary force on the satellite, but the satellite evidently exerted a momentary force back on the astronaut as well, for the astronaut moved away from the original position in the opposite direction. Newton did not have astronauts and satellites to think about, but this is the kind of reasoning he did when he concluded that forces always occur in matched pairs that are equal and opposite. Thus the **third law of motion** is as follows:

> Whenever two objects interact, the force exerted on one object is equal in size and opposite in direction to the force exerted on the other object.

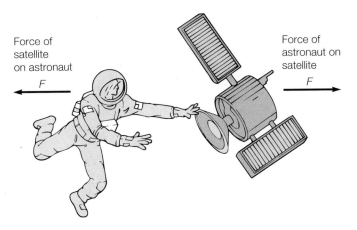

Force of satellite on astronaut

F

Force of astronaut on satellite

F

FIGURE 2.22
Forces occur in matched pairs that are equal in magnitude and opposite in direction.

FIGURE 2.23
The football player's foot is pushing against the ground, but it is the ground pushing against the foot that accelerates the player forward to catch a pass.

Perhaps it would be more common to move a satellite with a small rocket. A satellite is maneuvered in space by firing a rocket in the direction opposite to the direction someone wants to move the satellite. Exhaust gases (or compressed gases) are accelerated in one direction and exert an equal but opposite force on the satellite that accelerates it in the opposite direction. This is another example of the third law.

Consider how the pairs of forces work on the earth's surface. You walk by pushing your feet against the ground (Figure 2.23). Of course you could not do this if it were not for friction. You would slide as on slippery ice without friction. But since friction does exist, you exert a backward horizontal force on the ground, and, as the third law explains, the ground exerts an equal and opposite force on you. You accelerate forward from the net force as explained by the second law. If the earth had the same mass as you, however, it would accelerate backward at the same rate that you were accelerated forward. The earth is much more massive than you, however, so any acceleration of the earth is a vanishingly small amount. The overall effect is that you are accelerated forward by the force the ground exerts on you.

Return now to the example of riding a bicycle that was discussed previously. What is the source of the *external* force that accelerates you and the bike? Pushing against the pedals is not external to you and the bike, so that force will *not* accelerate you and the bicycle forward. This force is transmitted through the bike mechanism to the rear tire, which pushes against the ground. It is the ground exerting an equal and opposite force against the system of you and the bike that accelerates you forward. You must consider the forces that act on the system of the bike and you before you can apply $F = ma$. The only forces that will affect the forward motion of the bike system are the force of the ground pushing it forward and the frictional forces that oppose the forward motion. This is another example of the third law.

The third law states that forces always occur in matched pairs that act in opposite directions and on two *different* bodies. You could express this law with symbols as

$$F_{A \text{ due to } B} = F_{B \text{ due to } A}$$

equation 2.7

where the force on the astronaut, for example, would be "A due to B," and the force on the satellite would be "B due to A."

Sometimes the third law of motion is expressed as follows: "For every action there is an equal and opposite reaction," but this can be misleading. Neither force is the cause of the other. The forces are at every instant the cause of each other and they appear and disappear at the same time. If you are going to describe the force exerted on a satellite by an astronaut, then you must realize that there is a simultaneous force exerted on the astronaut by the satellite. The forces (astronaut on satellite and satellite on astronaut) are equal in magnitude but opposite in direction.

EXAMPLE 2.12

A 60.0 kg astronaut is freely floating in space and pushes on a freely floating 120.0 kg satellite with a force of 30.0 N for 1.50 s. (a) Compare the forces exerted on the astronaut and the satellite, and (b) compare the acceleration of the astronaut to the acceleration of the satellite.

Solution

(a) According to Newton's third law of motion (equation 2.7),

$$F_{A \text{ due to } B} = F_{B \text{ due to } A}$$
$$30.0 \text{ N} = 30.0 \text{ N}$$

Both feel a 30.0 N force for 1.50 s but in opposite directions.

(b) Newton's second law describes a relationship between force, mass, and acceleration, $F = ma$.

For the astronaut:

$m = 60.0 \text{ kg}$

$F = 30.0 \text{ N}$

$a = ?$

$F = ma \quad \therefore \quad a = \dfrac{F}{m}$

$$a = \dfrac{30.0 \, \frac{\text{kg·m}}{\text{s}^2}}{60.0 \text{ kg}}$$

$$= \dfrac{30.0}{60.0} \left(\dfrac{\text{kg·m}}{\text{s}^2} \right)\left(\dfrac{1}{\text{kg}} \right)$$

$$= 0.500 \dfrac{\text{kg·m}}{\text{kg·s}^2} = \boxed{0.500 \, \dfrac{\text{m}}{\text{s}^2}}$$

For the satellite:

$m = 120.0 \text{ kg}$

$F = 30.0 \text{ N}$

$a = ?$

$F = ma \quad \therefore \quad a = \dfrac{F}{m}$

$$a = \dfrac{30.0 \, \frac{\text{kg·m}}{\text{s}^2}}{120.0 \text{ kg}}$$

$$= \dfrac{30.0}{120.0} \left(\dfrac{\text{kg·m}}{\text{s}^2} \right)\left(\dfrac{1}{\text{kg}} \right)$$

$$= 0.250 \dfrac{\text{kg·m}}{\text{kg·s}^2} = \boxed{0.250 \, \dfrac{\text{m}}{\text{s}^2}}$$

EXAMPLE 2.13

After the interaction and acceleration between the astronaut and satellite described previously, they both move away from their original positions. What is the new speed for each? (Answer: Astronaut $v_f = 0.750 \text{ m/s}$. Satellite $v_f = 0.375 \text{ m/s}$) (Hint: $v_f = at + v_i$)

Concepts Applied

Laws of Motion in the Kitchen

Relationships between variables involved in the second and third laws of motion can be studied with some common materials. Obtain a plastic 35 mm film container, an Alka-Seltzer tablet, and a large metal cookie sheet.

Procedure: Add enough cool water to about half fill a plastic 35 mm film container. Place 1/4 Alka-Seltzer tablet in the water and quickly snap on the lid. Now quickly place the container on its side at the center of a large metal cookie sheet.

Analysis: Explain the result in terms of the third law of motion. Compare the mass and evidence of the acceleration of the lid and container, explaining the result in terms of the second law of motion.

Momentum

Sportscasters often refer to the *momentum* of a team, and newscasters sometimes refer to an election where one of the candidates has *momentum*. Both situations describe a competition where one side is moving toward victory and it is difficult to stop. It seems appropriate to borrow this term from the physical sciences because momentum is a property of movement. It takes a longer time to stop something from moving when it has a lot of momentum. The physical science concept of momentum is closely related to Newton's laws of motion. **Momentum** (p) is defined as the product of the mass (m) of an object and its velocity (v),

$$\text{momentum} = \text{mass} \times \text{velocity}$$

or

$$p = mv$$

equation 2.8

The astronaut in example 2.12 had a mass of 60.0 kg and a velocity of 0.750 m/s as a result of the interaction with the satellite. The resulting momentum was therefore (60.0 kg) (0.750 m/s), or 45.0 kg·m/s. As you can see, the momentum would be greater if the astronaut had acquired a greater velocity or if the astronaut had a greater mass and acquired the same velocity. Momentum involves both the inertia and the velocity of a moving object.

Conservation of Momentum

Notice that the momentum acquired by the satellite in example 2.12 is *also* 45.0 kg·m/s. The astronaut gained a certain momentum in one direction, and the satellite gained the *very same momentum in the opposite direction*. Newton originally defined the second law in terms of a rate of change of momentum being proportional to the net force acting on an object. Since the third law explains that the forces exerted on both the astronaut and satellite were equal and opposite, you would expect both objects to acquire equal momentum in the opposite direction. This result is observed any time objects in a system interact and the only forces involved are those between the interacting objects (Figure 2.24). This statement leads to a particular kind of relationship called a *law of conservation*. In this case, the law applies to momentum and is called the **law of conservation of momentum:**

> **The total momentum of a group of interacting objects remains the same in the absence of external forces.**

Conservation of momentum, energy, and charge are among examples of conservation laws that apply to everyday situations. These situations always illustrate two understandings, that (1) each conservation law is an expression of symmetry that describes a physical principle that can be observed; and, (2) each law holds regardless of the details of an interaction or how it took place. Since the conservation laws express symmetry that always occurs, they tell us what might be expected to happen, and what might be expected not to happen in a given situation. The symmetry also allows unknown quantities to be found by analysis.

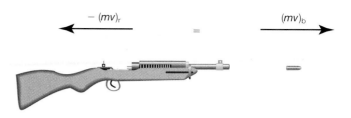

$-(mv)_r$ $=$ $(mv)_b$

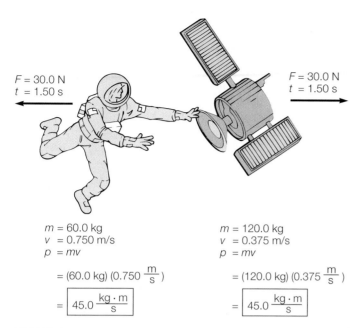

$F = 30.0$ N
$t = 1.50$ s

$F = 30.0$ N
$t = 1.50$ s

$m = 60.0$ kg
$v = 0.750$ m/s
$p = mv$

$= (60.0 \text{ kg}) (0.750 \frac{m}{s})$

$= \boxed{45.0 \frac{kg \cdot m}{s}}$

$m = 120.0$ kg
$v = 0.375$ m/s
$p = mv$

$= (120.0 \text{ kg}) (0.375 \frac{m}{s})$

$= \boxed{45.0 \frac{kg \cdot m}{s}}$

FIGURE 2.24

Both the astronaut and the satellite received a force of 30.0 N for 1.50 s when they pushed on each other. Both then have a momentum of 45.0 kg·m/s in the opposite direction. This is an example of the law of conservation of momentum.

The law of conservation of momentum, for example, is useful in analyzing motion in simple systems of collisions such as those of billiard balls, automobiles, or railroad cars. It is also useful in measuring action and reaction interactions, as in rocket propulsion, where the backward momentum of the exhaust gases equals the momentum given to the rocket in the opposite direction. When this is done, momentum is always found to be conserved.

The firing of a bullet from a rifle and the concurrent "kick" or recoil of the rifle is often used as an example of conservation of momentum where the interaction between objects results in momentum in opposite directions (Figure 2.25). When the rifle is fired, the exploding gunpowder propels the bullet with forward momentum. At the same time, the force from the exploding gunpowder pushes the rifle backward with a momentum opposite that of the bullet. The bullet moves forward with a momentum of $(mv)_b$ and the rifle moves in an opposite direction to the bullet, so its momentum is $-(mv)_r$. According to the law of conservation of momentum, the momentum of the bullet $(mv)_b$ must equal the momentum of the rifle $-(mv)_r$ in the opposite direction. If the bullet and rifle had the same mass, they would each move with equal velocities when the rifle was fired. The rifle is much more massive than a bullet, however, so the bullet has a much greater velocity than the rifle. The momentum of the rifle is nonetheless equal to the momentum of the bullet, and the recoil can be significant if the rifle is not held firmly against the shoulder. When held firmly against the shoulder, the rifle and the person's body are one object. The increased mass results in a proportionally smaller recoil velocity.

FIGURE 2.25

A rifle and bullet provide an example of conservation of momentum. Before being fired, a rifle and bullet have a total momentum ($p = mv$) of zero since there is no motion. When fired, the bullet is then propelled in one direction with a forward momentum $(mv)_b$. At the same time, the rifle is pushed backward with a momentum opposite to that of the bullet, so its momentum is shown with a minus sign, or $-(mv)_r$. Since $(mv)_b$ plus $-(mv)_r$ equals zero, the total momentum of the rifle and bullet is zero after as well as before the rifle is fired.

EXAMPLE 2.14

A 20,000 kg railroad car is coasting at 3 m/s when it collides and couples with a second, identical car at rest. What is the resulting speed of the combined cars?

Solution

Moving car → $m_1 = 20,000$ kg, $v_1 = 3$ m/s
Second car → $m_2 = 20,000$ kg, $v_2 = 0$
Combined cars → $v_{1\&2} = ?$ m/s

Since momentum is conserved, the total momentum of the cars should be the same before and after the collision. Thus

$$\text{momentum before} = \text{momentum after}$$
$$\text{car 1} + \text{car 2} = \text{coupled cars}$$
$$m_1 v_1 + m_2 v_2 = (m_1 + m_2) v_{1\&2}$$
$$v_{1\&2} = \frac{m_1 v_1}{(m_1 + m_2)}$$

$$v_{1\&2} = \frac{(20,000 \text{ kg})\left(3 \frac{m}{s}\right)}{(20,000 \text{ kg}) + (20,000 \text{ kg})}$$

$$= \frac{20,000 \text{ kg} \cdot 3 \frac{m}{s}}{40,000 \text{ kg}}$$

$$= 0.5 \times 3 \frac{kg \cdot m}{s} \times \frac{1}{kg}$$

$$= 1.5 \frac{m}{s}$$

$$= \boxed{2 \frac{m}{s}}$$

(Answer rounded to one significant figure.)

Car 2 had no momentum with a velocity of zero, so m_2v_2 on the left side of the equation equals zero. When the cars couple the mass is doubled $(m + m)$, and the velocity of the coupled cars will be 2 m/s.

EXAMPLE 2.15

A student and her rowboat have a combined mass of 100.0 kg. Standing in the motionless boat in calm water, she tosses a 5.0 kg rock out the back of the boat with a velocity of 5.0 m/s. What will be the resulting speed of the boat? (Answer: 0.25 m/s)

Impulse

Have you ever heard that you should "follow through" when hitting a ball? When you follow through, the bat is in contact with the ball for a longer period of time. The force of the hit is important, of course, but both the force and how long the force is applied determine the result. The product of the force and the time of application is called **impulse**. This quantity can be expressed as

$$impulse = Ft$$

where F is the force applied during the time of contact t. The impulse you give the ball determines how fast the ball will move, and thus how far it will travel.

Impulse is related to the change of motion of a ball of a given mass, so the change of momentum (mv) is brought about by the impulse. This can be expressed as

$$change\ of\ momentum = (applied\ force)(time\ of\ contact)$$
$$\Delta p = Ft$$

equation 2.9

where Δp is a change of momentum. You "follow through" while hitting a ball in order to increase the contact time. If the same force is used, a longer contact time will result in a greater impulse. A greater impulse means a greater change of momentum, and since the mass of the ball does not change, the overall result is a moving ball with a greater velocity. This means following through will result in more distance from hitting the ball with the same force. That's why it is important to follow through when you hit the ball.

Now consider bringing a moving object to a stop by catching it. In this case the mass and the velocity of the object are fixed at the time you catch it, and there is nothing you can do about these quantities. The change of momentum is equal to the impulse, and the force and time of force application *can* be manipulated. For example, consider how you would catch a raw egg that is tossed to you. You would probably move your hands with the egg as you caught it, increasing the contact time.

Increasing the contact time has the effect of reducing the force since $\Delta p = Ft$. You change the force applied by increasing the contact time, and hopefully you reduce the force sufficiently so the egg does not break.

Contact time is also important in safety. Automobile airbags, the padding in elbow and knee pads, and the plastic barrels off the highway in front of overpass supports are examples of designs intended to increase the contact time. Again, increasing the contact time reduces the force since $\Delta p = Ft$. The impact force is reduced and so are the injuries. Think about this the next time you see a car that was crumpled and bent by a collision. The driver and passengers were probably saved from more serious injuries since more time was involved in stopping the car that crumpled. A car that crumples is a safer car in a collision.

Concepts Applied

Momentum Experiment

The popular novelty item of a frame with five steel balls hanging from strings can be used to observe momentum exchanges during elastic collisions. When one ball is pulled back and released, it stops as it transfers its momentum to the ball it strikes and the momentum is transferred ball to ball until the end ball swings out. Make some predictions, then do the experiment for the following. What will happened when: (1) Two balls are released together on one side. (2) One ball on each side is released at the same time. (3) Two balls on one side are released together as two balls are simultaneously released together on the other side. (4) Two balls on one side are released together as a single ball is simultaneously released on the other side. Analyze the momentum transfers down the line for each situation.

As an alternative to the use of the swinging balls, consider a similar experiment using a line of marbles in contact with each other in a grooved ruler. Here, you could also vary the mass of marbles in collisions.

Forces and Circular Motion

Consider a communications satellite that is moving at a uniform speed around the earth in a circular orbit. According to the first law of motion there *must be* forces acting on the satellite, since it does *not* move off in a straight line. The second law of motion also indicates forces, since an unbalanced force is required to change the motion of an object.

Recall that acceleration is defined as a change in velocity, and that velocity has both magnitude and direction. The velocity is changed by a change in speed, direction, or both speed and direction. The satellite in a circular orbit is continuously being accelerated. This means that there is a continuously acting

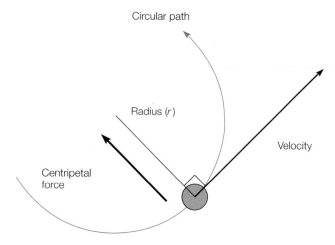

FIGURE 2.26

Centripetal force on the ball causes it to change direction continuously, or accelerate into a circular path. Without the unbalanced force acting on it, the ball would continue in a straight line.

unbalanced force on the satellite that pulls it out of a straight-line path.

The force that pulls an object out of its straight-line path and into a circular path is called a **centripetal** (center-seeking) **force**. Perhaps you have swung a ball on the end of a string in a horizontal circle over your head. Once you have the ball moving, the only unbalanced force (other than gravity) acting on the ball is the centripetal force your hand exerts on the ball through the string. This centripetal force pulls the ball from its natural straight-line path into a circular path. There are no outward forces acting on the ball. The force that you feel on the string is a consequence of the third law; the ball exerts an equal and opposite force on your hand. If you were to release the string, the ball would move away from the circular path in a *straight line* that has a right angle to the radius at the point of release (Figure 2.26). When you release the string, the centripetal force ceases, and the ball then follows its natural straight-line motion. If other forces were involved, it would follow some other path. Nonetheless, the apparent outward force has been given a name just as if it were a real force. The outward tug is called a **centrifugal force.**

The magnitude of the centripetal force required to keep an object in a circular path depends on the inertia, or mass, of the object and the acceleration of the object, just as you learned in the second law of motion. The acceleration of an object moving in a circle can be shown by geometry or calculus to be directly proportional to the square of the speed around the circle (v^2) and inversely proportional to the radius of the circle (r). (A smaller radius requires a greater acceleration.) Therefore, the acceleration of an object moving in uniform circular motion (a_c) is

$$a_c = \frac{v^2}{r}$$

equation 2.10

The magnitude of the centripetal force of an object with a mass (m) that is moving with a velocity (v) in a circular orbit of a radius (r) can be found by substituting equation 3.5 in $F = ma$, or

$$F = \frac{mv^2}{r}$$

equation 2.11

EXAMPLE 2.16

A 0.25 kg ball is attached to the end of a 0.5 m string and moved in a horizontal circle at 2.0 m/s. What net force is needed to keep the ball in its circular path?

Solution

$$m = 0.25 \text{ kg} \qquad F = \frac{mv^2}{r}$$

$$r = 0.5 \text{ m}$$

$$v = 2.0 \text{ m/s} \qquad = \frac{(0.25 \text{ kg})(2.0 \text{ m/s})^2}{0.5 \text{ m}}$$

$$F = ? \qquad = \frac{(0.25 \text{ kg})(4.0 \text{ m}^2/\text{s}^2)}{0.5 \text{ m}}$$

$$= \frac{(0.25)(4.0)}{0.5} \frac{\text{kg·m}^2}{\text{s}^2} \times \frac{1}{\text{m}}$$

$$= 2\frac{\text{kg·m}^2}{\text{m·s}^2}$$

$$= 2\frac{\text{kg·m}}{\text{s}^2}$$

$$= \boxed{2 \text{ N}}$$

EXAMPLE 2.17

Suppose you make the string in example 2.16 half as long, 0.25 m. What force is now needed? (Answer: 4.0 N)

Newton's Law of Gravitation

You know that if you drop an object, it always falls to the floor. You define *down* as the direction of the object's movement and *up* as the opposite direction. Objects fall because of the force of gravity, which accelerates objects at $g = 9.8$ m/s^2 (32 ft/s^2) and gives them weight, $w = mg$.

Gravity is an attractive force, a pull that exists between all objects in the universe. It is a mutual force that, just like all other forces, comes in matched pairs. Since the earth attracts you with a certain force, you must attract the earth with an exact opposite force. The magnitude of this force of mutual

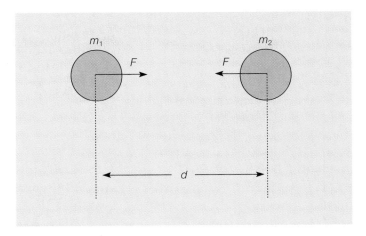

FIGURE 2.27
The variables involved in gravitational attraction. The force of attraction (F) is proportional to the product of the masses (m_1, m_2) and inversely proportional to the square of the distance (d) between the centers of the two masses.

attraction depends on several variables. These variables were first described by Newton in *Principia*, his famous book on motion that was printed in 1687. Newton had, however, worked out his ideas much earlier, by the age of twenty-four, along with ideas about his laws of motion and the formula for centripetal acceleration. In a biography written by a friend in 1752, Newton stated that the notion of gravitation came to mind during a time of thinking that "was occasioned by the fall of an apple." He was thinking about why the Moon stays in orbit around Earth rather than moving off in a straight line as would be predicted by the first law of motion. Perhaps the same force that attracts the moon toward the earth, he thought, attracts the apple to the earth. Newton developed a theoretical equation for gravitational force that explained not only the motion of the moon but the motion of the whole solar system. Today, this relationship is known as the **universal law of gravitation:**

> **Every object in the universe is attracted to every other object with a force that is directly proportional to the product of their masses and inversely proportional to the square of the distances between them.**

In symbols, m_1 and m_2 can be used to represent the masses of two objects, d the distance between their centers, and G a constant of proportionality. The equation for the law of universal gravitation is therefore

$$F = G\frac{m_1m_2}{d^2}$$

equation 2.12

This equation gives the magnitude of the attractive force that each object exerts on the other. The two forces are oppositely directed. The constant G is a universal constant, since the law applies to all objects in the universe. It was first measured experimentally by Henry Cavendish in 1798. The accepted value today is $G = 6.67 \times 10^{-11}\,\text{N}\cdot\text{m}^2/\text{kg}^2$. Do not confuse G, the universal constant, with g, the acceleration due to gravity on the surface of the earth.

Thus, the magnitude of the force of gravitational attraction is determined by the mass of the two objects and the distance between them (Figure 2.27). The law also states that *every* object is attracted to every other object. You are attracted to all the objects around you—chairs, tables, other people, and so forth. Why don't you notice the forces between you and other objects? The answer is in example 2.18.

EXAMPLE 2.18

What is the force of gravitational attraction between two 60.0 kg (132 lb) students who are standing 1.00 m apart?

Solution

$$G = 6.67 \times 10^{-11}\,\text{N}\cdot\text{m}^2/\text{kg}^2$$
$$m_1 = 60.0\,\text{kg}$$
$$m_2 = 60.0\,\text{kg}$$
$$d = 1.00\,\text{m}$$
$$F = ?$$

$$F = G\frac{m_1m_2}{d^2}$$

$$= \frac{(6.67 \times 10^{-11}\,\text{N}\cdot\text{m}^2/\text{kg}^2)(60.0\,\text{kg})(60.0\,\text{kg})}{(1.00\,\text{m})^2}$$

$$= (6.67 \times 10^{-11})(3.60 \times 10^3)\,\frac{\text{N}\cdot\text{m}^2\cdot\cancel{\text{kg}^2}}{\cancel{\text{kg}^2}}}{\text{m}^2}$$

$$= 2.40 \times 10^{-7}\,(\text{N}\cdot\text{m}^2)\left(\frac{1}{\text{m}^2}\right)$$

$$= 2.40 \times 10^{-7}\,\frac{\text{N}\cdot\cancel{\text{m}^2}}{\cancel{\text{m}^2}}$$

$$= \boxed{2.40 \times 10^{-7}\,\text{N}}$$

(Note: A force of 2.40×10^{-7} (0.00000024) N is equivalent to a force of 5.40×10^{-8} lb (0.00000005 lb), a force that you would not notice. In fact, it would be difficult to measure such a small force.)

EXAMPLE 2.19

What would be the value of g if the earth were less dense, with the same mass and double the radius? (Answer: $g = 2.45\,\text{m/s}^2$)

As you can see in example 2.18, one or both of the interacting objects must be quite massive before a noticeable force results from the interaction. That is why you do not notice the force of gravitational attraction between you and objects that are not very massive compared to the earth. The attraction between you and the earth overwhelmingly predominates, and that is all you notice.

Newton was able to show that the distance used in the equation is the distance from the center of one object to the center of the second object. This does not mean that the force originates at the center, but that the overall effect is the same as if you considered all the mass to be concentrated at a center point. The weight of an object, for example, can be calculated by using a form of Newton's second law, $F = ma$. This general law shows a relationship between *any* force acting *on* a body, the mass of a body, and the resulting acceleration. When the acceleration is due to gravity, the equation becomes $F = mg$. The law of gravitation deals *specifically with the force of gravity* and how it varies with distance and mass. Since weight is a force, then $F = mg$. You can write the two equations together,

$$mg = G\frac{mm_e}{d^2}$$

where m is the mass of some object on earth, m_e is the mass of the earth, g is the acceleration due to gravity, and d is the distance between the centers of the masses. Canceling the m's in the equation leaves

$$g = G\frac{m_e}{d^2}$$

which tells you that on the surface of the earth the acceleration due to gravity, 9.8 m/s², is a constant because the other two variables (mass of the earth and the distance to center of earth) are constant. Since the m's canceled, you also know that the mass of an object does not affect the rate of free fall; all objects fall at the same rate, with the same acceleration, no matter what their masses are.

Example 2.20 shows that the acceleration due to gravity, g, is about 9.8 m/s² and is practically a constant for relatively short distances above the surface. Notice, however, that Newton's law of gravitation is an inverse square law. This means if you double the distance, the force is $1/(2)^2$ or 1/4 as great. If you triple the distance, the force is $1/(3)^2$ or 1/9 as great. In other words, the force of gravitational attraction and g decrease inversely with the square of the distance from the earth's center. The weight of an object and the value of g are shown for several distances in Figure 2.28. If you have the time, a good calculator, and the inclination, you could check the values given in Figure 2.28 for a 70.0 kg person by doing problems similar to example 2.20. In fact, you could even calculate the mass of the earth, since you already have the value of g.

Using reasoning similar to that found in example 2.20, Newton was able to calculate the acceleration of the Moon toward Earth, about 0.0027 m/s². The Moon "falls" toward Earth because it is accelerated by the force of gravitational attraction. This attraction acts as a *centripetal force* that keeps the Moon from following a straight-line path as would be predicted from the first law. Thus, the acceleration of the

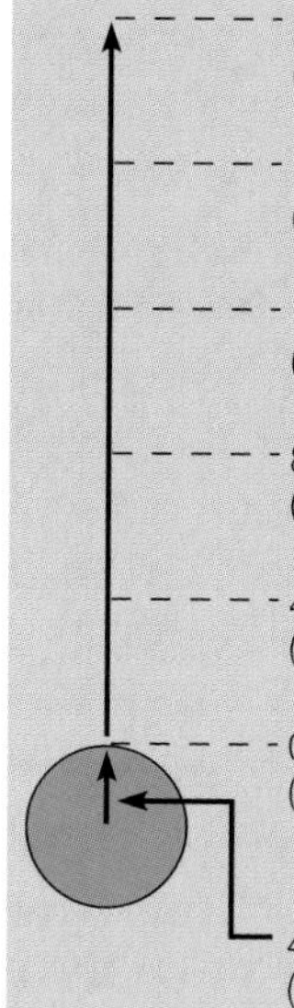

Distance above surface	Value of g	Mass	Weight
20,000 mi (38,400 km)	1 ft/s² (0.3 m/s²)	70.0 kg	4.7 lb (21 N)
16,000 mi (25,600 km)	1.3 ft/s² (0.4 m/s²)	70.0 kg	6.3 lb (28 N)
12,000 mi (19,200 km)	2 ft/s² (0.6 m/s²)	70.0 kg	9.5 lb (42 N)
8,000 mi (12,800 km)	3.6 ft/s² (1.1 m/s²)	70.0 kg	17 lb (77 N)
4,000 mi (6,400 km)	7.9 ft/s² (2.4 m/s²)	70.0 kg	37 lb (168 N)
0 mi (0 km)	32 ft/s² (9.80 m/s²)	70.0 kg	154 lb (686 N)

4,000 mi (6,400 km)

FIGURE 2.28

The force of gravitational attraction decreases inversely with the square of the distance from the earth's center. Note the weight of a 70.0 kg person at various distances above the earth's surface.

Moon keeps it in a somewhat circular orbit around Earth. Figure 2.29 shows that the Moon would be in position A if it followed a straight-line path instead of "falling" to position B as it does. The Moon thus "falls" around Earth. Newton was

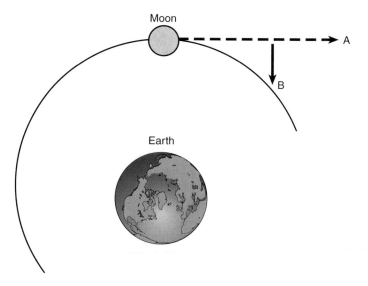

FIGURE 2.29

Gravitational attraction acts as a centripetal force that keeps the Moon from following the straight-line path shown by the dashed line to position A. It was pulled to position B by gravity (0.0027 m/s²) and thus "fell" toward Earth the distance from the dashed line to B, resulting in a somewhat circular path.

When do astronauts experience weightlessness, or "zero gravity"? Theoretically, the gravitational field of Earth extends to the whole universe. You know that it extends to the Moon, and indeed, even to the Sun some 93 million miles away. There is a distance, however, at which the gravitational force must become immeasurably small. But even at an altitude of 20,000 miles above the surface of Earth, gravity is measurable. At 20,000 miles, the value of g is about 1 ft/s^2 (0.3 m/s^2) compared to 32 ft/s^2 (9.8 m/s^2) on the surface. Since gravity does exist at these distances, how can an astronaut experience "zero gravity"?

Gravity does act on astronauts in spacecraft that are in orbit around Earth. The spacecraft stays in orbit, in fact, because of the gravitational attraction and because it has the correct speed. If the speed were less than 5 mi/s, the spacecraft would return to Earth. Astronauts fire their retro-rockets, which slow the speed, causing the spacecraft to fall down to the earth. If the speed were more than 7 mi/s, the spacecraft would fly off into space. The spacecraft stays in orbit because it has the right speed to continu-ously "fall" around and around the earth. Gravity provides the necessary centripetal force that causes the spacecraft to fall out of its natural straight-line motion.

Since gravity is acting on the astronaut and spacecraft, the term *zero gravity* is not an accurate description of what is happening. The astronaut, spacecraft, and everything in it are experiencing *apparent weightlessness* because they are continuously falling toward the earth. Everything seems to float because everything is falling together. But, strictly speaking, everything still has weight, be-cause weight is defined as a gravitational force acting on an object ($w = mg$).

Whether weightlessness is apparent or real, however, the effects on people are the same. Long-term orbital flights have provided evidence that the human body changes from the effect of weightlessness. Bones lose calcium and other minerals, the heart shrinks to a much smaller size, and leg muscles shrink so much on prolonged flights that astronauts cannot walk when they return to the earth. These changes oc-cur because on Earth humans are constantly subjected to the force of gravity. The nature of the skeleton and the strength of the muscles are determined by how the body reacts to this force. Metabolic pathways and physiological processes that maintain strong bones and muscles evolved having to cope with a specific gravitational force. When we are suddenly subjected to a place where gravity is signifi-cantly different, these processes result in weakened systems. If we had evolved on a planet with a different gravitational force, we would have muscles and bones that were adapted to the gravity on that planet. All organisms have evolved in a world with gravity. Many kinds of organisms have been used in experiments in space to try to de-velop a better understanding of how their systems work.

The problems related to prolonged weightlessness must be worked out before long-term weightless flights can take place. One solution to these problems might be a large, uniformly spinning spacecraft. The as-tronauts tend to move in a straight line, and the side of the turning spacecraft (now the "floor") exerts a force on them to make them go in a curved path. This force would act as an artificial gravity.

---

able to analyze the motion of the Moon quantitatively as evidence that it is gravitational force that keeps the Moon in its orbit. The law of gravitation was extended to the Sun, other planets, and eventually the universe. The quantitative predictions of observed relationships among the planets were strong evidence that all objects obey the same law of gravita-tion. In addition, the law provided a means to calculate the mass of Earth, the Moon, the planets, and the Sun. Newton's law of gravitation, laws of motion, and work with mathemat-ics formed the basis of most physics and technology for the next two centuries, as well as accurately describing the world of everyday experience.

EXAMPLE 2.20

The surface of the earth is approximately 6,400 km from its center. If the mass of the earth is 6.0×10^{24} kg, what is the acceleration due to gravity, g, near the surface?

$$G = 6.67 \times 10^{-11} \, \text{N·m}^2/\text{kg}^2$$
$$m_e = 6.0 \times 10^{24} \, \text{kg}$$
$$d = 6,400 \, \text{km} \, (6.4 \times 10^6 \, \text{m})$$
$$g = ?$$

$$g = \frac{Gm_e}{d^2}$$

$$= \frac{(6.67 \times 10^{-11} \, \text{N·m}^2/\text{kg}^2)(6.0 \times 10^{24} \, \text{kg})}{(6.4 \times 10^6 \, \text{m})^2}$$

$$= \frac{(6.67 \times 10^{-11})(6.0 \times 10^{24})}{4.1 \times 10^{13}} \frac{\text{N·m}^2 \cdot \text{kg}}{\text{kg}^2}$$

$$= \frac{4.0 \times 10^{14}}{4.1 \times 10^{13}} \frac{\frac{\text{kg·m}}{\text{s}^2}}{\text{kg}}$$

$$= \boxed{9.8 \, \text{m/s}^2}$$

(Note: In the unit calculation, remember that a newton is a kg·m/s^2.)

Isaac Newton (1642–1727)

Isaac Newton was a British physicist and mathematician who is regarded as one of the greatest scientists ever to have lived. In physics, he discovered the three laws of motion that bear his name and was the first to explain gravitation, clearly defining the nature of mass, weight, force, inertia, and acceleration. In his honor, the SI unit of force is called the newton. Newton also made fundamental discoveries in optics, finding that white light is composed of a spectrum of colors and inventing the reflecting telescope. In mathematics, Newton's principal contribution was to formulate the calculus and the binomial theorem.

Newton was born at Woolsthorpe, Lincolnshire, on December 25, 1642 by the old Julian calendar, but on January 4, 1643 by modern reckoning. His birthplace, Woolsthorpe Manor, is now preserved. Newton's was an inauspicious beginning for he was a premature, sickly baby born after his father's death, and his survival was not expected. When he was three, his mother remarried and the young Newton was left in his grandmother's care. He soon began to take refuge in things mechanical, reputedly making water-clocks, kites bearing fiery lanterns aloft, and a model mill powered by a mouse, as well as innumerable drawings and diagrams. When Newton was 12, he began to attend the King's School, Grantham, but his schooling was not to last. His mother, widowed again, returned to Woolsthorpe in 1658 and withdrew him from school with the intention of making him into a farmer. Fortunately, his uncle recognized Newton's ability and managed to get him back to school to prepare for university entrance. This Newton achieved in 1661, when he went to Trinity College, Cambridge, and began to delve widely and deeply into the scholarship of the day.

In 1665, the year that he became a Bachelor of Arts, the university was closed because of the plague and Newton spent eighteen months at Woolsthorpe, with only the occasional visit to Cambridge. Such seclusion was a prominent feature of Newton's creative life and, during this period, he laid the foundations of his work in mathematics, optics, dynamics, and celestial mechanics, performing his first prism experiments and reflecting on motion and gravitation.

Newton returned to Cambridge in 1666 and became a minor Fellow of Trinity in 1667 and a major Fellow the following year. He also received his Master of Arts degree in 1668 and became Lucasian Professor of Mathematics—at the age of only 26. It is said that the previous incumbent, Isaac Barrow (1630–1677), resigned the post to make way for Newton. Newton remained at Cambridge almost 30 years, studying alone for the most part, though in frequent contact with other leading scientists by letter and through the Royal Society in London, which elected him a Fellow in 1672. These were Newton's most fertile years. He labored day and night in his chemical laboratory, at his calculations, or immersed in theological and mystical speculations. In Cambridge, he completed what may be described as his greatest single work, the *Philosophae Naturalis Principia Mathematica* (*Mathematical Principles of Natural Philosophy*). This was presented to the Royal Society in 1686, who subsequently withdrew from publishing it through shortage of funds. The astronomer Edmund Halley (1656–1742), a wealthy man and friend of Newton, paid for the publication of the *Principia* in 1687. In it, Newton revealed his laws of motion and the law of universal gravitation.

Newton presented his conclusions on dynamics in the *Principia*. Although he had already developed the calculus, he did not use it in the *Principia*, preferring to prove all his results geometrically. In this great work, Newton's plan was first to develop the subject of general dynamics from a mathematical point of view and then to apply the results in the solution of important astronomical and physical problems. It included

a synthesis of Kepler's laws of planetary motion and Galileo's laws of falling bodies, developing the system of mechanics we know today, including the three famous laws of motion. The first law states that every body remains at rest or in constant motion in a straight line unless it is acted upon by a force. This defines inertia, finally disproving the idea which had been prevalent since Aristotle (384–322 B.C.) had mooted it, that force is required to keep anything moving. The second law states that a force accelerates a body by an amount proportional to its mass. This was the first clear definition of force and it also distinguished mass from weight. The third law states that action and reaction are equal and opposite, which showed how things could be made to move.

Newton also developed his general theory of gravitation as a universal law of attraction between any two objects, stating that the force of gravity is proportional to the masses of the objects and decreases in proportion to the square of the distance between the two bodies. Though, in the years before, there had been considerable correspondence between Newton, Hooke, Halley, and Kepler on the mathematical formulation of these laws, Newton did not complete the work until the writing of the *Principia*.

"I was in the prime of my age for invention" said Newton of those two years 1665 and 1666, and it was in that period that he performed his fundamental work in optics. Again it should be pointed out that the study of Newton's optics has been limited to his published letters and the *Opticks* of 1704, its publication delayed until after Hooke's death to avoid yet another controversy over originality. No adequate edition or full translation of the voluminous *Lectiones Opticae* exists. Newton began those first, crucial experiments by passing sunlight through a prism, finding that it dispersed the white light into a spectrum of colors. He then took a second

—*Continued top of next page*

prism and showed that it could combine the colors in the spectrum and form white light again. In this way, Newton proved that the colors are a property of light and not of the prism. An interesting by-product of these early speculations was the development of the reflecting telescope. Newton held the erroneous opinion that optical dispersion was independent of the medium through which the light was refracted and, therefore, that nothing could be done to correct the chromatic aberration caused by lenses. He therefore set about building a telescope in which the objective lens is replaced by a curved mirror, in which aberration could not occur. In 1668 Newton succeeded in making the first reflecting telescope, a tiny instrument only 15 cm/6 in long, but the direct ancestor of today's huge astronomical reflecting telescopes. In this invention, Newton was anticipated to some degree by James Gregory (1638–1675) who had produced a design for a reflecting telescope five years earlier but had not succeeded in constructing one.

Other scientists, Hooke especially, were critical of Newton's early reports, seeing too little connection between experimental result and theory, so that, in the course of a debate lasting several years, Newton was forced to refine his theories with considerable subtlety. He performed further experiments in which he investigated many other optical phenomena, including thin film interference effects, one of which, "Newton's rings," is named for him.

The *Opticks* presented a highly systematized and organized account of Newton's work and his theory of the nature of light and the effects that light produces. In fact, although he held that light rays were corpuscular in nature, he integrated into his ideas the concept of periodicity, holding that "ether waves" were associated with light corpuscles, a remarkable conceptual leap, for Hooke and Huygens, the founder of the wave theory, both denied periodicity to light waves. The corpuscle concept lent itself to an analysis by forces and established an analogy between the action of gross bodies and that of light, reinforcing the universalizing tendency of the *Principia.*

However, Newton's prestige was such that the corpuscular theory held sway for much longer than it deserved, not being finally overthrown until early in the 1800s. Ironically, it was the investigation of interference effects by Thomas Young (1773–1829) that led to the establishment of the wave theory of light.

Although comparatively little is known of the bulk of Newton's complete writings in chemistry and physics, we know even less about his chemistry and alchemy, chronology, prophecy, and theology. The vast number of documents he wrote on these matters have never yet been properly analyzed, but what is certain is that he took great interest in alchemy, performing many chemical experiments in his own laboratory and being in contact with Robert Boyle (1627–1691). He also wrote much on ancient chronology and the authenticity of certain biblical texts.

Newton's greatest achievement was to demonstrate that scientific principles are of universal application. In the *Principia Mathematica,* he built logically and analytically from mathematical premises and the evidence of experiment and observation to develop a model of the universe that is still of general validity. "If I have seen further than other men," he once said with perhaps assumed modesty, "it is because I have stood on the shoulders of giants"; and Newton was certainly able to bring together the knowledge of his forebears in a brilliant synthesis. Newton's life marked the first great flowering of the scientific method, which had been evolving in fits and starts since the time of the ancient Greeks. But Newton really established it, completing a scientific revolution in Europe that had begun with Nicolaus Copernicus (1473–1543) and ushering in the Age of Reason, in which the scientific method was expected to yield complete knowledge by the elucidation of the basic laws that govern the universe. No knowledge can ever be total, but Newton's example brought about an explosion of investigation and discovery that has never really abated. He perhaps foresaw this when he remarked "To myself, I seem to have been only like a boy playing on the seashore, and diverting myself in now and then finding a smoother pebble or a prettier shell than ordinary, whilst the great ocean of truth lay all undiscovered before me."

With his extraordinary insight into the workings of nature and rare tenacity in wresting its secrets and revealing them in as fundamental and concise a way as possible, Newton stands as a colossus of science. In physics, only Archimedes (287–212 B.C.) and Albert Einstein (1879–1955), who also possessed these qualities, may be compared to him.

Summary

Motion can be measured by speed, velocity, and acceleration. *Speed* is a measure of how fast something is moving. It is a ratio of the distance covered between two locations to the time that elapsed while moving between the two locations. The *average speed* considers the distance covered during some period of time, while the *instantaneous speed* is the speed at some specific instant. *Velocity* is a measure of the speed and direction of a moving object. *Acceleration* is the change of velocity during some period of time.

A *force* is a push or a pull that can change the motion of an object. The *net force* is the sum of all the forces acting on an object.

Galileo determined that a continuously applied force is not necessary for motion and defined the concept of *inertia:* an object remains in unchanging motion in the absence of a net force. Galileo also determined that falling objects accelerate toward the earth's surface independent of the weight of the object. He found the acceleration due to gravity, g, to be 9.8 m/s^2 (32 ft/s^2), and the distance an object falls is proportional to the square of the time of free fall ($d \propto t^2$).

Compound motion occurs when an object is projected into the air. Compound motion can be described by splitting the motion into vertical and horizontal parts. The acceleration due to gravity, g, is a constant that is acting at all times and acts independently of any motion that an object has. The path of an object that is projected at some angle to the horizon is therefore a parabola.

Newton's *first law of motion* is concerned with the motion of an object and the lack of a net force. Also known as the *law of inertia,* the first law states that an object will retain its state of straight-line motion (or state of rest) unless a net force acts on it.

The *second law of motion* describes a relationship between net force, mass, and acceleration. A *newton* of force is the force needed to give a 1.0 kg mass an acceleration of 1.0 m/s^2.

Weight is the downward force that results from the earth's gravity acting on the mass of an object. Weight is measured in *newtons* in the metric system and *pounds* in the English system.

Newton's *third law of motion* states that forces are produced by the interaction of *two different* objects. These forces always occur in matched pairs that are equal in size and opposite in direction.

Momentum is the product of the mass of an object and its velocity. In the absence of external forces, the momentum of a group of interacting objects always remains the same. This relationship is the *law of conservation of momentum. Impulse* is a change of momentum equal to a force times the time of application.

An object moving in a circular path must have a force acting on it, since it does not move in a straight line. The force that pulls an object out of its straight-line path is called a *centripetal force.* The centripetal force needed to keep an object in a circular path depends on the mass of the object, its velocity, and the radius of the circle.

The *universal law of gravitation* is a relationship between the masses of two objects, the distance between the objects, and a proportionality constant. Newton was able to use this relationship to show that gravitational attraction provides the centripetal force that keeps the moon in its orbit.

Summary of Equations

2.1

$$\text{average speed} = \frac{\text{distance}}{\text{time}}$$

$$\bar{v} = \frac{d}{t}$$

2.2

$$\text{acceleration} = \frac{\text{change of velocity}}{\text{time}}$$

$$= \frac{\text{final velocity} - \text{initial velocity}}{\text{time}}$$

$$a = \frac{v_f - v_i}{t}$$

2.3

$$\text{average velocity} = \frac{\text{final velocity} + \text{initial velocity}}{2}$$

$$\bar{v} = \frac{v_f + v_i}{2}$$

2.4

$$\text{distance} = \frac{1}{2}(\text{acceleration})(\text{time})^2$$

$$d = \frac{1}{2}at^2$$

2.5

$$\text{force} = \text{mass} \times \text{acceleration}$$

$$F = ma$$

2.6

$$\text{weight} = \text{mass} \times \text{acceleration due to gravity}$$

$$w = mg$$

2.7

$$\text{force on object A} = \text{force on object B}$$

$$F_{\text{A due to B}} = F_{\text{B due to A}}$$

2.8

$$\text{momentum} = \text{mass} \times \text{velocity}$$

$$p = mv$$

2.9

$$\text{change of momentum} = \text{force} \times \text{time}$$

$$\Delta p = Ft$$

2.10

$$\text{centripetal acceleration} = \frac{\text{velocity squared}}{\text{radius of circle}}$$

$$a_c = \frac{v^2}{r}$$

2.11

$$\text{centripetal force} = \frac{\text{mass} \times \text{velocity squared}}{\text{radius of circle}}$$

$$F = \frac{mv^2}{r}$$

2.12

$$\text{gravitational force} = \text{constant} \times \frac{\text{one mass} \times \text{another mass}}{\text{distance squared}}$$

$$F = G\frac{m_1 m_2}{d^2}$$

KEY TERMS

acceleration (p. **29**)
centrifugal force (p. **50**)
centripetal force (p. **50**)
first law of motion (p. **41**)
force (p. **32**)
free fall (p. **35**)
g (p. **37**)
impulse (p. **49**)
inertia (p. **34**)
law of conservation of momentum (p. **47**)

mass (p. **44**)
momentum (p. **47**)
net force (p. **32**)
newton (p. **43**)
second law of motion (p. **43**)
speed (p. **27**)
third law of motion (p. **45**)
universal law of gravitation (p. **51**)
velocity (p. **29**)

1. A quantity of 5 m/s^2 is a measure of
   a. metric area.
   b. acceleration.
   c. speed.
   d. velocity.

2. An automobile has how many different devices that can cause it to undergo acceleration?
   a. none
   b. one
   c. two
   d. three or more

3. Ignoring air resistance, an object falling toward the surface of the earth has a velocity that is
   a. constant.
   b. increasing.
   c. decreasing.
   d. acquired instantaneously, but dependent on the weight of the object.

4. Ignoring air resistance, an object falling near the surface of the earth has an *acceleration* that is
   a. constant.
   b. increasing.
   c. decreasing.
   d. dependent on the weight of the object.

5. Two objects are released from the same height at the same time, and one has twice the weight of the other. Ignoring air resistance,
   a. the heavier object hits the ground first.
   b. the lighter object hits the ground first.
   c. they both hit at the same time.
   d. whichever hits first depends on the distance dropped.

6. A ball rolling across the floor slows to a stop because
   a. there is a net force acting on it.
   b. the force that started it moving wears out.
   c. the forces are balanced.
   d. the net force equals zero.

7. Considering the forces on the system of you and a bicycle as you pedal the bike at a constant velocity in a horizontal straight line,
   a. the force you are exerting on the pedal is greater than the resisting forces.
   b. all forces are in balance, with the net force equal to zero.
   c. the resisting forces of air and tire friction are less than the force you are exerting.
   d. the resisting forces are greater than the force you are exerting.

8. If you double the unbalanced force on an object of a given mass, the acceleration will be
   a. doubled.
   b. increased fourfold.
   c. increased by one-half.
   d. increased by one-fourth.

9. If you double the mass of a cart while it is undergoing a constant unbalanced force, the acceleration will be
   a. doubled.
   b. increased fourfold.
   c. half as much.
   d. one-fourth as much.

10. Doubling the distance between the center of an orbiting satellite and the center of the earth will result in what change in the gravitational attraction of the earth for the satellite?
    a. one-half as much
    b. one-fourth as much
    c. twice as much
    d. four times as much

11. If a ball swinging in a circle on a string is moved twice as fast, the force on the string will be
    a. twice as great.
    b. four times as great.
    c. one-half as much.
    d. one-fourth as much.

12. A ball is swinging in a circle on a string when the string length is doubled. At the same velocity, the force on the string will be
    a. twice as great.
    b. four times as great.
    c. one-half as much.
    d. one-fourth as much.

Answers
1. b **2.** d **3.** b **4.** a **5.** c **6.** a **7.** b **8.** a **9.** c **10.** b **11.** b **12.** c

1. An insect inside a bus flies from the back toward the front at 5.0 mi/h. The bus is moving in a straight line at 50.0 mi/h. What is the speed of the insect?

2. Disregarding air friction, describe all the forces acting on a bullet shot from a rifle into the air.

3. Can gravity act in a vacuum? Explain.

4. Is it possible for a small car to have the same momentum as a large truck? Explain.

5. What net force is needed to maintain the constant velocity of a car moving in a straight line? Explain.

6. How can there ever be an unbalanced force on an object if every action has an equal and opposite reaction?

7. Why should you bend your knees as you hit the ground after jumping from a roof?

8. Is it possible for your weight to change as your mass remains constant? Explain.

9. What maintains the speed of the earth as it moves in its orbit around the sun?

10. Suppose you are standing on the ice of a frozen lake and there is no friction whatsoever. How can you get off the ice? (Hint: Friction is necessary to crawl or walk, so that will not get you off the ice.)

11. A rocket blasts off from a platform on a space station. An identical rocket blasts off from free space. Considering everything else to be equal, will the two rockets have the same acceleration? Explain.

12. An astronaut leaves a spaceship that is moving through free space to adjust an antenna. Will the spaceship move off and leave the astronaut behind? Explain.

PARALLEL EXERCISES

The exercises in groups A and B cover the same concepts. Solutions to group A exercises are located in appendix D.

Note: *Neglect all frictional forces in all exercises.*

Group A

1. What is the average speed, in km/h, for a car that travels 22 km in exactly 15 min?

2. Suppose a radio signal travels from earth and through space at a speed of 3.0×10^8 m/s. How far into space did the signal travel during the first 20.0 minutes?

3. How far away was a lightning strike if thunder is heard 5.00 seconds after seeing the flash? Assume that sound traveled at 350.0 m/s during the storm.

4. A car is driven at an average speed of 100.0 km/h for two hours, then at an average speed of 50.0 km/h for the next hour. What was the average speed for the three-hour trip?

5. What is the acceleration of a car that moves from rest to 15.0 m/s in 10.0 s?

6. How long will be required for a car to go from a speed of 20.0 m/s to a speed of 25.0 m/s if the acceleration is 3.0 m/s²?

7. A bullet leaves a rifle with a speed of 2,360 ft/s. How much time elapses before it strikes a target 1 mile (5,280 ft) away?

8. A pitcher throws a ball at 40.0 m/s, and the ball is electronically timed to arrive at home plate 0.4625 s later. What is the distance from the pitcher to the home plate?

9. The Sun is 1.50×10^8 km from Earth, and the speed of light is 3.00×10^8 m/s. How many minutes elapse as light travels from the Sun to Earth?

10. An archer shoots an arrow straight up with an initial velocity magnitude of 100.0 m/s. After 5.00 s, the velocity is 51.0 m/s. At what rate is the arrow decelerated?

11. A ball thrown straight up climbs for 3.0 s before falling. Neglecting air resistance, with what velocity was the ball thrown?

12. A ball dropped from a building falls for 4.00 s before it hits the ground. (a) What was its final velocity just as it hit the ground? (b) What was the average velocity during the fall? (c) How high was the building?

13. You drop a rock from a cliff, and 5.00 s later you see it hit the ground. How high is the cliff?

14. What is the resulting acceleration when an unbalanced force of 100 N is applied to a 5 kg object?

15. What is the momentum of a 100 kg football player who is moving at 6 m/s?

16. A car weighing 13,720 N is speeding down a highway with a velocity of 91 km/h. What is the momentum of this car?

Group B

1. A boat moves 15.0 km across a lake in 45 min. What was the average speed of the boat in kilometers per hour?

2. If the Sun is a distance of 1.5×10^8 km from Earth, how long does it take sunlight to reach Earth if it moves at 3.0×10^8 m/s?

3. How many meters away is a cliff if an echo is heard 0.500 s after the original sound? Assume that sound traveled at 343 m/s on that day.

4. A car has an average speed of 80.0 km/h for one hour, then an average speed of 90.0 km/h for two hours during a three-hour trip. What was the average speed for the three-hour trip?

5. What is the acceleration of a car that moves from a speed of 5.0 m/s to a speed of 15 m/s during a time of 6.0 s?

6. How much time is needed for a car to accelerate from 8.0 m/s to a speed of 22 m/s if the acceleration is 3.0 m/s²?

7. A rocket moves through outer space at 11,000 m/s. At this rate, how much time would be required to travel the distance from Earth to the Moon, which is 380,000 km?

8. Sound travels at 1,140 ft/s in the warm air surrounding a thunderstorm. How far away was the place of discharge if thunder is heard 4.63 s after a lightning flash?

9. How many hours are required for a radio signal from a space probe near the planet Pluto, 6.00×10^9 km away, to reach Earth? Assume that the radio signal travels at the speed of light, 3.00×10^8 m/s.

10. A rifle is fired straight up, and the bullet leaves the rifle with an initial velocity magnitude of 724 m/s. After 5.00 s the velocity is 675 m/s. At what rate is the bullet decelerated?

11. A rock thrown straight up climbs for 2.50 s, then falls to the ground. Neglecting air resistance, with what velocity did the rock strike the ground?

12. An object is observed to fall from a bridge, striking the water below 2.50 s later. (a) With what velocity did it strike the water? (b) What was its average velocity during the fall? (c) How high is the bridge?

13. A ball dropped from a window strikes the ground 2.00 seconds later. How high is the window above the ground?

14. Find the resulting acceleration from a 300 N force that acts on an object with a mass of 3,000 kg.

15. What is the momentum of a 30.0 kg shell fired from a cannon with a velocity of 500 m/s?

16. What is the momentum of a 39.2 N bowling ball with a velocity of 7.00 m/s?

17. A 15 g bullet is fired with a velocity of 200 m/s from a 6 kg rifle. What is the recoil velocity of the rifle?

18. An astronaut and equipment weigh 2,156 N on Earth. Weightless in space, the astronaut throws away a 5.0 kg wrench with a velocity of 5.0 m/s. What is the resulting velocity of the astronaut in the opposite direction?

19. (a) What is the weight of a 1.25 kg book? (b) What is the acceleration when a net force of 10.0 N is applied to the book?

20. What net force is needed to accelerate a 1.25 kg book 5.00 m/s²?

21. What net force does the road exert on a 70.0 kg bicycle and rider to give them an acceleration of 2.0 m/s²?

22. A 1,500 kg car accelerates uniformly from 44.0 km/h to 80.0 km/h in 10.0 s. What was the net force exerted on the car?

23. A net force of 5,000.0 N accelerates a car from rest to 90.0 km/h in 5.0 s. (a) What is the mass of the car? (b) What is the weight of the car?

24. What is the weight of a 70.0 kg person?

25. How much centripetal force is needed to keep a 0.20 kg ball on a 1.50 m string moving in a circular path with a speed of 3.0 m/s?

26. On Earth, an astronaut and equipment weigh 1,960.0 N. While weightless in space, the astronaut fires a 100 N rocket backpack for 2.0 s. What is the resulting velocity of the astronaut and equipment?

17. A 30.0 kg shell is fired from a 2,000 kg cannon with a velocity of 500 m/s. What is the resulting velocity of the cannon?

18. An 80.0 kg man is standing on a frictionless ice surface when he throws a 4.00 kg book at 20.0 m/s. With what velocity does the man move across the ice?

19. (a) What is the weight of a 5.00 kg backpack? (b) What is the acceleration of the backpack if a net force of 10.0 N is applied?

20. What net force is required to accelerate a 20.0 kg object to 10.0 m/s²?

21. What forward force must the ground apply to the foot of a 60.0 kg person to result in an acceleration of 1.00 m/s²?

22. A 1,000.0 kg car accelerates uniformly to double its speed from 36.0 km/h in 5.00 s. What net force acted on this car?

23. A net force of 3,000.0 N accelerates a car from rest to 36.0 km/h in 5.00 s. (a) What is the mass of the car? (b) What is the weight of the car?

24. How much does a 60.0 kg person weigh?

25. What tension must a 50.0 cm length of string support in order to whirl an attached 1,000.0 g stone in a circular path at 5.00 m/s?

26. A 200.0 kg astronaut and equipment move with a velocity of 2.00 m/s toward an orbiting spacecraft. How long will the astronaut need to fire a 100.0 N rocket backpack to stop the motion relative to the spacecraft?

The wind can be used as a source of energy. All you need is a way to capture the energy—such as these wind turbines in California—and to live somewhere where the wind blows enough to make it worthwhile.

CHAPTER

3

Energy

The term *energy* is closely associated with the concepts of force and motion. Naturally moving matter, such as the wind or moving water, exerts forces. You have felt these forces if you have ever tried to walk against a strong wind or stand in one place in a stream of rapidly moving water. The motion and forces of moving air and moving water are used as *energy sources* (Figure 3.1). The wind is an energy source as it moves the blades of a windmill, performing useful work. Moving water is an energy source as it forces the blades of a water turbine to spin, turning an electric generator. Thus, moving matter exerts a force on objects in its path, and objects moved by the force can also be used as an energy source.

Matter does not have to be moving to supply energy; matter *contains* energy. Food supplied the energy for the muscular exertion of the humans and animals that accomplished most of the work before this century. Today, machines do the work that was formerly accomplished by muscular exertion. Machines also use the energy contained in matter. They use gasoline, for example, as they supply the forces and motion to accomplish work.

Moving matter and matter that contains energy can be used as energy sources to perform work. The concepts of work and energy and the relationship to matter are the topics of this chapter. You will learn how energy flows in and out of your surroundings as well as a broad, conceptual view of energy that will be developed more fully throughout the course.

Work

You learned earlier that the term *force* has a special meaning in science that is different from your everyday concept of force. In everyday use, you use the term in a variety of associations such as police force, economic force, or the force of an argument. Earlier, force was discussed in a general way as a push or pull. Then a more precise scientific definition of force was developed from Newton's laws of motion—a force is a result of an interaction that is capable of changing the state of motion of an object.

The word *work* represents another one of those concepts that has a special meaning in science that is different from your everyday concept. In everyday use, work is associated with a task to be accomplished or the time spent in performing the task. You might work at understanding physical science, for example, or you might tell someone that physical science is a lot of work. You also probably associate physical work, such as lifting or moving boxes, with how tired you become from the effort. The definition of mechanical work is not concerned with tasks, time, or how tired you become from doing a task. It is concerned with the application of a force to an object and the distance the object moves as a result of the force. The **work** done on the object is defined as *the product of the applied force and the parallel distance through which the force acts:*

$$\text{work} = \text{force} \times \text{distance}$$
$$W = Fd$$

equation 3.1

Mechanical work is the product of a force and the distance an object moves as a result of the force. There are two important considerations to remember about this definition: (1) something *must move* whenever work is done, and (2) the movement must be in the *same direction* as the direction of the force. When you move a book to a higher shelf in a bookcase you are doing work on the book. You apply a vertically upward force equal to the weight of the book as you move it in the same direction as the direction of the applied force. The work done on the book can therefore be calculated by multiplying the weight of the book by the distance it was moved (Figure 3.2).

Units of Work

The units of work can be obtained from the definition of work, $W = Fd$. In the metric system a force is measured in newtons (N), and distance is measured in meters (m), so the unit of work is

$$W = Fd$$
$$W = (\text{newton})(\text{meter})$$
$$W = (\text{N})(\text{m})$$

The newton-meter is therefore the unit of work. This derived unit has a name. The newton-meter is called a **joule** (J) (pronounced "jool").

$$1 \text{ joule} = 1 \text{ newton meter}$$

The units for a newton are kg·m/s^2, and the unit for a meter is m. It therefore follows that the units for a joule are $\text{kg·m}^2/\text{s}^2$.

In the English system, the force is measured in pounds (lb), and the distance is measured in feet (ft). The unit of work in the English system is therefore the ft·lb. The ft·lb does not have a

FIGURE 3.1

This is Glen Canyon Dam on the Colorado River between Utah and Arizona. The 216 m (about 710 ft) wall of concrete holds Lake Powell, which is 170 m (about 560 ft) deep at the dam when full. Water falling through 5 m (about 15 ft) diameter penstocks to generators at the bottom of the dam produce up to 1,300 MW of electrical power.

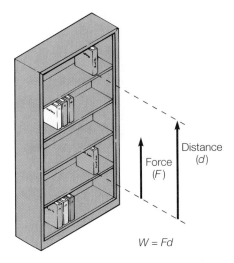

FIGURE 3.2

The force on the book moves it through a vertical distance from the second shelf to the fifth shelf, and work is done, $W = Fd$.

name of its own as the N·m does (Figure 3.3). [Note that although the equation is $W = Fd$, and this means = (pounds) (feet), the unit is called the ft·lb.]

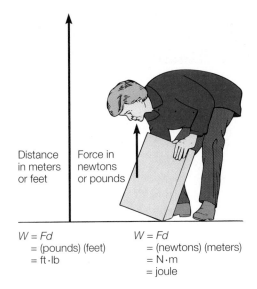

$W = Fd$
= (pounds) (feet)
= ft·lb

$W = Fd$
= (newtons) (meters)
= N·m
= joule

FIGURE 3.3

Work is done against gravity when lifting an object. Work is measured in joules or in foot-pounds.

EXAMPLE 3.1

How much work is needed to lift a 5.0 kg backpack to a shelf 1.0 m above the floor?

Solution

The backpack has a mass (m) of 5.0 kg, and the distance (d) is 1.0 m. To lift the backpack requires a vertically upward force equal to the weight of the backpack. Weight can be calculated from $w = mg$:

$$m = 5.0 \text{ kg} \qquad w = mg$$
$$g = 9.8 \text{ m/s}^2 \qquad\quad = (5.0 \text{ kg})\left(9.8 \frac{\text{m}}{\text{s}^2}\right)$$
$$W = ?$$
$$= (5.0 \times 9.8) \text{ kg} \times \frac{\text{m}}{\text{s}^2}$$
$$= 49 \frac{\text{kg·m}}{\text{s}^2}$$
$$= \boxed{49 \text{ N}}$$

The definition of work is found in equation 3.1,

$$F = 49 \text{ N} \qquad W = Fd$$
$$d = 1.0 \text{ m} \qquad\quad = (49 \text{ N})(1.0 \text{ m})$$
$$W = ? \qquad\qquad = (49 \times 1.0)(\text{N·m})$$
$$= 49 \text{ N·m}$$
$$= \boxed{49 \text{ J}}$$

Simple machines are tools that people use to help them do work. Recall that work is a force times a distance and you can see that the simple machine helps you do work by changing a force or a distance that something is moved. The force or distance advantage you gain by using the machine is called the *mechanical advantage*. The larger the mechanical advantage, the greater the effort that you would save by using the machine.

A lever is a simple machine, and Box Figure 3.1 shows what is involved when a lever reduces a force needed to do work. First, note there are two forces involved. The force that you provide by using the machine is called the *effort force*. You and the machine are working against the second force, called the *resistance force*. In the illustration, a 60 lb effort force is used to move a resistance force of 300 lb.

There are also two distances involved in using the lever. The distance over which your effort force acts is called the *effort distance*, and the distance the resistance moves is called the *resistance distance*. You would push down with an effort force of 60 lb through an effort distance of 2.5 feet. The 300 lb rock, on the other hand, was raised a resistance distance of 0.5 ft.

You did 60 × 2.5 ft, or 150 ft·lb of work on the lever. The work done on the rock by the lever was 300 lb × 0.5 ft, or 150 ft·lb of work. The work done by you on the lever is the same as the work done by the lever on the rock, so

> work input = work output

Since work is force times distance, we can write this concept as

> effort force × effort distance =
>
> resistance force × resistance distance

Ignoring friction, the work you get out of any simple machine is the same as the work you put into it. The lever enabled you to trade force for distance, and the mechanical advantage (*MA*) can be found from a ratio of the resistance force (F_R)

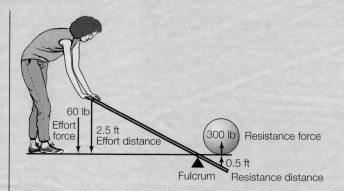

BOX FIGURE 3.1

The lever is one of six simple machines.

divided by the effort force (F_E):

$$MA = \frac{F_R}{F_E}$$

Therefore, the example lever in Box Figure 3.1 had a mechanical advantage of

$$MA = \frac{F_R}{F_E}$$

$$= \frac{300 \text{ lb}}{60 \text{ lb}}$$

$$= 5$$

You can also find the mechanical advantage by dividing the effort distance (d_E) by the resistance distance (d_R):

$$MA = \frac{d_E}{d_R}$$

For the example lever we find

$$MA = \frac{d_E}{d_R}$$

$$= \frac{2.5 \text{ ft}}{0.5 \text{ ft}}$$

$$= 5$$

So, we can use either the forces or distances involved in simple machines to calculate the mechanical advantage. In summary, a simple machine works for you by making it possible to apply a small force over a large distance to get a large force working over a small distance.

There are six kinds of simple machines: inclined plane, wedge, screw, lever, wheel and axle, and pulley. As you will see, the screw and wedge can be considered types of inclined planes; the wheel and axle and the pulley can be considered types of levers.

1. The *inclined plane* is a stationary ramp that is used to trade distance for force. You are using an inclined plane when you climb a stairway, drive a road that switches back and forth when going up a mountainside, or use a board to slide a heavy box up to a loading dock. Each use gives a large mechanical advantage by trading distance for force. For example, sliding a heavy box up a 10 m ramp to a 2 m-high loading dock raises the box with less force through a greater distance. The mechanical advantage of this inclined plane would be

$$MA = \frac{d_E}{d_R}$$

$$= \frac{10 \text{ m}}{2 \text{ m}}$$

$$= 5$$

Ignoring friction, a mechanical advantage of 5 means that a force of only 20 newtons would be needed to

—*Continued top of next page*

Continued—

push a box weighing 100 newtons up the ramp.

2. The *wedge* is an inclined plane that moves. An ax is two back-to-back inclined planes that move through the wood it is used to split. Wedges are found in knives, axes, hatchets, and nails.

3. The *screw* is an inclined plane that has been wrapped around a cylinder, with the threads playing the role of the incline. A finely threaded screw has a higher mechanical advantage and requires less force to turn, but it also requires a greater effort distance.

4. The *lever* is a bar or board that is free to pivot about a fixed point called a fulcrum. There are three classes of levers based on the location of the fulcrum, effort force, and resistance force (Box Figure 3.2). A first-class lever has the fulcrum between the effort force and the resistance force. Examples are a seesaw, pliers, scissors, crowbars, and shovels. A second-class lever has the effort resistance between the fulcrum and the effort force. Examples are nutcrackers and twist-type jar openers. A third-class lever has the effort force between the resistance force and the fulcrum. Examples are fishing rods and tweezers.

   A claw hammer can be used as a first-class lever to remove nails from a

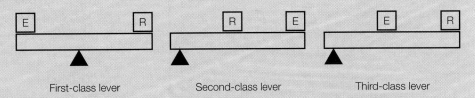

First-class lever Second-class lever Third-class lever

BOX FIGURE 3.2
The three classes of levers are defined by the relative locations of the fulcrum, effort, and resistance.

board. If the hammer handle is 30 cm and the distance from the nail slot to the fulcrum is 5 cm, the mechanical advantage will be

$$MA = \frac{d_E}{d_R}$$

$$= \frac{30 \text{ cm}}{5 \text{ cm}}$$

$$= 6$$

5. A *wheel and axle* has two circles, with the smaller circle called the axle and the larger circle called the wheel. The wheel and axle can be considered to be a lever that can move in a circle. Examples are a screwdriver, doorknob, steering wheel, and any application of a turning crank. The mechanical advantage is found from the radius of the wheel, where the effort is applied, to the radius of the axle, which is the distance over which the resistance

moves. For example, a large screwdriver has a radius of 15 cm in the handle (the wheel) and 0.5 in the bit (the axle). The mechanical advantage of this screwdriver is

$$MA = \frac{d_E}{d_R}$$

$$= \frac{3 \text{ cm}}{0.5 \text{ cm}}$$

$$= 6$$

6. A *pulley* is a movable lever that rotates around a fulcrum. A single fixed pulley can only change the direction of a force. To gain a mechanical advantage you need a fixed pulley and a movable pulley such as those found in a block and tackle. The mechanical advantage of a block and tackle can be found by comparing the length of rope or chain pulled to the distance the resistance has moved.

EXAMPLE 3.2

How much work is required to lift a 50 lb box vertically a distance of 2 ft? (Answer: 100 ft·lb)

Power

You are doing work when you walk up a stairway, since you are lifting yourself through a distance. You are lifting your weight (force exerted) the *vertical* height of the stairs (distance through which the force is exerted). Consider a person who weighs 120 lb and climbs a stairway with a vertical distance of

Concepts Applied

Book Work

Place a tied loop of string between the center pages of a small book. Pick up the loop so the string lifts the book, supporting it with open pages down. Use a spring scale to find the weight of the book in newtons. Measure the done work in lifting the book 1 meter. Use the spring scale to measure the work done in pulling the book along a tabletop for 1 meter. Is the amount of work done lifting the book the same as the amount of work done pulling the book along the tabletop? Why or why not?

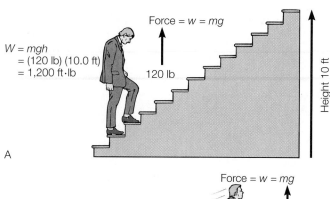

$$W = mgh$$
$$= (120 \text{ lb}) (10.0 \text{ ft})$$
$$= 1,200 \text{ ft·lb}$$

Force = w = mg

120 lb

Height 10 ft

A

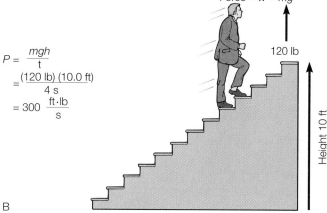

Force = w = mg

120 lb

$$P = \frac{mgh}{t}$$
$$= \frac{(120 \text{ lb}) (10.0 \text{ ft})}{4 \text{ s}}$$
$$= 300 \frac{\text{ft·lb}}{\text{s}}$$

Height 10 ft

B

FIGURE 3.4

(A) The work accomplished in climbing a stairway is the person's weight times the vertical distance. (B) The power level is the work accomplished per unit of time.

10 ft. This person will do (120 lb)(10 ft) or 1,200 ft·lb of work. Will the amount of work change if the person were to run up the stairs? The answer is no; the same amount of work is accomplished. Running up the stairs, however, is more tiring than walking up the stairs. You use the same amount of energy but at a greater *rate* when running. The rate at which energy is transformed or the rate at which work is done is called **power** (Figure 3.4). Power is measured as work per unit of time,

$$\text{power} = \frac{\text{work}}{\text{time}}$$

$$P = \frac{W}{t}$$

equation 3.2

Considering just the work and time factors, the 120 lb person who ran up the 10 ft height of stairs in 4 seconds would have a power rating of

$$P = \frac{W}{t} = \frac{(120 \text{ lb})(10 \text{ ft})}{4 \text{ s}} = 300 \frac{\text{ft·lb}}{\text{s}}$$

If the person had a time of 3 s on the same stairs, the power rating would be greater, 400 ft·lb/s. This is a greater *rate* of energy use, or greater power.

When the steam engine was first invented, there was a need to describe the rate at which the engine could do work. Since people at this time were familiar with using horses to do their work, the steam engines were compared to horses. James Watt, who designed a workable steam engine, defined **horsepower** as a power rating of 550 ft·lb/s (Figure 3.5A). To convert a power rating in the English units of ft·lb/s to horsepower, divide the power rating by 550 ft·lb/s/hp. For example, the 120 lb person who had a power rating of 400 ft·lb/s had a horsepower of 400 ft·lb/s ÷ 550 ft·lb/s/hp, or 0.7 hp.

In the metric system, power is measured in joules per second. The unit J/s, however, has a name. A J/s is called a **watt** (W). The watt (Figure 3.5B) is used with metric prefixes for large numbers: 1,000 W = 1 kilowatt (kW) and 1,000,000 W = 1 megawatt (MW). It takes 746 W to equal 1 horsepower. One kilowatt is equal to about 1 1/3 horsepower.

The electric utility company charges you for how much electrical energy you have used. Electrical energy is measured by power (kW) times the time of use (h). The kWh is a unit of *work*, not power. Since power is

$$P = \frac{W}{t}$$

then it follows that

$$W = Pt$$

So power times time equals a unit of work, kWh. We will return to kilowatts and kilowatt-hours later when we discuss electricity.

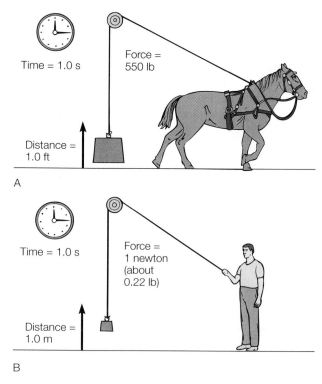

Time = 1.0 s

Force = 550 lb

Distance = 1.0 ft

A

Time = 1.0 s

Force = 1 newton (about 0.22 lb)

Distance = 1.0 m

B

FIGURE 3.5

(A) A horsepower is defined as a power rating of 550 ft·lb/s. (B) A watt is defined as a newton-meter per second, or joule per second.

EXAMPLE **3.3**

An electric lift can raise a 500.0 kg mass a distance of 10.0 m in 5.0 s. What is the power of the lift?

Solution

Power is work per unit time ($P = W/t$), and work is force times distance ($W = Fd$). The vertical force required is the weight lifted, and $w = mg$. Therefore the work accomplished would be $W = mgh$, and the power would be $P = mgh/t$. Note that h is for height, a vertical distance (d).

$m = 500.0 \text{ kg}$

$g = 9.8 \text{ m/s}^2$

$h = 10.0 \text{ m}$

$t = 5.0 \text{ s}$

$P = ?$

$$P = \frac{mgh}{t}$$

$$= \frac{(500.0 \text{ kg})(9.8 \text{ m/s}^2)(10.0 \text{ m})}{5.0 \text{ s}}$$

$$= \frac{(500.0)(9.8)(10.0)}{5.0} \frac{\text{kg} \cdot \frac{\text{m}}{\text{s}^2} \cdot \text{m}}{\text{s}}$$

$$= 9,800 \frac{\text{N} \cdot \text{m}}{\text{s}}$$

$$= 9,800 \frac{\text{J}}{\text{s}}$$

$$= 9,800 \text{ W}$$

$$= 9.8 \text{ kW}$$

The power in horsepower (hp) units would be

$$9,800 \text{ W} \times \frac{\text{hp}}{746 \text{ W}} = \boxed{13 \text{ hp}}$$

EXAMPLE **3.4**

A 150 lb person runs up a 15 ft stairway in 10.0 s. What is the horsepower rating of the person? (Answer: 0.41 horsepower)

Concepts Applied

Work and Power

Power is the rate of expending energy or of doing work. You can find your power output by taking a few measurements.

First, let's find how much work you do in walking up a flight of stairs. Your work output will be approximately equal to the change in your potential energy (mgh), so you will need (1) to measure the vertical height of a flight of stairs in metric units, and (2) to calculate or measure your mass (conversion factors are located inside the front cover). Record your findings in your report.

Second, find your power output by climbing the stairs as fast as you can while someone measures the time with a stopwatch. Find your power output in watts. Convert this to horsepower by consulting the conversion factors inside the front cover. Did you develop at least one horsepower? Does a faster person always have more horsepower?

$$\text{walking power} = \frac{(\quad \text{kg})(9.8 \text{m/s}^2)(\quad \text{m})}{(\quad \text{s})}$$

$$\text{running power} = \frac{(\quad \text{kg})(9.8 \text{m/s}^2)(\quad \text{m})}{(\quad \text{s})}$$

Motion, Position, and Energy

Closely related to the concept of work is the concept of **energy**. Energy can be defined as the *ability to do work*. This definition of energy seems consistent with everyday ideas about energy and physical work. After all, it takes a lot of energy to do a lot of work. In fact, one way of measuring the energy of something is to see how much work it can do. Likewise, when work is done *on* something, a change occurs in its energy level. The following examples will help clarify this close relationship between work and energy.

Potential Energy

Consider a book on the floor next to a bookcase. You can do work on the book by vertically raising it to a shelf. You can measure this work by multiplying the vertical upward force applied times the distance that the book is moved. You might find, for example, that you did an amount of work equal to 10 J on the book (see example 3.1).

Suppose that the book has a string attached to it, as shown in Figure 3.6. The string is threaded over a frictionless pulley and attached to an object on the floor. If the book is caused to fall from the shelf, the object on the floor will be vertically lifted through some distance by the string. The falling book exerts a force on the object through the string, and the object is moved through a distance. In other words, the *book* did work on the object through the string, $W = Fd$.

The book can do more work on the object if it falls from a higher shelf, since it will move the object a greater distance. The higher the shelf, the greater the *potential* for the book to do work. The ability to do work is defined as energy. The energy that an object has because of its position is called **potential energy** (PE). Potential energy is defined as *energy due to position*. This type of potential energy is called *gravitational potential energy,* since it is a result of gravitational attraction. There are other types of potential energy, such as that in a compressed or stretched spring.

Note the relationship between work and energy in the example. You did 10 J of work to raise the book to a higher shelf. In so doing, you increased the potential energy of the book

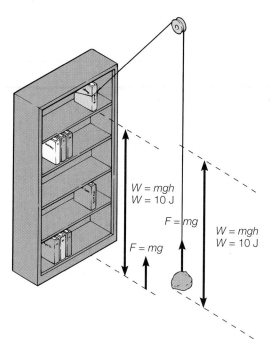

FIGURE 3.6

If moving a book from the floor to a high shelf requires 10 J of work, then the book will do 10 J of work on an object of the same mass when the book falls from the shelf.

by 10 J. The book now has the *potential* of doing 10 J of additional work on something else, therefore,

Work done on an object to change position	=	Increase in potential energy	=	Increase in work the object can do
work on book	=	potential energy of book	=	work by book
(10 J)		(10 J)		(10 J)

As you can see, a joule is a measure of work accomplished on an object. A joule is also a measure of potential energy. And, a joule is a measure of how much work an object can do. Both work and energy are measured in joules (or ft·lbs).

The potential energy of an object can be calculated, as described previously, from the work done *on* the object to change its position. You exert a force equal to its weight as you lift it some height above the floor, and the work you do is the product of the weight and height. Likewise, the amount of work the object *could* do because of its position is the product of its weight and height. For the metric unit of mass, weight is the product of the mass of an object times *g*, the acceleration due to gravity, so

$$\text{potential energy} = \text{weight} \times \text{height}$$

$$PE = mgh$$

equation 3.3

For English units, the pound *is* the gravitational unit of force, or weight, so equation 3.3 becomes $PE = (w)(h)$.

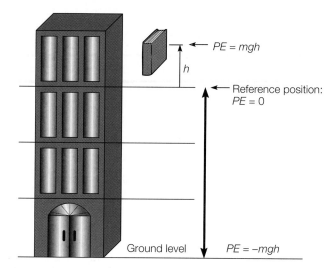

FIGURE 3.7

The zero reference level for potential energy is chosen for convenience. Here the reference position chosen is the third floor, so the book will have a negative potential energy at ground level.

Under what conditions does an object have zero potential energy? Considering the book in the bookcase, you could say that the book has zero potential energy when it is flat on the floor. It can do no work when it is on the floor. But what if that floor happens to be the third floor of a building? You could, after all, drop the book out of a window. The answer is that it makes no difference. The same results would be obtained in either case since it is the *change of position* that is important in potential energy. The zero reference position for potential energy is therefore arbitrary. A zero reference point is chosen as a matter of convenience. Note that if the third floor of a building is chosen as the zero reference position, a book on ground level would have negative potential energy. This means that you would have to do work on the book to bring it back to the zero potential energy position (Figure 3.7). You will learn more about negative energy levels later in the chapters on chemistry.

EXAMPLE 3.5

What is the potential energy of a 2.0 lb book that is on a bookshelf 4.0 ft above the floor?

Solution

Equation 3.3, *PE = mgh,* shows the relationship between potential energy (*PE*), weight (*mg*), and height (*h*).

$w = 2.0$ lb	$PE = mgh = (w)(h)$
$h = 4.0$ ft	$= (2.0 \text{ lb})(4.0 \text{ ft})$
$PE = ?$	$= (2.0)(4.0) \text{ ft·lb}$
	$= \boxed{8.0 \text{ ft·lb}}$

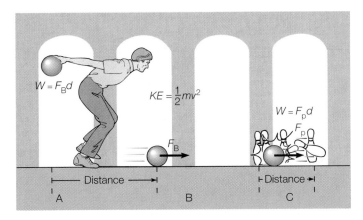

FIGURE 3.8

(*A*) Work is done on the bowling ball as a force (F_B) moves it through a distance. (*B*) This gives the ball a kinetic energy equal in amount to the work done on it. (*C*) The ball does work on the pins and has enough remaining energy to crash into the wall behind the pins.

EXAMPLE 3.6

How much work can a 5.00 kg mass do if it is 5.00 m above the ground? (Answer: 245 J)

Kinetic Energy

Moving objects have the ability to do work on other objects because of their motion. A rolling bowling ball exerts a force on the bowling pins and moves them through a distance, but the ball loses speed as a result of the interaction (Figure 3.8). A moving car has the ability to exert a force on a small tree and knock it down, again with a corresponding loss of speed. Objects in motion have the ability to do work, so they have energy. The energy of motion is known as **kinetic energy.** Kinetic energy can be measured (1) in terms of the work done to put the object in motion or (2) in terms of the work the moving object will do in coming to rest. Consider objects that you put into motion by throwing. You exert a force on a football as you accelerate it through a distance before it leaves your hand. The kinetic energy that the ball now has is equal to the work (force times distance) that you did on the ball. You exert a force on a baseball through a distance as the ball increases its speed before it leaves your hand. The kinetic energy that the ball now has is equal to the work that you did on the ball. The ball exerts a force on the hand of the person catching the ball and moves it through a distance. The net work done on the hand is equal to the kinetic energy that the ball had. Therefore,

Work done to put an object in motion	=	Increase in kinetic energy	=	Increase in work the object can do

A baseball and a bowling ball moving with the same velocity do not have the same kinetic energy. You cannot knock down many bowling pins with a slowly rolling baseball. Obviously the more massive bowling ball can do much more work than a less massive baseball with the same velocity. Is it possible for the bowling ball and the baseball to have the same kinetic energy? The answer is yes, if you can give the baseball sufficient velocity. This might require shooting the baseball from a cannon, however. Kinetic energy is proportional to the mass of a moving object, but velocity has a greater influence. Consider two balls of the same mass, but one is moving twice as fast as the other. The ball with twice the velocity will do *four* times as much work as the slower ball. A ball with three times the velocity will do *nine* times as much work as the slower ball. Kinetic energy is proportional to the square of the velocity ($2^2 = 4$; $3^2 = 9$). The kinetic energy (*KE*) of an object is

$$\text{kinetic energy} = \frac{1}{2}(\text{mass})(\text{velocity})^2$$

$$KE = \frac{1}{2}mv^2$$

equation 3.4

The unit of mass is the kg, and the unit of velocity is m/s. Therefore the unit of kinetic energy is

$$KE = (\text{kg})\left(\frac{\text{m}}{\text{s}}\right)^2$$

$$= (\text{kg})\left(\frac{\text{m}^2}{\text{s}^2}\right)$$

$$= \frac{\text{kg·m}^2}{\text{s}^2}$$

which is the same thing as

$$\left(\frac{\text{kg·m}}{\text{s}^2}\right)(\text{m})$$

or

N·m

or

joule (J)

Kinetic energy is measured in joules.

EXAMPLE 3.7

A 7.00 kg bowling ball is moving in a bowling lane with a velocity of 5.00 m/s. What is the kinetic energy of the ball?

Solution

The relationship between kinetic energy (KE), mass (m), and velocity (v) is found in equation 3.4, $KE = 1/2\, mv^2$:

$$m = 7.00 \text{ kg} \qquad KE = \frac{1}{2} mv^2$$

$$v = 5.00 \text{ m/s}$$

$$KE = ?$$

$$= \frac{1}{2}(7.00 \text{ kg})\left(5.00\,\frac{\text{m}}{\text{s}}\right)^2$$

$$= \frac{1}{2}(7.00 \times 25.0)\text{ kg} \times \frac{\text{m}^2}{\text{s}^2}$$

$$= \frac{1}{2}175\,\frac{\text{kg·m}^2}{\text{s}^2}$$

$$= 87.5\,\frac{\text{kg·m}}{\text{s}^2}\cdot\text{m}$$

$$= 87.5 \text{ N·m}$$

$$= \boxed{87.5 \text{ J}}$$

EXAMPLE **3.8**

A 100.0 kg football player moving with a velocity of 6.0 m/s tackles a stationary quarterback. How much work was done on the quarterback? (Answer: 1,800 J)

Concepts Applied

Energy Converter Coaster

A roller coaster is an energy converter that swaps kinetic and potential energy back and forth. The only time an outside energy source is used is to move the cars to the top of the first hill. The first hill is always the highest hill above the ground, and here the cars will have the most potential energy they will have for the entire ride.

When the coaster reaches the top of the first hill, it begins to move down a sloping track with increasing speed as potential energy is converted to kinetic energy. At the bottom of this hill the track then starts up a second hill, and the cars this time convert kinetic energy back to potential energy. Ideally, all the potential energy is converted to kinetic energy as the cars move down a hill, and all the kinetic energy is converted to potential energy as the cars move up a hill. There is no perfect energy conversion, and some energy is lost to air resistance and friction. The roller coaster design allows for these losses and leaves room

for a slight surplus of kinetic energy. You know this is true because the operator must apply brakes at the end of the ride.

Is the speed of the moving roller coaster at the bottom of a hill directly proportional to the height of the hill? Ignoring friction, the speed of a coaster at the bottom of a hill is proportional to the *square root of the height* of the hill. This means that to double the speed of the coaster at the bottom of the hill, you would need to build the hill more than four times higher. It would need to be more than four times higher since the coaster does not drop straight down. Doubling the height increases the theoretical speed only 40 percent.

What is the relationship between the weight of the people on a roller coaster and the speed achieved at the bottom of the hill? The answer is that the weight of the people does not matter. Ignoring air resistance and friction, a heavy roller coaster full of people and a lighter one with just a few people will both have the same speed at the bottom of the hill.

Draw a profile of a roller coaster ride to find out about potential and kinetic energy exchanges and the height of each hill. The profile should show the relative differences in height between the crest of each hill and the bottom of each upcoming dip. From such a profile you could find where in a ride the maximum speed would occur, as well as what speed to expect. As you study a profile, keep in mind that the roller coaster is designed to produce many changes of speed—accelerations—rather than high speed alone.

Energy Flow

The key to understanding the individual concepts of work and energy is to understand the close relationship between the two. When you do work on something you give it energy of position (potential energy) or you give it energy of motion (kinetic energy). In turn, objects that have kinetic or potential energy can now do work on something else as the transfer of energy continues. Where does all this energy come from and where does it go? The answer to this question is the subject of this section on energy flow.

Work and Energy

Energy is used to do work on an object, exerting a force through a distance. This force is usually *against* something (Figure 3.9), and there are five main groups of resistance:

1. *Work against inertia.* A net force that changes the state of motion of an object is working against inertia. According to the laws of motion, a net force acting through a distance is needed to change the velocity of an object. For uniform circular motion, work done by centripetal force is zero.

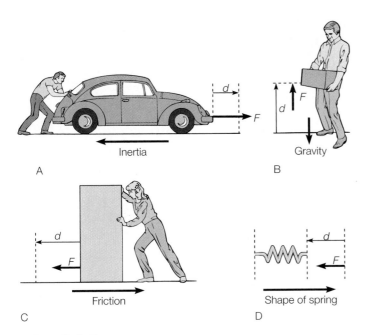

FIGURE 3.9

Examples of working against (A) inertia, (B) gravity, (C) friction, and (D) shape.

2. *Work against gravity.* Consider the force from gravitational attraction. A net force that changes the position of an object is a downward force from the acceleration due to gravity acting on a mass, $w = mg$. To change the position of an object, a force opposite to mg is needed to act through the distance of the position change. Thus lifting an object requires doing work against the force of gravity.

3. *Work against friction.* The force that is needed to maintain the motion of an object is working against friction. Friction is always present when two surfaces in contact move over each other. Friction resists motion.

4. *Work against shape.* The force that is needed to stretch or compress a spring is working against the shape of the spring. Other examples of work against shape include compressing or stretching elastic materials. If the elastic limit is reached, then the work goes into deforming or breaking the material.

5. *Work against any combination of inertia, fundamental forces, friction, and/or shape.* It is a rare occurrence on the earth that work is against only one type of resistance. Pushing on the back of a stalled automobile to start it moving up a slope would involve many resistances. This is complicated, however, so a single resistance is usually singled out for discussion.

Work is done against the main groups of resistances, but what is the result? The result is that some kind of *energy change* has taken place. Among the possible energy changes are the following:

1. *Increased kinetic energy.* Work against inertia results in an increase of kinetic energy, the energy of motion.
2. *Increased potential energy.* Work against gravity and work against shape result in an increase of potential energy, the energy of position.
3. *Increased temperature.* Work against friction results in an increase in the temperature. Temperature is a manifestation of the kinetic energy of the particles making up an object, as you will learn in the next chapter.

FIGURE 3.10

Mechanical energy is the energy of motion, or the energy of position, of many familiar objects. This boat has energy of motion.

4. *Increased combinations of kinetic energy, potential energy, and/or temperature.* Again, isolated occurrences are more the exception than the rule. In all cases, however, the sum of the total energy changes will be equal to the work done.

Work was done *against* various resistances, and energy was *increased* as a result. The object with increased energy can now do work on some other object or objects. A moving object has kinetic energy, so it has the ability to do work. An object with potential energy has energy of position, and it, too, has the ability to do work. You could say that energy *flowed* into and out of an object during the entire process. The following energy scheme is intended to give an overall conceptual picture of energy flow. Use it to develop a broad view of energy. You will learn the details later throughout the course.

Energy Forms

Energy comes in various forms, and different terms are used to distinguish one form from another. Although energy comes in various *forms,* this does not mean that there are different *kinds* of energy. The forms are the result of the more common fundamental forces—gravitational, electromagnetic, and nuclear—and objects that are interacting. Energy can be categorized into five forms: (1) *mechanical,* (2) *chemical,* (3) *radiant,* (4) *electrical,* and (5) *nuclear.* The following is a brief discussion of each of the five forms of energy.

Mechanical energy is the form of energy of familiar objects and machines (Figure 3.10). A car moving on a highway has kinetic mechanical energy. Water behind a dam has potential mechanical energy. The spinning blades of a steam turbine have kinetic mechanical energy. The form of mechanical energy is usually associated with the kinetic energy of everyday-sized objects and the potential energy that results from gravity. There are other possibilities (e.g., sound), but this description will serve the need for now.

A

B

FIGURE 3.11
Chemical energy is a form of potential energy that is released during a chemical reaction. Both (*A*) wood and (*B*) coal have chemical energy that has been stored through the process of photosynthesis. The pile of wood might provide fuel for a small fireplace for several days. The pile of coal might provide fuel for a power plant for a hundred days.

Chemical energy is the form of energy involved in chemical reactions (Figure 3.11). Chemical energy is released in the chemical reaction known as *oxidation*. The fire of burning wood is an example of rapid oxidation. A slower oxidation releases energy from food units in your body. As you will learn in the chemistry unit, chemical energy involves electromagnetic forces between the parts of atoms. Until then, consider the following comparison. Photosynthesis is carried on in green plants. The plants use the energy of sunlight to rearrange carbon dioxide and water into plant materials and oxygen. Leaving out many steps and generalizing, this reaction could be represented by the following word equation:

energy + carbon dioxide + water = wood + oxygen

FIGURE 3.12
This demonstration solar cell array converts radiant energy from the sun to electrical energy, producing an average of 200,000 watts of electric power (after conversion).

The plant took energy and two substances and made two different substances. This is similar to raising a book to a higher shelf in a bookcase. That is, the new substances have more energy than the original ones did. Consider a word equation for the burning of wood:

wood + oxygen = carbon dioxide + water + energy

Notice that this equation is exactly the reverse of photosynthesis. In other words, the energy used in photosynthesis was released during oxidation. Chemical energy is a kind of potential energy that is stored and later released during a chemical reaction.

Radiant energy is energy that travels through space (Figure 3.12). Most people think of light or sunlight when considering this form of energy. Visible light, however, occupies only a small part of the complete electromagnetic spectrum, as shown in Figure 3.13. Radiant energy includes light and all other parts of the spectrum. Infrared radiation is sometimes called "heat radiation" because of the association with heating when this type of radiation is absorbed. For example, you feel the interaction of infrared radiation when you hold your hand near a warm range element. However, infrared radiation is another type of radiant energy. In fact, some snakes, such as rattlesnakes, copperheads, and water mocassins, have pits between their eyes that can detect infrared radiation emitted from warm animals where you see total darkness. Microwaves are another type of radiant energy that are used in cooking. As with other forms of energy, light, infrared, and microwaves will be considered in more detail later. For now, consider all types of radiant energy to be forms of energy that travel through space.

Electrical energy is another form of energy from electromagnetic interactions that will be considered in detail later. You are familiar with electrical energy that travels through wires to your home from a power plant (Figure 3.14), electrical energy that is generated by chemical cells in a flashlight, and electrical energy that can be "stored" in a car battery.

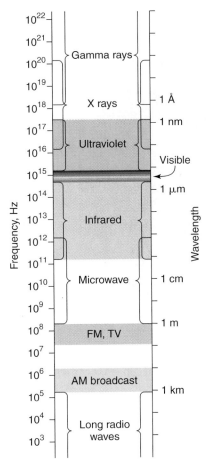

FIGURE 3.13

The electromagnetic spectrum. Notice that in some regions the names of the electromagnetic radiation overlap.

Nuclear energy is a form of energy often discussed because of its use as an energy source in power plants. Nuclear energy is another form of energy from the atom, but this time the energy involves the nucleus, the innermost part of an atom, and nuclear interactions.

Energy Conversion

Potential energy can be converted to kinetic energy and vice versa. The simple pendulum offers a good example of this conversion. A simple pendulum is an object, called a bob, suspended by a string or wire from a support. If the bob is moved to one side and then released, it will swing back and forth in an arc. At the moment that the bob reaches the top of its swing, it stops for an instant, then begins another swing. At the instant of stopping, the bob has 100 percent potential energy and no kinetic energy. As the bob starts back down through the swing, it is gaining kinetic energy and losing potential energy. At the instant the bob is at the bottom of the swing, it has 100 percent kinetic energy and no potential energy. As the bob now climbs through the other half of the arc, it is gaining potential energy and losing kinetic energy until it again reaches an instantaneous stop at the top, and the process

starts over. The kinetic energy of the bob at the bottom of the arc is equal to the potential energy it had at the top of the arc (Figure 3.15). Disregarding friction, the sum of the potential energy and the kinetic energy remains constant throughout the swing.

The potential energy lost during a fall equals the kinetic energy gained (Figure 3.16). In other words,

$$PE_{lost} = KE_{gained}$$

Substituting the values from equations 3.3 and 3.4,

$$mgh = \frac{1}{2}mv^2$$

Canceling the m and solving for v_f,

$$v_f = \sqrt{2gh}$$

equation 3.5

Equation 3.5 tells you the final speed of a falling object after its potential energy is converted to kinetic energy. This assumes, however, that the object is in free fall, since the effect of air resistance is ignored. Note that the m's cancel, showing again that the mass of an object has no effect on its final speed.

EXAMPLE 3.9

A 1.0 kg book falls from a height of 1.0 m. What is its velocity just as it hits the floor?

Solution

The relationships involved in the velocity of a falling object are given in equation 3.5.

$$h = 1.0 \text{ m}$$
$$g = 9.8 \text{ m/s}^2$$
$$v_f = ?$$

$$v_f = \sqrt{2gh}$$
$$= \sqrt{(2)(9.8 \text{ m/s}^2)(1.0 \text{ m})}$$
$$= \sqrt{2 \times 9.8 \times 1.0 \frac{m}{s^2} \cdot m}$$
$$= \sqrt{19.6 \frac{m^2}{s^2}}$$
$$= \boxed{4.4 \text{ m/s}}$$

EXAMPLE 3.10

What is the kinetic energy of a 1.0 kg book just before it hits the floor after a 1.0 m fall? (Answer: 9.8 J)

Any *form* of energy can be converted to another form. In fact, most technological devices that you use are nothing more

FIGURE 3.14

The blades of a steam turbine. In a power plant, chemical or nuclear energy is used to heat water to steam, which is directed against the turbine blades. The mechanical energy of the turbine turns an electric generator. Thus, a power plant converts chemical or nuclear energy to mechanical energy, which is then converted to electrical energy.

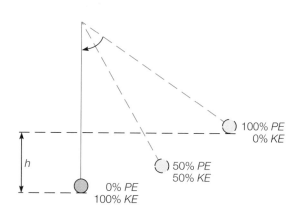

100% *PE*
0% *KE*

50% *PE*
50% *KE*

h

0% *PE*
100% *KE*

FIGURE 3.15

This pendulum bob loses potential energy (*PE*) and gains an equal amount of kinetic energy (*KE*) as it falls through a distance *h*. The process reverses as the bob moves up the other side of its swing.

10 m
(height of release)

$PE = mgh = 98$ J
$v = \sqrt{2gh} = 0$ (at time of release)
$KE = 1/2mv^2 = 0$

5 m

$PE = mgh = 49$ J
$v = \sqrt{2gh} = 9.9$ m/s
$KE = 1/2mv^2 = 49$ J

0 m

$PE = mgh = 0$ (as it hits)
$v = \sqrt{2gh} = 14$ m/s
$KE = 1/2mv^2 = 98$ J

FIGURE 3.16

The ball trades potential energy for kinetic energy as it falls. Notice that the ball had 98 J of potential energy when dropped and has a kinetic energy of 98 J just as it hits the ground.

than *energy-form converters* (Figure 3.17). A lightbulb, for example, converts electrical energy to radiant energy. A car converts chemical energy to mechanical energy. A solar cell converts radiant energy to electrical energy, and an electric motor converts electrical energy to mechanical energy. Each technological device converts some form of energy (usually chemical or electrical) to another form that you desire (usually mechanical or radiant).

It is interesting to trace the *flow of energy* that takes place in your surroundings. Suppose, for example, that you are riding a bicycle. The bicycle has kinetic mechanical energy as it moves along.

Where did the bicycle get this energy? From you, as you use the chemical energy of food units to contract your muscles and move the bicycle along. But where did your chemical energy come from? It came from your food, which consists of plants, animals who eat plants, or both plants and animals. In any case, plants are at the bottom of your food chain. Plants convert radiant energy from the sun into chemical energy. Radiant energy comes to the plants from the sun because of the nuclear reactions that took place in the core of the sun. Your bicycle is therefore powered by nuclear energy that has undergone a number of form conversions!

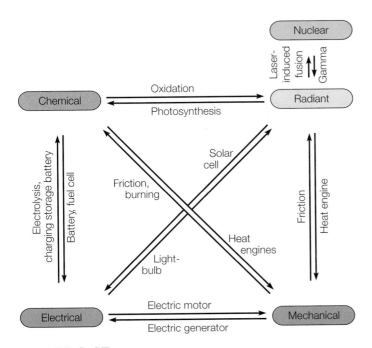

FIGURE 3.17
The energy forms and some conversion pathways.

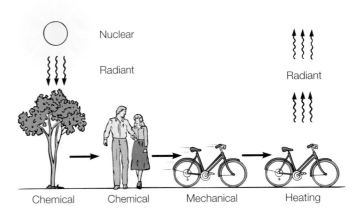

FIGURE 3.18
Energy arrives from the sun, goes through a number of conversions, then radiates back into space. The total sum leaving eventually equals the original amount that arrived.

Energy Conservation

Energy can be transferred from one object to another, and it can be converted from one form to another form. If you make a detailed accounting of all forms of energy before and after a transfer or conversion, the total energy will be *constant*. Consider your bicycle coasting along over level ground when you apply the brakes. What happened to the kinetic mechanical energy of the bicycle? It went into heating the rim and brakes of your bicycle, then eventually radiated to space as infrared radiation. All radiant energy that reaches the earth is eventually radiated back to space (Figure 3.18). Thus, throughout all the form conversions and energy transfers that take place, the total sum of energy remains constant.

The total energy is constant in every situation that has been measured. This consistency leads to another one of the conservation laws of science, the **law of conservation of energy:**

> **Energy is never created or destroyed. Energy can be converted from one form to another but the total energy remains constant.**

You may be wondering about the source of nuclear energy. Does a nuclear reaction create energy? Albert Einstein answered this question back in the early 1900s when he formulated his now-famous relationship between mass and energy, $E = mc^2$. This relationship will be discussed in detail in chapter 14. Basically, the relationship states that mass *is* a form of energy, and this has been experimentally verified many times.

Energy Transfer

Earlier it was stated that when you do work on something, you give it energy. The result of work could be increased kinetic mechanical energy, increased gravitational potential energy, or an increase in the temperature of an object. You could summarize this by stating that either *working* or *heating* is always involved any time energy is transformed. This is not unlike your financial situation. In order to increase or decrease your financial status, you need some mode of transfer, such as cash or checks, as a means of conveying assets. Just as with cash flow from one individual to another, energy flow from one object to another requires a mode of transfer. In energy matters the mode of transfer is working or heating. Any time you see working or heating occurring you know that an energy transfer is taking place. The next time you see heating, think about what energy form is being converted to what new energy form. (The final form is usually radiant energy.) Heating is the topic of chapter 4, where you will consider the role of heat in energy matters.

Energy Sources Today

Prometheus, according to ancient Greek mythology, stole fire from heaven and gave it to humankind. Fire has propelled human advancement ever since. All that was needed was something to burn—fuel for Prometheus's fire.

Any substance that burns can be used to fuel a fire, and various fuels have been used over the centuries as humans advanced. First, wood was used as a primary source for heating. Then coal fueled the industrial revolution. Eventually, humankind roared into the twentieth century burning petroleum. According to a 1999 report on primary energy consumed in the United States, petroleum was the most widely used source of energy (Figure 3.19). It provided about 40 percent of the total energy used, and natural gas contributed about 23 percent of the total. The use of coal provided about 21 percent of the total. Biomass, which is any material formed by photosynthesis, contributed about 3 percent of the total. Note that petroleum, coal, biomass, and natural gas are all chemical sources of energy, sources that are mostly burned for their energy. These chemical sources supplied about 89 percent of the total energy consumed.

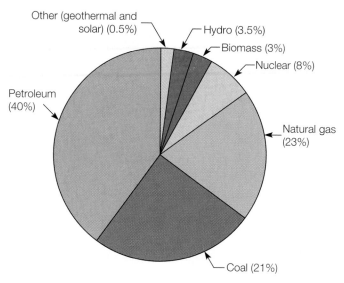

FIGURE 3.19

Primary energy consumed in the United States by source, 1999.

*Source: Energy Information Administration (www.eia.doe.gov/emea/sep/us/frame.html)

About a third of this was burned for heating, and the rest was burned to drive engines or generators.

Nuclear energy and hydropower are the nonchemical sources of energy. These sources are used to generate electrical energy. The alternative sources of energy, such as solar and geothermal, provided about 0.5 percent of the total energy consumed.

The energy-source mix has changed from past years, and it will change in the future. Wood supplied 90 percent of the energy until the 1850s, when the use of coal increased. Then, by 1910, coal was supplying about 75 percent of the total energy needs. Then petroleum began making increased contributions to the energy supply. Now increased economic and environmental constraints and a decreasing supply of petroleum are producing another supply shift. The present petroleum-based energy era is about to shift to a new energy era.

About 95 percent of the total energy consumed today is provided by four sources, (1) petroleum (including natural gas), (2) coal, (3) hydropower, and (4) nuclear. The following is a brief introduction to these four sources.

Petroleum

The word *petroleum* is derived from the word "petra," meaning rock, and the word "oleum," meaning oil. Petroleum is oil that comes from oil-bearing rock. Natural gas is universally associated with petroleum and has similar origins. Both petroleum and natural gas form from organic sediments, materials that have settled out of bodies of water. Sometimes a local condition permits the accumulation of sediments that are exceptionally rich in organic material. This could occur under special conditions in a freshwater lake, or it could occur on shallow ocean basins. In either case, most of the organic material is from plankton—tiny free-floating animals and plants such as algae. It is from such accumulations of buried organic material that

petroleum and natural gas are formed. The exact process by which these materials become petroleum and gas is not understood. It is believed that bacteria, pressure, appropriate temperatures, and time are all important. Natural gas is formed at higher temperatures than is petroleum. Varying temperatures over time may produce a mixture of petroleum and gas or natural gas alone.

Petroleum forms a thin film around the grains of the rock where it formed. Pressure from the overlying rock and water move the petroleum and gas through the rock until it reaches a rock type or structure that stops it. If natural gas is present it occupies space above the accumulating petroleum. Such accumulations of petroleum and natural gas are the sources of supply for these energy sources.

Discussions about the petroleum supply and the cost of petroleum usually refer to a "barrel of oil." The *barrel* is an accounting device of 42 United States gallons. Such a 42-gallon barrel does not exist. When or if oil is shipped in barrels, each drum holds 55 United States gallons. The various uses of petroleum products are discussed in chapter 13.

The supply of petroleum and natural gas is limited. Most of the continental drilling prospects appear to be exhausted, and the search for new petroleum supplies is now offshore. In general, over 25 percent of our nation's petroleum is estimated to come from offshore wells. Imported petroleum accounts for more than half of the oil consumed, with most imported oil coming from Mexico, Canada, Venezuela, Nigeria, and Saudi Arabia.

Petroleum is used for gasoline (about 45 percent), diesel (about 40 percent), and heating oil (about 15 percent). Petroleum is also used in making medicine, clothing fabrics, plastics, and ink.

Coal

Petroleum and natural gas formed from the remains of tiny organisms that lived millions of years ago. Coal, on the other hand, formed from an accumulation of plant materials that collected under special conditions millions of years ago. Thus, petroleum, natural gas, and coal are called **fossil fuels.** Fossil fuels contain the stored radiant energy of organisms that lived millions of years ago.

The first thing to happen in the formation of coal was that plants in swamps died and sank. Stagnant swamp water protected the plants and plant materials from consumption by animals and decomposition by microorganisms. Over time, chemically altered plant materials collected at the bottom of pools of water in the swamp. This carbon-rich material is *peat* (not to be confused with peat moss). Peat is used as a fuel in many places in the world. The flavor of Scotch (whisky) is the result of the peat fires used to brew the liquor. Peat is still being produced naturally in swampy areas today. Under pressure and at high temperatures peat will eventually be converted to coal. There are several stages, or *ranks*, in the formation of coal. The lowest rank is lignite (brown coal), and then subbituminous, then bituminous (soft coal), and the highest rank is anthracite (hard coal).

Each rank of coal has different burning properties and a different energy content. Coal also contains impurities of clay, silt, iron oxide, and sulfur. The mineral impurities leave an ash

An alternative source of energy is one that is different from the typical sources used today. The sources used today are the fossil fuels (coal, petroleum, and natural gas), nuclear, and falling water. Alternative sources could be solar, geothermal, hydrogen gas, fusion, or any other energy source that a new technology could utilize.

The term *solar energy* is used to describe a number of technologies that directly or indirectly utilize sunlight as an alternative energy source (Box Figure 3.3). There are eight main categories of these solar technologies:

1. *Solar cells.* A solar cell is a thin crystal of silicon, gallium, or some polycrystalline compound that generates electricity when exposed to light. Also called *photovoltaic devices,* solar cells have no moving parts and produce electricity directly, without the need for hot fluids or intermediate conversion states. Solar cells have been used extensively in space vehicles and satellites. Here on the earth, however, use has been limited to demonstration projects, remote site applications, and consumer specialty items such as solar-powered watches and calculators. The problem with solar cells today is that the manufacturing cost is too high (they are essentially handmade). Research is continuing on the development of highly efficient, affordable solar cells that could someday produce electricity for the home.

2. *Power tower.* This is another solar technology designed to generate electricity. One type of planned power tower will have a 171 m (560 ft) tower surrounded by some 9,000 special mirrors called heliostats. The heliostats will focus sunlight on a boiler at the top of the tower where salt (a mixture of sodium nitrate and potassium nitrate) will be heated to about 566°C (about 1,050°F). This molten salt will be pumped to a steam generator,

BOX FIGURE 3.3

Wind is another form of solar energy. This wind turbine generates electrical energy for this sailboat, charging batteries for backup power when the wind is not blowing. In case you are wondering, the turbine cannot be used to make a wind to push the boat along. In accord with Newton's laws of motion, this would not produce a net force on the boat.

and the steam will be used to drive a generator, just like other power plants. Water could be heated directly in the power tower boiler. Molten salt is used because it can be stored in an insulated storage tank for use when the sun is not shining, perhaps for up to twenty hours.

3. *Passive application.* In passive applications energy flows by natural means, without mechanical devices such as motors, pumps, and so forth. A passive solar house would include considerations like the orientation of a house to the sun, the size and positioning of windows, and a roof overhang that lets sunlight in during the winter but keeps it out during the summer. There are different design plans to capture, store, and distribute solar energy throughout a house.

4. *Active application.* An active solar application requires a solar collector in which sunlight heats air, water, or some liquid. The liquid or air is pumped through pipes in a house to generate electricity, or it is used directly for hot water. Solar water heating makes more economic sense today than the other applications.

5. *Wind energy.* The wind has been used for centuries to move ships, grind grain into flour, and pump water. The wind blows, however, because radiant energy from the sun heats some parts of the earth's surface more than other parts. This differential heating results in pressure differences and the horizontal movement of air, which is called *wind.* Thus, wind is another form of solar

—Continued top of next page

energy. Wind turbines are used to generate electrical energy or mechanical energy. The biggest problem with wind energy is the inconsistency of the wind. Sometimes the wind speed is too great, and other times it is not great enough. Several methods of solving this problem are being researched.

6. *Biomass.* Biomass is any material formed by photosynthesis, including small plants, trees, and crops, and any garbage, crop residue, or animal waste. Biomass can be burned directly as a fuel, converted into a gas fuel (methane), or converted into liquid fuels such as alcohol. The problems with using biomass include the energy expended in gathering the biomass, as well as the energy used to convert it to a gaseous or liquid fuel.

7. *Agriculture and industrial heating.* This is a technology that simply uses sunlight to dry grains, cure paint, or do anything that can be done with sunlight rather than using traditional energy sources.

8. *Ocean thermal energy conversion (OTEC).* This is an electric generating plant that would take advantage of the approximately 22°C (about 40°F) temperature difference between the surface and the depths of tropical, subtropical, and equatorial ocean waters.

Basically, warm water is drawn into the system to vaporize a fluid, which expands through a turbine generator. Cold water from the depths condenses the vapor back to a liquid form, which is then cycled back to the warm-water side. The concept has been tested several times and was found to be technically successful. The greatest interest seems to be islands that have warm surface waters (and cold depths) such as Hawaii, Puerto Rico, Guam, the Virgin Islands, and others.

---

when the coal is burned, and the sulfur produces sulfur dioxide, a pollutant.

Most of the coal mined today is burned by utilities to generate electricity (about 80 percent). The coal is ground to a face-powder consistency and blown into furnaces. This greatly increases efficiency but produces *fly ash,* ash that "flies" up the chimney. Industries and utilities are required by the Federal Clean Air Act to remove sulfur dioxide and fly ash from plant emissions. About 20 percent of the cost of a new coal-fired power plant goes into air pollution control equipment. Coal is an abundant but dirty energy source.

Moving Water

Moving water has been used as a source of energy for thousands of years. It is considered a renewable energy source, inexhaustible as long as the rain falls. Today, hydroelectric plants generate about 3 percent of the nation's *total* energy consumption at about 2,400 power-generating dams across the nation. Hydropower furnished about 40 percent of the United States' electric power in 1940. Today, dams furnish 9 percent of the electric power. It is projected that this will drop even lower, perhaps to 7 percent in the near future. Energy consumption has increased, but hydropower production has not kept pace because geography limits the number of sites that can be built.

Water from a reservoir is conducted through large pipes called penstocks to a powerhouse, where it is directed against turbine blades that turn a shaft on an electric generator. A rough approximation of the power that can be extracted from the falling water can be made by multiplying the depth of the

Concepts Applied

City Power

Compare amounts of energy sources needed to produce electric power. Generally, 1 MW (1,000,000 W) will supply the electrical needs of 1,000 people.

1. Use the population of your city to find how many megawatts of electricity are required for your city.
2. Use the following equivalencies to find out how much coal, oil, gas, or uranium would be consumed in one day to supply the electrical needs.

$$\frac{1 \text{ kWh}}{\text{of electricity}} = \begin{array}{l} 1 \text{ lb of coal} \\ 0.08 \text{ gal of oil} \\ 9 \text{ cubic ft of gas} \\ 0.00013 \text{ g of uranium} \end{array}$$

Example

Assume your city has 36,000 people. Then 36 MW of electricity will be needed. How much oil is needed to produce this electricity?

$$36\,\text{MW} \times \frac{1{,}000 \text{ kW}}{\text{MW}} \times \frac{24 \text{ h}}{\text{day}} \times \frac{0.08 \text{ gal}}{\text{kWh}} = \begin{array}{l} 69{,}120 \text{ or about} \\ 70{,}000 \text{ gal/day} \end{array}$$

Since there are 42 gallons in a barrel,

$$\frac{70{,}000 \text{ gal/day}}{42 \text{ gal/barrel}} = \frac{70{,}000}{42} \times \frac{\text{gal}}{\text{day}} \times \frac{\text{barrel}}{\text{gal}} = \begin{array}{l} 1{,}666, \text{ or about} \\ 2{,}000 \text{ barrel/day} \end{array}$$

James Prescott Joule (1818–1889)

James Joule was a British physicist who verified the principle of conservation of energy by making the first accurate determination of the mechanical equivalent of heat. He also discovered Joule's law, which defines the relation between heat and electricity, and with Lord Kelvin (1824–1907) the Joule-Thomson effect. In recognition of Joule's pioneering work on energy, the SI unit of energy is named the joule.

Joule was born at Salford on December 24, 1818, into a wealthy brewing family. He and his brother were educated at home between 1833 and 1837 in elementary mathematics, natural philosophy, and chemistry, partly by John Dalton (1766–1844). Joule was a delicate child and very shy, and apart from his early education he was entirely self-taught in science. He does not seem to have played any part in the family brewing business, although some of his first experiments were done in a laboratory at the brewery.

Joule had great dexterity as an experimenter, and he could measure temperatures very precisely. At first, other scientists could not credit such accuracy and were disinclined to believe the theories that Joule developed to explain his results. The encouragement of Lord Kelvin from 1847 changed these attitudes, however, and Kelvin subsequently used Joule's practical ability to great advantage. By 1850, Joule was highly thought of among scientists and became a Fellow of the Royal Society. He was awarded the Society's Copley Medal in 1866 and was president of the British Association for the Advancement of Science in 1872 and again in 1887. Joule's own wealth was able to fund his scientific career, and he never took an academic post. His funds eventually ran out, however. He was awarded a pension in 1878 by Queen Victoria, but by that time his mental powers were going. He suffered a long illness and died in Sale, Cheshire, on October 11, 1889.

Joule realized the importance of accurate measurement very early on, and exact quantitative data became his hallmark. His most active research period was between 1837 and 1847 and led to the establishment of the principle of conservation of energy and the equivalence of heat and other forms of energy. In a long series of experiments, he studied the quantitative relationship between electrical, mechanical, and chemical effects and heat, and in 1843 he was able to announce his determination of the amount of work required to produce a unit of heat. This is called the mechanical equivalent of heat (currently accepted value 4.1868 joules per calorie).

Joule's first experiments related the chemical and electrical energy expended to the heat produced in metallic conductors and voltaic and electrolytic cells. These results were published between 1840 and 1843. He proved the relationship, known as Joule's law, that the heat produced in a conductor of resistance R by a current I is proportional to I^2R per second. He went on to discuss the relationship between heat and mechanical power in 1843. Joule first measured the rise in temperature and the current and the mechanical work involved when a small electromagnet rotated in water between the poles of another magnet, his training for these experiments having

been provided by early research with William Sturgeon (1783–1850), a pioneer of electromagnetism. Joule then checked the rise in temperature by a more accurate experiment, forcing water through capillary tubes. The third method depended on the compression of air, and the fourth produced heat from friction in water using paddles that rotated under the action of a falling weight. This has become the best-known method for the determination of the mechanical equivalent of heat. Joule showed that the results obtained using different liquids (water, mercury, and sperm oil) were the same. In the case of water, 772 ft·lb of work produced a rise of 1°F. This value was universally accepted as the mechanical equivalent of heat.

The great value of Joule's work in demonstrating the conservation of energy lay in the variety and completeness of his experimental evidence. He showed that the same relationship held in all cases that could be examined experimentally and that the ratio of equivalence of the different forms of energy did not depend on how one form was converted into another or on the materials involved. The principle that Joule had established is in fact the first law of thermodynamics: that energy cannot be created or destroyed, but only transformed.

Because he had not received any formal mathematical training, Joule was unable to keep up with the new science of thermodynamics to which he had made such a fundamental and important contribution. However, the presentation of his final work on the mechanical equivalent of heat in 1847 attracted great interest and support from William Thomson, then only 22 and later to become Lord Kelvin. Much of Joule's later work was carried out with him, for Kelvin had need of Joule's experimental prowess to put his ideas on

—*Continued top of next page*

Continued—

thermodynamics into practice. This led in 1852 to the discovery of the Joule-Thomson effect, which produces cooling in a gas when the gas expands freely. The effect is caused by the conversion of heat into work done by the molecules in overcoming attractive forces between them as they move apart. It was to prove vital to techniques in the liquefaction of gases and low-temperature physics.

Joule lives on in the use of his name to measure energy, supplanting earlier units such as the erg and calorie. It is an appropriate reflection of his great experimental ability and his tenacity in establishing a basic law of science.

water (in feet) by the amount of water flowing (in cubic feet per second), then dividing by 10. The result is roughly equal to the horsepower.

Nuclear

Nuclear power plants use nuclear energy to produce electricity. Energy is released as the nuclei of uranium and plutonium atoms split, or undergo fission. The fissioning takes place in a large steel vessel called a *reactor*. Water is pumped through the reactor to produce steam, which is used to produce electrical energy, just as in the fossil fuel power plants. The nuclear processes are described in detail in chapter 14, and the process of producing electrical energy is described in detail in chapter 6. Nuclear power plants use nuclear energy to produce electricity, but some people oppose the use of this process. The electric utility companies view nuclear energy as *one* energy source used to produce electricity. They state that they have no allegiance to any one energy source but are seeking to utilize the most reliable and dependable of several energy sources. Petroleum, coal, and hydropower are also used as energy sources for electric power production. The electric utility companies are concerned that petroleum and natural gas are becoming increasingly expensive, and there are questions about long-term supplies. Hydropower has limited potential for growth, and solar energy is prohibitively expensive today. Utility companies see two major energy sources that are available for growth: coal and nuclear. There are problems and advantages to each, but the utility companies feel they must use coal and nuclear power until the new technologies, such as solar power, are economically feasible.

Summary

Work is defined as the product of an applied force and the distance through which the force acts. Work is measured in newton-meters, a metric unit called a *joule*. *Power* is work per unit of time. Power is measured in *watts*. One watt is 1 joule per second. Power is also measured in *horsepower*. One horsepower is 550 ft·lb/s.

Energy is defined as the ability to do work. An object that is elevated against gravity has a potential to do work. The object is said to have *potential energy*, or *energy of position*. Moving objects have the ability to do work on other objects because of their motion. The *energy of motion* is called *kinetic energy*.

Work is usually done *against inertia, gravity, friction, shape,* or *combinations of these.* As a result, there is a gain of *kinetic energy, potential energy, an increased temperature,* or *any combination of these.* Energy comes in the *forms of mechanical, chemical, radiant, electrical,* or *nuclear.* Potential energy can be converted to kinetic and kinetic can be *converted* to potential. Any form of energy can be *converted* to any other form. Most technological devices are *energy-form converters* that do work for you. Energy flows into and out of the surroundings, but the amount of energy is always constant. The *law of conservation of energy* states that *energy is never created or destroyed.* Energy conversion always takes place through *heating* or *working.*

The basic energy sources today are the chemical *fossil fuels* (petroleum, natural gas, and coal), *nuclear energy,* and *hydropower. Petroleum* and *natural gas* were formed from organic material of plankton, tiny free-floating plants and animals. A barrel of petroleum is 42 United States gallons, but such a container does not actually exist. *Coal* formed from plants that were protected from consumption by falling into a swamp. The decayed plant material, *peat,* was changed into the various *ranks* of coal by pressure and heating over some period of time. Coal is a dirty fuel that contains impurities and sulfur. Controlling air pollution from burning coal is costly. Water power and nuclear energy are used for the generation of electricity.

Summary of Equations

3.1

$$\text{work} = \text{force} \times \text{distance}$$

$$W = Fd$$

3.2

$$\text{power} = \frac{\text{work}}{\text{time}}$$

$$P = \frac{W}{t}$$

3.3

$$\text{potential energy} = \text{weight} \times \text{height}$$

$$PE = mgh$$

3.4

$$\text{kinetic energy} = \frac{1}{2}(\text{mass})(\text{velocity})^2$$

$$KE = \frac{1}{2}mv^2$$

3.5

$$\text{final velocity} = \text{square root of } (2 \times \begin{array}{c}\text{acceleration due}\\\text{to gravity}\end{array} \times \text{height of fall})$$

$$v_f = \sqrt{2gh}$$

KEY TERMS

chemical energy (p. **72**)

electrical energy (p. **72**)

energy (p. **67**)

fossil fuels (p. **76**)

horsepower (p. **66**)

joule (p. **62**)

kinetic energy (p. **69**)

law of conservation
of energy (p. **75**)

mechanical energy (p. **71**)

nuclear energy (p. **73**)

potential energy (p. **67**)

power (p. **66**)

radiant energy (p. **72**)

watt (p. **66**)

work (p. **62**)

APPLYING THE CONCEPTS

1. According to the scientific definition of work, pushing on a rock accomplishes no work unless there is
   a. movement.
   b. a net force.
   c. an opposing force.
   d. movement in the same direction as the direction of the force.

2. The metric unit of a joule (J) is a unit of
   a. potential energy.
   b. work.
   c. kinetic energy.
   d. any of the above.

3. Which of the following is the combination of units called a joule?
   a. $kg \cdot m/s^2$
   b. kg/s^2
   c. $kg \cdot m^2/s^2$
   d. $kg \cdot m/s$

4. The newton-meter is called a joule. The similar unit in the English system, the foot-pound, is called a
   a. horsepower.
   b. slug.
   c. gem.
   d. foot-pound, with no other name.

5. Power is
   a. the rate at which work is done.
   b. the rate at which energy is expended.
   c. work per unit time.
   d. any of the above.

6. A power rating of 600 ft·lb/s is
   a. more than 1 horsepower.
   b. exactly 1 horsepower.
   c. less than 1 horsepower.

7. A N·m/s is a unit of
   a. work.
   b. power.
   c. energy.
   d. none of the above.

8. Which of the following is a combination of units called a watt?
   a. N·m/s
   b. $kg \cdot m^2/s^2/s$
   c. J/s
   d. all of the above.

9. About how many watts are equivalent to 1 horsepower?
   a. 7.5
   b. 75
   c. 750
   d. 7,500

10. A kilowatt-hour is actually a unit of
    a. power.
    b. work.
    c. time.
    d. electrical charge.

11. In calculating the upward force required to lift an object, it is necessary to use *g* if the mass is given in kg. The quantity of *g* is not needed if the weight is given in lb because
    a. the rules of measurement are different in the English system.
    b. the symbol for metric mass has the letter "g" in it, and the symbol for pound does not.
    c. a pound is defined as a measure of force, and a kilogram is not.
    d. a kilogram is a unit of weight.

12. The potential energy of a box on a shelf, relative to the floor, is a measure of
    a. the work that was required to put the box on the shelf from the floor.
    b. the weight of the box times the distance above the floor.
    c. the energy the box has because of its position above the floor.
    d. all of the above.

13. A rock on the ground is considered to have zero potential energy. In the bottom of a well, then, the rock would be considered to have
    a. zero potential energy, as before.
    b. negative potential energy.
    c. positive potential energy.
    d. zero potential energy, but will require work to bring it back to ground level.

14. Which quantity has the greatest influence on the amount of kinetic energy that a large truck has while moving down the highway?
    a. mass
    b. weight
    c. velocity
    d. size

15. Which of the following is a form of energy that is a kind of potential energy?
    a. radiant
    b. electrical
    c. chemical
    d. None of the above is correct.

16. Electrical energy can be converted to
    a. chemical energy.
    b. mechanical energy.
    c. radiant energy.
    d. any of the above.

17. Most all energy comes to and leaves the earth in the form of
    a. nuclear energy.
    b. chemical energy.
    c. radiant energy.
    d. kinetic energy.

18. The law of conservation of energy is basically that
    a. energy must not be used up faster than it is created or the supply will run out.
    b. energy should be saved because it is easily destroyed.
    c. energy is never created or destroyed.
    d. you are breaking a law if you needlessly destroy energy.

19. The most widely used source of energy today is
    a. coal.
    b. petroleum.
    c. nuclear.
    d. water power.

20. The accounting device of a "barrel of oil" is defined to hold how many U.S. gallons of petroleum?
    a. 24
    b. 42
    c. 55
    d. 100

Answers

1. d 2. d 3. c 4. d 5. d 6. a 7. b 8. d 9. c 10. b 11. c 12. d 13. b 14. c 15. c 16. d 17. c 18. c 19. b 20. b

QUESTIONS FOR THOUGHT

1. How is work related to energy?
2. What is the relationship between the work done while moving a book to a higher bookshelf and the potential energy that the book has on the higher shelf?
3. Does a person standing motionless in the aisle of a moving bus have kinetic energy? Explain.
4. A lamp bulb is rated at 100 W. Why is a time factor not included in the rating?
5. Is a kWh a unit of work, energy, power, or more than one of these? Explain.
6. If energy cannot be destroyed, why do some people worry about the energy supplies?
7. A spring clamp exerts a force on a stack of papers it is holding together. Is the spring clamp doing work on the papers? Explain.
8. Why are petroleum, natural gas, and coal called *fossil fuels?*
9. From time to time, people claim to have invented a machine that will run forever without energy input and develops more energy than it uses (perpetual motion). Why would you have reason to question such a machine?
10. Define a joule. What is the difference between a joule of work and a joule of energy?
11. Compare the energy needed to raise a mass 10 m on Earth to the energy needed to raise the same mass 10 m on the Moon. Explain the difference, if any.
12. What happens to the kinetic energy of a falling book when the book hits the floor?

PARALLEL EXERCISES

The exercises in groups A and B cover the same concepts. Solutions to group A exercises are located in appendix D.

Note: *Neglect all frictional forces in all exercises.*

Group A

1. A force of 200 N is needed to push a table across a level classroom floor for a distance of 3 m. How much work was done on the table?
2. An 880 N box is pushed across a level floor for a distance of 5.0 m with a force of 440 N. How much work was done on the box?

Group B

1. How much work is done when a force of 800.0 N is exerted while pushing a crate across a level floor for a distance of 1.5 m?
2. A force of 400.0 N is exerted on a 1,250 N car while moving it a distance of 3.0 m. How much work was done on the car?

3. How much work is done in raising a 10.0 kg backpack from the floor to a shelf 1.5 m above the floor?

4. If 5,000 J of work is used to raise a 102 kg crate to a shelf in a warehouse, how high was the crate raised?

5. A 60.0 kg student runs up a 5.00-meter-high stairway in a time of 3.92 seconds. How many watts of power did she develop?

6. (a) How many horsepower is a 1,400 W blow dryer? (b) How many watts is a 3.5 hp lawnmower?

7. What is the kinetic energy of a 2,000 kg car moving at 72 km/h?

8. How much work is needed to stop a 1,000.0 kg car that is moving straight down the highway at 54.0 km/h?

9. A horizontal force of 10.0 lb is needed to push a bookcase 5 ft across the floor. (a) How much work was done on the bookcase? (b) How much did the gravitational potential energy change as a result?

10. (a) How much work is done in moving a 2.0 kg book to a shelf 2.00 m high? (b) What is the potential energy of the book as a result? (c) How much kinetic energy will the book have as it hits the ground when it falls?

11. A 150 g baseball has a velocity of 30.0 m/s. What is its kinetic energy in J?

12. (a) What is the kinetic energy of a 1,000.0 kg car that is traveling at 90.0 km/h? (b) How much work was done to give the car this kinetic energy? (c) How much work must be done to stop the car?

13. A 60.0 kg jogger moving at 2.0 m/s decides to double the jogging speed. How did this change in speed change the kinetic energy?

14. A bicycle and rider have a combined mass of 70.0 kg and are moving at 6.00 m/s. A 70.0 kg person is now given a ride on the bicycle. (Total mass is 140.0 kg.) How did the addition of the new rider change the kinetic energy at the same speed?

15. A 170.0 lb student runs up a stairway to a classroom 25.0 ft above ground level in 10.0 s. (a) How much work did the student do? (b) What was the average power output in hp?

16. (a) How many seconds will it take a 20.0 hp motor to lift a 2,000.0 lb elevator a distance of 20.0 ft? (b) What was the average velocity of the elevator?

17. A ball is dropped from 9.8 ft above the ground. Using energy considerations only, find the velocity of the ball just as it hits the ground.

18. What is the velocity of a 1,000.0 kg car if its kinetic energy is 200 kJ?

19. A Foucault pendulum swings to 3.0 in above the ground at the highest points and is practically touching the ground at the lowest point. What is the maximum velocity of the pendulum?

20. An electric hoist is used to lift a 250.0 kg load to a height of 80.0 m in 39.2 s. (a) What is the power of the hoist motor in kW? (b) in hp?

3. A 5.0 kg textbook is raised a distance of 30.0 cm as a student prepares to leave for school. How much work did the student do on the book?

4. An electric hoist does 196,000 J of work in raising a 250.0 kg load. How high was the load lifted?

5. What is the horsepower of a 1,500.0 kg car that can go to the top of a 360.0 m high hill in exactly 1.00 minute?

6. (a) How many horsepower is a 250 W lightbulb? (b) How many watts is a 230 hp car?

7. What is the kinetic energy of a 30-gram bullet that is traveling at 200 m/s?

8. How much work will be done by a 30-gram bullet traveling at 200 m/s?

9. A force of 50.0 lb is used to push a box 10.0 ft across a level floor. (a) How much work was done on the box? (b) What is the change of potential energy as a result of this move?

10. (a) How much work is done in raising a 50.0 kg crate a distance of 1.5 m above a storeroom floor? (b) What is the change of potential energy as a result of this move? (c) How much kinetic energy will the crate have as it falls and hits the floor?

11. What is the kinetic energy in J of a 60.0 g tennis ball approaching a tennis racket at 20.0 m/s?

12. (a) What is the kinetic energy of a 1,500.0 kg car with a velocity of 72.0 km/h? (b) How much work must be done on this car to bring it to a complete stop?

13. The driver of an 800.0 kg car decides to double the speed from 20.0 m/s to 40.0 m/s. What effect would this have on the amount of work required to stop the car, that is, on the kinetic energy of the car?

14. Compare the kinetic energy of an 800.0 kg car moving at 20.0 m/s to the kinetic energy of a 1,600.0 kg car moving at an identical speed.

15. A 175.0 lb hiker is able to ascend a 1,980.0 ft high slope in 1 hour and 45 minutes. (a) How much work did the hiker do? (b) What was the average power output in hp?

16. (a) How many seconds will it take a 10.0 hp motor to lift a 2,000.0 lb elevator a distance of 20.0 feet? (b) What was the average velocity of the elevator?

17. A ball is dropped from 20.0 ft above the ground. (a) At what height is half of its energy kinetic and half potential? (b) Using energy considerations only, what is the velocity of the ball just as it hits the ground?

18. What is the velocity of a 60.0 kg jogger with a kinetic energy of 1,080.0 J?

19. A small sports car and a pickup truck start coasting down a 10.0 m hill together, side by side. Assuming no friction, what is the velocity of each vehicle at the bottom of the hill?

20. A 70.0 kg student runs up the stairs of a football stadium to a height of 10.0 m above the ground in 10.0 s. (a) What is the power of the student in kW? (b) in hp?

Sparks fly from a plate of steel as it is cut by an infrared laser. Today, lasers are commonly used to cut as well as weld metals, so the cutting and welding are done by light, not by a flame or electric current.

CHAPTER

4

Heat and Temperature

Heat has been closely associated with the comfort and support of people throughout history. You can imagine the appreciation when your earliest ancestors first discovered fire and learned to keep themselves warm and cook their food. You can also imagine the wonder and excitement about 3000 B.C., when people put certain earthlike substances on the hot, glowing coals of a fire and later found metallic copper, lead, or iron. The use of these metals for simple tools followed soon afterward. Today, metals are used to produce complicated engines that use heat for transportation and that do the work of moving soil and rock, construction, and agriculture. Devices made of heat-extracted metals are also used to control the temperature of structures, heating or cooling the air as necessary. Thus, the production and control of heat gradually built the basis of civilization today (Figure 4.1).

The sources of heat are the energy forms that you learned about in chapter 3. The fossil fuels are *chemical* sources of heat. Heat is released when oxygen is combined with these fuels. Heat also results when *mechanical* energy does work against friction, such as in the brakes of a car coming to a stop. Heat also appears when *radiant* energy is absorbed. This is apparent when solar energy heats water in a solar collector or when sunlight melts snow. The transformation of *electrical* energy to heat is apparent in toasters, heaters, and ranges. *Nuclear* energy provides the heat to make steam in a nuclear power plant. Thus, all energy forms can be converted to heat.

The relationship between energy forms and heat appears to give an order to nature, revealing patterns that you will want to understand. All that you need is some kind of explanation for the relationships—a model or theory that helps make sense of it all. This chapter is concerned with heat and temperature and their relationship to energy. It begins with a simple theory about the structure of matter, and then uses the theory to explain the concepts of heat, energy, and temperature changes.

The Kinetic Molecular Theory

The idea that substances are composed of very small particles can be traced back to certain early Greek philosophers. The earliest record of this idea was written by Democritus during the fifth century B.C. He wrote that matter was empty space filled with tremendous numbers of tiny, indivisible particles called *atoms*. This idea, however, was not acceptable to most of the ancient Greeks, because matter seemed continuous, and empty space was simply not believable. The idea of atoms was rejected by Aristotle as he formalized his belief in continuous matter composed of the earth, air, fire, and water elements. Aristotle's belief about matter, like his beliefs about motion, predominated through the 1600s. Some people, such as Galileo and Newton, believed the ideas about matter being composed of tiny particles, or atoms, since this theory seemed to explain the behavior of matter. Widespread acceptance of the particle model did not occur, however, until strong evidence was developed through chemistry in the late 1700s and early 1800s. The experiments finally led to a collection of assumptions about the small particles of matter and the space around them. Collectively, the assumptions could be called the **kinetic molecular theory**. The following is a general description of some of these assumptions.

Molecules

The basic assumption of the kinetic molecular theory is that all matter is made up of tiny, basic units of structure called *atoms*. Atoms are neither divided, created, nor destroyed during any type of chemical or physical change. There are similar groups of atoms that make up the pure substances known as chemical *elements*. Each element has its own kind of atom, which is different from the atoms of other elements. For example, hydrogen, oxygen, carbon, iron, and gold are chemical elements, and each has its own kind of atom.

In addition to the chemical elements, there are pure substances called *compounds* that have more complex units of structure (Figure 4.2). Pure substances, such as water, sugar, and alcohol, are composed of atoms of two or more elements that join together in definite proportions. Water, for example, has structural units that are made up of two atoms of hydrogen tightly bound to one atom of oxygen (H_2O). These units are not easily broken apart and stay together as small physical particles of which water is composed. Each is the smallest particle of water that can exist, a molecule of water. A *molecule* is generally defined as a tightly bound group of atoms in which the atoms maintain their identity. How atoms become bound together to form molecules is discussed in chapters 8–10.

Some elements exist as gases at ordinary temperatures, and all elements are gases at sufficiently high temperatures. At ordinary temperatures, the atoms of oxygen, nitrogen, and other gases are paired in groups of two to form *diatomic molecules*. Other gases, such as helium, exist as single, unpaired atoms at ordinary temperatures. At sufficiently high temperatures, iron, gold, and other metals vaporize to form gaseous, single, unpaired atoms. In the kinetic molecular theory the term *molecule* has the additional meaning of the smallest, ultimate particle of matter that can exist. Thus, the ultimate particle of a gas, whether it is made up of two or more atoms

FIGURE 4.1

Heat and modern technology are inseparable. These glowing steel slabs, at over 1,100°C (about 2,000°F), are cut by an automatic flame torch. The slab caster converts 300 tons of molten steel into slabs in about 45 minutes. The slabs are converted to sheet steel for use in the automotive, appliance, and building industries.

FIGURE 4.2

Metal atoms appear in the micrograph of a crystal of titanium niobium oxide, magnified 7,800,000 times with the help of an electron microscope.

bound together or of a single atom, is conceived of as a molecule. A single atom of helium, for example, is known as a *monatomic molecule*. For now, a **molecule** is defined as the smallest particle of a compound, or a gaseous element, that can exist and still retain the characteristic properties of that substance.

Molecules Interact

Some molecules of solids and liquids interact, strongly attracting and clinging to each other. When this attractive force is between the same kind of molecules, it is called *cohesion*. It is a stronger cohesion that makes solids and liquids different from gases, and without cohesion all matter would be in the form of gases. Sometimes one kind of molecule attracts and clings to a different kind of molecule. The attractive force between unlike molecules is called *adhesion*. Water wets your skin because the adhesion of water molecules and skin is stronger than the cohesion of water molecules. Some substances, such as glue, have a strong force of adhesion when they harden from a liquid state, and they are called adhesives.

Phases of Matter

Three phases of matter are common on earth under conditions of ordinary temperature and pressure. These phases—or forms of existence—are solid, liquid, and gas. Each of these has a different molecular arrangement (Figure 4.3). The different characteristics of each phase can be attributed to the molecular

arrangements and the strength of attraction between the molecules (Table 4.1).

Solids have definite shapes and volumes because they have molecules that are fixed distances apart and bound by relatively strong cohesive forces. Each molecule is a nearly fixed distance

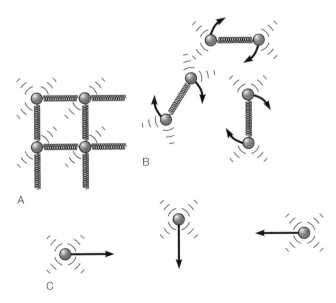

FIGURE 4.3

(*A*) In a solid, molecules vibrate around a fixed equilibrium position and are held in place by strong molecular forces. (*B*) In a liquid, molecules can rotate and roll over each other because the molecular forces are not as strong. (*C*) In a gas, molecules move rapidly in random, free paths.

4-3   **Chapter Four:** Heat and Temperature   **87**

TABLE 4.1

The shape and volume characteristics of solids, liquids, and gases are reflections of their molecular arrangements*

	Solids	Liquids	Gases
Shape	Fixed	Variable	Variable
Volume	Fixed	Fixed	Variable

*These characteristics are what would be expected under ordinary temperature and pressure conditions on the surface of the earth.

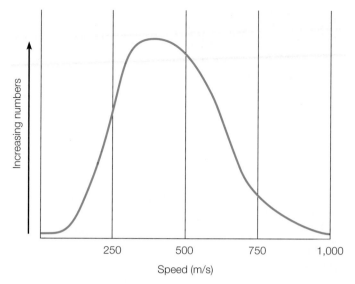

FIGURE 4.4

The number of oxygen molecules with certain speeds that you might find in a sample of air at room temperature. Notice that a few are barely moving and some have speeds over 1,000 m/s at a given time, but the *average* speed is somewhere around 500 m/s.

from the next, but it does vibrate and move around an equilibrium position. The masses of these molecules and the spacing between them determine the density of the solid. The hardness of a solid is the resistance of a solid to forces that tend to push its molecules further apart.

Liquids have molecules that are not confined to an equilibrium position as in a solid. The molecules of a liquid are close together and bound by cohesive forces that are not as strong as in a solid. This permits the molecules to move from place to place within the liquid. The molecular forces are strong enough to give the liquid a definite volume but not strong enough to give it a definite shape. Thus, a liter of water is always a liter of water (unless it is under tremendous pressure) and takes the shape of the container holding it. Because the forces between the molecules of a liquid are weaker than the forces between the molecules of a solid, a liquid cannot support the stress of a rock placed on it as a solid does. The liquid molecules *flow*, rolling over each other as the rock pushes its way between the molecules. Yet, the molecular forces are strong enough to hold the liquid together, so it keeps the same volume.

Gases are composed of molecules with weak cohesive forces acting between them. The gas molecules are relatively far apart and move freely in a constant, random motion that is changed often by collisions with other molecules. Gases therefore have neither fixed shapes nor fixed volumes.

There are other distinctions between the phases of matter. The term *vapor* is sometimes used to describe a gas that is usually in the liquid phase. Water vapor, for example, is the gaseous form of liquid water. Liquids and gases are collectively called *fluids* because of their ability to flow, a property that is lacking in most solids. A not-so-ordinary phase of matter on the earth is called *plasma*. Plasmas occur at extremely high temperatures and are made up of charged parts of atoms.

Molecules Move

Suppose you are in an evenly heated room with no air currents. If you open a bottle of ammonia, the odor of ammonia is soon noticeable everywhere in the room. According to the kinetic molecular theory, molecules of ammonia leave the bottle and bounce around among the other molecules making up the air until they are everywhere in the room, slowly becoming more evenly distributed. The ammonia molecules *diffuse*, or spread, throughout the room. The ammonia odor diffuses throughout the room faster if the air temperature is higher and slower if the air temperature is lower. This would imply a relationship between the temperature and the speed at which molecules move about.

The relationship between the temperature of a gas and the motion of molecules was formulated in 1857 by Rudolf Clausius. He showed that the temperature of a gas is proportional to the average kinetic energy of the gas molecules. This means that ammonia molecules have a greater average velocity at a higher temperature and a slower average velocity at a lower temperature. This explains why gases diffuse at a greater rate at higher temperatures. Recall, however, that kinetic energy involves the mass of the molecules as well as their velocity ($KE = 1/2\ mv^2$). It is the *average kinetic energy* that is proportional to the temperature, which involves the molecular mass as well as the molecular velocity. Whether the kinetic energy is jiggling, vibrating, rotating, or moving from place to place, the **temperature** of a substance is *a measure of the average kinetic energy of the molecules making up the substance* (Figure 4.4).

The kinetic molecular theory explains why matter generally expands with increased temperatures and contracts with decreased temperatures. At higher temperatures the molecules of a substance move faster, with increased agitation, and therefore they move a little farther apart, thus expanding the substance. As the substance cools, the motion slows, and the molecular forces are able to pull the molecules closer together, thus contracting the substance.

Temperature

If you ask people about the temperature, they usually respond with a referent ("hotter than the summer of '89") or a number ("68°F or 20°C"). Your response, or feeling, about the referent or number depends on a number of factors, including a *relative* comparison. A temperature of 20°C (68°F), for example, might seem cold during the month of July but warm during the month of January. The 20°C temperature is compared to what is expected at the time, even though 20°C is 20°C, no matter what month it is.

When people ask about the temperature, they are really asking *how hot or how cold something is*. Without a thermometer, however, most people can do no better than *hot* or *cold*, or perhaps *warm* or *cool*, in describing a relative temperature. Even then, there are other factors that confuse people about temperature. Your body judges temperature on the basis of the net *direction* of energy flow. You sense situations in which heat is flowing into your body as *warm* and situations in which heat is flowing from your body as *cool*. Perhaps you have experienced having your hands in snow for some time, then washing your hands in cold water. The cold water feels warm. Your hands are colder than the water, energy flows into your hands, and they communicate "warm."

Thermometers

The human body is a poor sensor of temperature, so a device called a *thermometer* is used to measure the hotness or coldness of something. Most thermometers are based on the relationship between some property of matter and changes in temperature. Almost all materials expand with increasing temperatures. A strip of metal is slightly longer when hotter and slightly shorter when cooler, but the change of length is too small to be useful in a thermometer. A more useful, larger change is obtained when two metals that have different expansion rates are bonded together in a strip. The bimetallic ("bi" = two; "metallic" = metal) strip will bend toward the metal with less expansion when the strip is heated (Figure 4.5). Such a bimetallic strip is formed into a coil and used in thermostats and dial thermometers (Figure 4.6).

The common glass thermometer is a glass tube with a bulb containing a liquid, usually mercury or colored alcohol, that

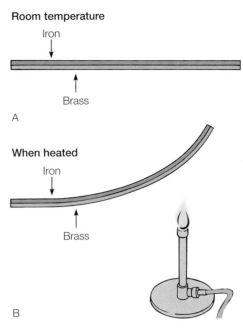

FIGURE 4.5

(*A*) A bimetallic strip is two different metals, such as iron and brass, bonded together as a single unit, shown here at room temperature. (*B*) Since one metal expands more than the other, the strip will bend when it is heated. In this example, the brass expands more than the iron, so the bimetallic strip bends away from the brass.

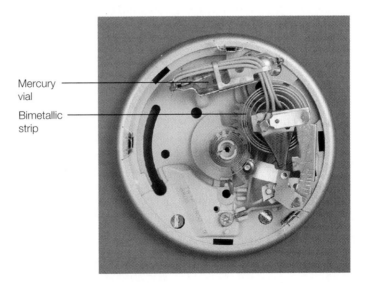

FIGURE 4.6

This thermostat has a coiled bimetallic strip that expands and contracts with changes in the room temperature. The attached vial of mercury is tilted one way or the other, and the mercury completes or breaks an electric circuit that turns the heating or cooling system on or off.

expands up the tube with increases in temperature and contracts back toward the bulb with decreases in temperature. The height of this liquid column is used with a referent scale to measure temperature. Some thermometers, such as a fever thermometer, have a small constriction in the bore so the liquid cannot normally return to the bulb. Thus, the thermometer shows the highest reading, even if the temperature it measures has fluctuated up and down during the reading. The liquid must be forced back into the bulb by a small swinging motion, bulb-end down, then sharply stopping the swing with a snap of the wrist. The inertia of the mercury in the bore forces it past the constriction and into the bulb. The fever thermometer is then ready to use again.

Today, scientists have developed a different type of thermometer and a way around the problems of using a glass mercury fever thermometer. This new approach measures the internal core temperature by quickly reading infrared radiation from the eardrum. All bodies with a temperature above absolute zero emit radiation, including your body. The intensity of the radiation is a sensitive function of body temperature, so reading the radiation emitted will tell you about the temperature of that body.

The human eardrum is close to the hypothalamus, the body's thermostat, so a temperature reading taken here must be close to the temperature of the internal core. You cannot use a mercury thermometer in the ear because of the very real danger of puncturing the eardrum, along with obtaining doubtful readings from a mercury bulb. You can use a pyroelectric material to measure the infrared radiation coming from the entrance to the ear canal, however, to quickly obtain a temperature reading. A pyroelectric material is a polarized crystal that generates an electric charge in proportion to a temperature change. The infrared fever thermometer has a short barrel, which is inserted in the ear canal opening. A button opens a shutter inside the battery-powered device, admitting infrared radiation for about 300 milliseconds. Infrared radiation from the ear canal increases the temperature of a thin pyroelectric crystal, which develops an electric charge. A current spike from the pyroelectric sensor now moves through some filters and converters and into a microprocessor chip. The chip is programmed with the relationship between the body temperature and the infrared radiation emitted. Using this information, it calculates the temperature by measuring the current spike produced by the infrared radiation falling on the pyroelectric crystal. The microprocessor now sends the temperature reading to an LCD display on the outside of the device, where it can be read almost immediately.

Concepts Applied

Human Thermometer?

Here is a way to find out how well the human body senses temperature. Obtain three containers that are large enough to submerge your hand in water. In one container place enough ice water, including ice cubes, to cover a hand. In a second container place enough water as hot as you can tolerate (without burning yourself) to cover a hand. Fill the third container with enough moderately warm water to cover a hand.

Submerge your right hand in the hot water and your left hand in the ice water for one minute. Dry your hands quickly, then submerge both in the warm water. How does the water feel to your right hand? How does it feel to your left hand? How well do your hands sense temperature?

Thermometer Scales

There are several referent scales used to define numerical values for measuring temperatures (Figure 4.7). The **Fahrenheit scale** was developed by the German physicist Gabriel D. Fahrenheit in about 1715. Fahrenheit invented a mercury-in-glass thermometer with a scale based on two arbitrarily chosen reference points. The original Fahrenheit scale was based on the temperature of an ice and salt mixture for the lower reference point (0°) and the temperature of the human body as the upper reference point (about 100°). Thus, the original Fahrenheit scale was a centigrade scale with 100 divisions between the high and the low reference points. The distance between the two reference points was then divided into equal intervals called *degrees*. There were problems with identifying a "normal" human body temperature as a reference point, since body temperature naturally changes during a given day and from day to day. However, some people "normally" have a higher body temperature than others. Some may have a normal body temperature of 99.1°F while others have a temperature of 97°F. The average for a large population is 98.6°F. The only consistent thing about the human body temperature is constant change. The standards for the Fahrenheit scale were eventually changed to something more consistent, the freezing point and the boiling point of water at normal atmospheric pressure. The original scale was retained with the new reference points, however, so the "odd" numbers of 32°F (freezing point of water) and 212°F (boiling point of water

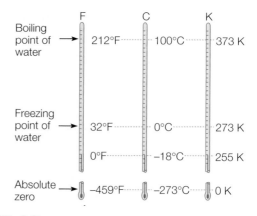

FIGURE 4.7

The Fahrenheit, Celsius, and Kelvin temperature scales.

under normal pressure) came to be the reference points. There are 180 equal intervals, or degrees, between the freezing and boiling points on the Fahrenheit scale.

The **Celsius scale** was invented by Anders C. Celsius, a Swedish astronomer, in about 1735. The Celsius scale uses the freezing point and the boiling point of water at normal atmospheric pressure, but it has different arbitrarily assigned values. The Celsius scale identifies the freezing point of water as 0°C and the boiling point as 100°C. There are 100 equal intervals, or degrees, between these two reference points, so the Celsius scale is sometimes called the *centigrade* scale.

There is nothing special about either the Celsius scale or the Fahrenheit scale. Both have arbitrarily assigned numbers, and neither is more accurate than the other. The Celsius scale is more convenient because it is a decimal scale and because it has a direct relationship with a third scale to be described shortly, the Kelvin scale. Both scales have arbitrarily assigned reference points and an arbitrary number line that indicates *relative* temperature changes. Zero is simply one of the points on each number line and does *not* mean that there is no temperature. Likewise, since the numbers are relative measures of temperature change, 2° is not twice as hot as a temperature of 1° and 10° is not twice as hot as a temperature of 5°. The numbers simply mean some measure of temperature *relative to* the freezing and boiling points of water under normal conditions.

You can convert from one temperature to the other by considering two differences in the scales: (1) the difference in the degree size between the freezing and boiling points on the two scales, and (2) the difference in the values of the lower reference points.

The Fahrenheit scale has 180° between the boiling and freezing points (212°F − 32°F) and the Celsius scale has 100° between the same two points. Therefore each Celsius degree is 180/100 or 9/5 as large as a Fahrenheit degree. Each Fahrenheit degree is 100/180 or 5/9 of a Celsius degree. You know that this is correct because there are more Fahrenheit degrees than Celsius degrees between freezing and boiling. The relationship between the degree sizes is 1°C = 9/5°F and 1°F = 5/9°C. In addition, considering the difference in the values of the lower reference points (0°C and 32°F) gives the equations for temperature conversion. (For a review of the sequence of mathematical operations used with equations, refer to the "Working with Equations" section in the Mathematical Review of appendix A.)

$$T_F = \frac{9}{5} T_C + 32°$$

equation 4.1

$$T_C = \frac{5}{9}(T_F - 32°)$$

equation 4.2

EXAMPLE 4.1

The average human body temperature is 98.6°F. What is the equivalent temperature on the Celsius scale?

Solution

$$T_C = \frac{5}{9}(T_F - 32°)$$

$$= \frac{5}{9}(98.6° - 32°)$$

$$= \frac{5}{9}(66.6°)$$

$$= \frac{333°}{9}$$

$$= \boxed{37° \text{ C}}$$

EXAMPLE 4.2

A bank temperature display indicates 20°C (room temperature). What is the equivalent temperature on the Fahrenheit scale? (Answer: 68°F)

There is a temperature scale that does not have arbitrarily assigned reference points, and zero *does* mean nothing. This is not a relative scale but an absolute temperature scale called the **absolute scale,** or *Kelvin scale.* The zero point on the absolute scale is thought to be the lowest limit of temperature. *Absolute zero is the lowest temperature possible,* occurring when all random motion of molecules was historically projected to cease. Absolute zero is written as 0 K. A degree symbol is not used, and the K stands for the SI standard scale unit, Kelvin. The absolute scale uses the same degree size as the Celsius scale, and −273°C = 0 K. Note in Figure 4.7 that 273 K is the freezing point of water, and 373 K is the boiling point. You could think of the absolute scale as a Celsius scale with the zero point shifted by 273°. Thus, the relationship between the absolute and Celsius scales is

$$T_K = T_C + 273$$

equation 4.3

A temperature of absolute zero has never been reached, but scientists have cooled a sample of sodium to 700 nanokelvins, or 700 billionths of a kelvin above absolute zero.

EXAMPLE 4.3

A science article refers to a temperature of 300.0 K. (a) What is the equivalent Celsius temperature? (b) the equivalent Fahrenheit temperature?

Solution

(a) The relationship between the absolute scale and Celsius scale is found in equation 4.3, $T_K = T_C + 273$. Solving this equation for Celsius yields $T_C = T_K - 273$.

$$T_C = T_K - 273$$
$$= 300.0 - 273$$
$$= \boxed{27.0°C}$$

(b)
$$T_F = \frac{9}{5}T_C + 32°$$
$$= \frac{9}{5}27.0° + 32°$$
$$= \frac{243°}{5} + 32°$$
$$= 48.6° + 32°$$
$$= \boxed{81°F}$$

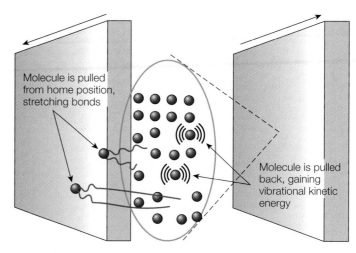

FIGURE 4.8

Here is how friction results in increased temperatures: Molecules on one moving surface will catch on another surface, stretching the molecular forces that are holding it. They are pulled back to their home position with a snap, resulting in a gain of vibrational kinetic energy.

(labels in figure:) Molecule is pulled from home position, stretching bonds

Molecule is pulled back, gaining vibrational kinetic energy

Heat

Suppose you have a bowl of hot soup or a cup of hot coffee that is too hot. What can you do to cool it? You can blow across the surface, which speeds evaporation and therefore results in cooling, but this is a slow process. If you were in a hurry, you would probably add something cooler, such as ice. Adding a cooler substance will cool the hot liquid.

You know what happens when you mix fluids or objects with a higher temperature with fluids or objects with a lower temperature. The warmer-temperature object becomes cooler and the cooler-temperature object becomes warmer. Eventually, both will have a temperature somewhere between the warmer and the cooler. This might suggest that something is moving between the warmer and cooler objects, changing the temperature. What is doing the moving?

The term **heat** is used to describe the "something" that moves between objects when two objects of different temperature are brought together. As you will learn in the next section, heat flow represents a form of energy transfer that takes place between objects. For now, we will continue to think of heat as "something" that moves between objects of different temperatures, such as your bowl of hot soup and a cold ice cube.

The relationship that exists between energy and temperature will help explain the concept of heat, so we will consider it first. If you rub your hands together a few times, they will feel a little warmer. If you rub them together vigorously for a while, they will feel a lot warmer, maybe hot. A temperature increase takes place anytime mechanical energy causes one surface to rub against another (Figure 4.8). The two surfaces could be solids, such as the two blocks, but they can also be the surface of a solid and a fluid, such as air. A solid object moving through the air encounters fluid friction, which results in a higher temperature of the surface. A high velocity meteor enters the earth's atmosphere and is heated so much from the friction that it begins to glow, resulting in the fireball and smoke trail of a "falling star."

To distinguish between the energy of the object and the energy of its molecules, we use the terms "external" and "internal" energy. **External energy** is the total potential and kinetic energy of an everyday-sized object. All the kinetic and potential energy considerations discussed in previous chapters were about the external energy of an object.

Internal energy is the total kinetic and potential energy of the *molecules* of an object. The kinetic energy of a molecule can be much more complicated than straight-line velocity might suggest, however, because a molecule can have many different types of motion at the same time (pulsing, twisting, turning, etc.). Overall, internal energy is characterized by properties such as temperature, density, heat, volume, pressure of a gas, and so forth.

When you push a table across the floor, the observable *external* kinetic energy of the table is transferred to the *internal* kinetic energy of the molecules between the table legs and the floor, resulting in a temperature increase (Figure 4.9). The relationship between external and internal kinetic energy explains why the heating is proportional to the amount of mechanical energy used.

Heat as Energy Transfer

Temperature is a measure of the degree of hotness or coldness of a body, a measure that is based on the average molecular kinetic energy. Heat, on the other hand, is based on the *total internal energy* of the molecules of a body. You can see one difference in heat and temperature by considering a cup of water and a large

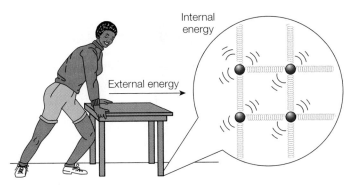

FIGURE 4.9

External energy is the kinetic and potential energy that you can see. *Internal energy* is the total kinetic and potential energy of molecules. When you push a table across the floor, you do work against friction. Some of the external mechanical energy goes into internal kinetic and potential energy, and the bottom surfaces of the legs become warmer.

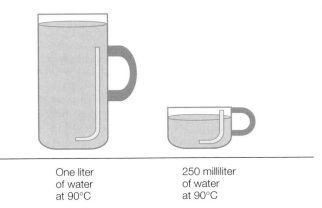

One liter of water at 90°C   250 milliliter of water at 90°C

FIGURE 4.10

Heat and temperature are different concepts, as shown by a liter of water (1,000 mL) and a 250 mL cup of water, both at the same temperature. You know the liter of water contains more internal energy because it will require more ice cubes to cool it to, say, 25°C than will be required for the cup of water. In fact, you will have to remove 48,750 *additional* calories to cool the liter of water.

tub of water. If both the small and the large amount of water have the same temperature, both must have the same average molecular kinetic energy. Now, suppose you wish to cool both by, say, 20°. The large tub of water would take much longer to cool, so it must be that the large amount of water has more internal energy (Figure 4.10). Heat is a measure based on the *total* internal energy of the molecules of a body, and there is more total energy in a large tub of water than in a cup of water at the same temperature.

Heat Defined

How can we measure heat? Since it is difficult to see molecules, internal energy is difficult to measure directly. Thus heat is nearly always measured during the process of a body gaining or losing energy. This measurement procedure will also give us a working definition of heat:

> **Heat is a measure of the internal energy that has been absorbed or transferred from one body to another.**

The *process* of increasing the internal energy is called "heating" and the *process* of decreasing internal energy is called "cooling." The word "process" is italicized to emphasize that heat is energy in transit, not a material thing you can add or take away. Heat is understood to be a measure of internal energy that can be measured as energy flows into or out of an object.

Two Heating Methods

There are two general ways that heating can occur. These are (1) from a temperature difference, with energy moving from the region of higher temperature, and (2) from an object gaining energy by way of an energy-form conversion.

When a *temperature difference* occurs, energy is transferred from a region of higher temperature to a region of lower temperature. Energy flows from a hot range element, for example, to a pot of cold water on a range. It is a natural process for

energy to flow from a region of higher temperature to a region of a lower temperature just as it is natural for a ball to roll downhill. The temperature of an object and the temperature of the surroundings determines if heat will be transferred to or from an object. The terms "heating" and "cooling" describe the direction of energy flow, naturally moving from a region of higher energy to one of lower energy.

The internal energy of an object can be increased during an *energy-form conversion* (mechanical, radiant, electrical, etc.), so we say that heating is taking place. The classic experiments by Joule showed an equivalence between mechanical energy and heating, electrical energy and heating, and other conversions. On a molecular level, the energy forms are doing work on the molecules, which can result in an increase of internal energy. Thus heating by energy-form conversion is actually a transfer of energy by *working*. This brings us back to the definition that "energy is the ability to do work." We can mentally note that this includes the ability to do work at the molecular level.

Heating that takes place because of a temperature difference will be considered in more detail after we consider how heat is measured.

Measures of Heat

Since heating is a method of energy transfer, a quantity of heat can be measured just like any quantity of energy. The metric unit for measuring work, energy, or heat is the *joule*. However, the separate historical development of the concepts of heat and the concepts of motion resulted in separate units, some based on temperature differences.

The metric unit of heat is called the **calorie** (cal). A calorie is defined as the *amount of energy (or heat) needed to increase the temperature of 1 gram of water 1 degree Celsius*. A more precise definition specifies the degree interval from 14.5°C to 15.5°C

Chapter Four: Heat and Temperature   **93**

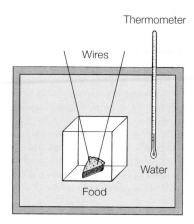

FIGURE 4.11

The Calorie value of food is determined by measuring the heat released from burning the food. If there is 10.0 kg of water and the temperature increased from 10° to 20°C, the food contained 100 Calories (100,000 calories). The food illustrated here would release much more energy than this.

because the energy required varies slightly at different temperatures. This precise definition is not needed for a general discussion. A **kilocalorie** (kcal) is the *amount of energy (or heat) needed to increase the temperature of 1 kilogram of water 1 degree Celsius.* The measure of the energy released by the oxidation of food is the kilocalorie, but it is called the Calorie (with a capital C) by nutritionists (Figure 4.11). Confusion can be avoided by making sure that the scientific calorie is never capitalized (cal) and the dieter's Calorie is always capitalized. The best solution would be to call the Calorie what it is, a kilocalorie (kcal).

The English system's measure of heating is called the **British thermal unit** (Btu). A Btu is *the amount of energy (or heat) needed to increase the temperature of 1 pound of water 1 degree Fahrenheit.* The Btu is commonly used to measure the heating or cooling rates of furnaces, air conditioners, water heaters, and so forth. The rate is usually expressed or understood to be in Btu per hour. A much larger unit is sometimes mentioned in news reports and articles about the national energy consumption. This unit is the *quad,* which is 1 quadrillion Btu (a million billion or 10^{15} Btu).

Heat is increased by an energy-form conversion, and the equivalence between energy and heating was first measured by James Joule. He found that the relationship between the energy form (mechanical, electrical radiant, etc.) and the resulting heating was always the same. For example, the relationship between mechanical work done and the resulting heating is always

$$4.184 \text{ J} = 1 \text{ cal}$$

or

$$4,184 \text{ J} = 1 \text{ kcal}$$

The establishment of this precise proportionality means that, fundamentally, mechanical work and heat are different forms of the same thing.

EXAMPLE **4.4**

A 1,000.0 kg car is moving at 90.0 km/h (25.0 m/s). How many kilocalories are generated when the car brakes to a stop?

Solution

The kinetic energy of the car is

$$KE = \frac{1}{2}mv^2$$

$$= \frac{1}{2}(1,000.0 \text{ kg})(25.0 \text{ m/s})^2$$

$$= (500.0)(625)\frac{\text{kg·m}^2}{\text{s}^2}$$

$$= 312,500 \text{ J} = 313,000 \text{ J or } 3.13 \times 10^5 \text{ J}$$

You can convert this to kcal by using the relationship between mechanical energy and heat:

$$(313,000 \text{ J})\left(\frac{1 \text{ kcal}}{4,184 \text{ J}}\right)$$

$$\frac{313,000}{4,184} \frac{\text{J·kcal}}{\text{J}}$$

$$\boxed{74.8 \text{ kcal}}$$

(Note: The temperature increase from this amount of heating could be calculated from equation 4.4.)

Specific Heat

You can observe a relationship between heat and different substances by doing an experiment in "kitchen physics." Imagine that you have a large pot of liquid to boil in preparing a meal. Three variables influence how much heat you need:

1. the initial temperature of the liquid,
2. how much liquid is in the pot, and,
3. the nature of the liquid (water or soup?).

What this means specifically, is

1. *Temperature change.* The amount of heat needed is proportional to the temperature change. It takes more heat to raise the temperature of cool water, so this relationship could be written as $Q \propto \Delta T$.
2. *Mass.* The amount of heat needed is also proportional to the amount of the substance being heated. A larger mass requires more heat to go through the same temperature change than a smaller mass. In symbols, $Q \propto m$.
3. *Substance.* Different materials require different amounts of heat to go through the same temperature range when their masses are equal (Figure 4.12). This property is called the **specific heat** of a material, which is defined as the amount of heat needed to increase the temperature of 1 gram of a substance 1 degree Celsius.

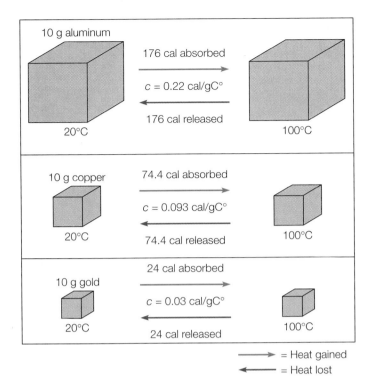

FIGURE 4.12

Of these three metals, aluminum needs the most heat per gram per degree when warmed, and releases the most heat when cooled. Why are the cubes different sizes?

TABLE 4.2

The specific heat of selected substances

Substance	Specific Heat (cal/gC° or kcal/kgC°)
Air	0.17
Aluminum	0.22
Concrete	0.16
Copper	0.093
Glass (average)	0.160
Gold	0.03
Ice	0.500
Iron	0.11
Lead	0.0305
Mercury	0.033
Seawater	0.93
Silver	0.056
Soil (average)	0.200
Steam	0.480
Water	1.00

Note: To convert to specific heat in J/kgC°, multiply each value by 4,184. Also note that 1 cal/gC° = 1 kcal/kgC°.

Considering all the variables involved in our kitchen physics cooking experience, we find the heat (Q) needed is described by the relationship

$$Q = mc\Delta T$$

equation 4.4

where c is the symbol for specific heat. Specific heat is related to the internal structure of a substance; some of the energy goes into the internal potential energy of the molecules and some goes into the internal kinetic energy of the molecules. The difference in values for the specific heat of different substances is related to the number of molecules in a 1-gram sample of each and to the way they form a molecular structure.

Specific heat is responsible for the fact that air temperatures vary more over land than over a large body of water. Table 4.2 gives the specific heat of soil as 0.200 cal/gC° and the specific heat of water as 1.00 cal/gC°. Since specific heat is defined as the amount of heat needed to increase the temperature of 1 gram of a substance 1 degree, this means 1 gram of water exposed to 1 calorie of sunlight will warm 1°C. One gram of soil exposed to 1 calorie of sunlight, on the other hand, will be warmed to 5°C because it only takes 0.2 calories to warm the soil 1°C. Thus the temperature is more even near large bodies of water because it is harder to change the temperature of the water.

EXAMPLE 4.5

How much heat must be supplied to a 500.0 g pan to raise its temperature from 20.0°C to 100.0°C if the pan is made of (a) iron and (b) aluminum?

Solution

The relationship between the heat supplied (Q), the mass (m), and the temperature change (ΔT) is found in equation 4.4. The specific heats (c) of iron and aluminum can be found in Table 4.2.

(a) Iron:

$m = 500.0$ g
$c = 0.11$ cal/gC°
$T_f = 100.0$°C
$T_i = 20.0$°C
$Q = ?$

$$Q = mc\Delta T$$

$$= (500.0 \text{ g})\left(0.11 \frac{\text{cal}}{\text{gC}°}\right)(80.0°\text{C})$$

$$= (500.0)(0.11)(80.0) \ \text{g} \times \frac{\text{cal}}{\text{gC}°} \times °\text{C}$$

$$= 4,400 \ \frac{\text{g·cal·C}°}{\text{gC}°}$$

$$= 4,400 \text{ cal}$$

$$= \boxed{4.4 \text{ kcal}}$$

(b) Aluminum:

$m = 500.0 \text{ g}$

$c = 0.22 \text{ cal/gC°}$

$T_f = 100.0°C$

$T_i = 20.0°C$

$Q = ?$

$Q = mc\Delta T$

$$= (500.0 \text{ g})\left(0.22 \frac{\text{cal}}{\text{gC°}}\right)(80.0°C)$$

$$= (500.0)(0.22)(80.0) \ \text{g} \times \frac{\text{cal}}{\text{gC°}} \times °C$$

$$= 8{,}800 \ \frac{\text{g·cal·}\cancel{C°}}{\cancel{\text{gC°}}}$$

$$= 8{,}800 \text{ cal}$$

$$= \boxed{8.8 \text{ kcal}}$$

It takes twice as much heat energy to warm the aluminum pan through the same temperature range as an iron pan. Thus, with equal rates of energy input, the iron pan will warm twice as fast as an aluminum pan.

EXAMPLE **4.6**

What is the specific heat of a 2 kg metal sample if 1.2 kcal are needed to increase the temperature from 20.0°C to 40.0°C? (Answer: 0.03 kcal/kgC°)

Concepts Applied

More Kitchen Physics

Consider the following information as it relates to the metals of cooking pots and pans.

1. It is easier to change the temperature of metals with low specific heats.
2. It is harder to change the temperature of metals with high specific heats.

   Look at the list of metals and specific heats in Table 4.2 and answer the following questions:

1. Considering specific heat alone, which metal could be used for making practical pots and pans that are the most energy efficient to use?
2. Again considering specific heat alone, would certain combinations of metals provide any advantages for rapid temperature changes?

Heat Flow

In a previous section you learned that heating is a transfer of energy that involves (1) a temperature difference or (2) energy-form conversions. Heat transfer that takes place because of a temperature difference takes place in three different ways: by conduction, convection, or radiation.

Conduction

Anytime there is a temperature difference, there is a natural transfer of heat from the region of higher temperature to the region of lower temperature. In solids, this transfer takes place as heat is *conducted* from a warmer place to a cooler one. Recall that the molecules in a solid vibrate in a fixed equilibrium position and that molecules in a higher temperature region have more kinetic energy, on the average, than those in a lower temperature region. When a solid, such as a metal rod, is held in a flame, the molecules in the warmed end vibrate violently. Through molecular interaction, this increased energy of vibration is passed on to the adjacent, slower-moving molecules, which also begin to vibrate more violently. They, in turn, pass on more vibrational energy to the molecules next to them. The increase in activity thus moves from molecule to molecule, causing the region of increased activity to extend along the rod. This is called **conduction,** the transfer of energy from molecule to molecule (Figure 4.13).

Most insulating materials are good insulators because they contain many small air spaces (Figure 4.14). The small air spaces are poor conductors because the molecules of air are far apart, compared to a solid, making it more difficult to pass the increased vibrating motion from molecule to molecule. Styrofoam, glass wool, and wool cloth are good insulators because they have many small air spaces, not because of the material they are made of. The best insulator is a vacuum, since there are no molecules to pass on the vibrating motion (Table 4.3).

Wooden and metal parts of your desk have the same temperature, but the metal parts will feel cooler if you touch them. Metal is a better conductor of heat than wood and feels cooler because it conducts heat from your finger faster. This is the same reason that a wood or tile floor feels cold to your bare feet. You use an insulating rug to slow the conduction of heat from your feet.

Concepts Applied

Touch Temperature

Objects that have been in a room with a constant temperature for some time should all have the same temperature. Touch metal, plastic, and wooden parts of a desk or chair to sense their temperature. Explain your findings.

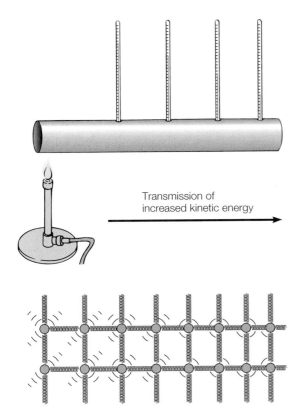

FIGURE 4.13

Thermometers placed in holes drilled in a metal rod will show that heat is conducted from a region of higher temperature to a region of lower temperature. The increased molecular activity is passed from molecule to molecule in the process of conduction.

TABLE 4.3	
Rate of conduction of materials*	

	Material	Value
	Silver	0.97
	Copper	0.92
	Aluminum	0.50
	Iron	0.11
	Lead	0.08
	Concrete	4.0×10^{-3}
	Glass	2.5×10^{-3}
	Tile	1.6×10^{-3}
	Brick	1.5×10^{-3}
	Water	1.3×10^{-3}
	Wood	3.0×10^{-4}
	Cotton	1.8×10^{-4}
	Styrofoam	1.0×10^{-4}
	Glass wool	9.0×10^{-5}
	Air	6.0×10^{-5}
	Vacuum	0

Better Conductor ↑ / Better Insulator ↓

*Based on temperature difference of 1°C per cm. Values are cal/s through a square centimeter of the material.

FIGURE 4.14

Fiberglass insulation is rated in terms of R-value, a ratio of the conductivity of the material to its thickness.

Convection

Convection is the transfer of heat by a large-scale displacement of groups of molecules with relatively higher kinetic energy. In conduction, increased kinetic energy is passed from molecule to molecule. In convection, molecules with higher kinetic energy are moved from one place to another place. Conduction happens primarily in solids, but convection happens only in liquids and gases, where fluid motion can carry molecules with higher kinetic energy over a distance. When molecules gain energy, they move more rapidly and push more vigorously against their surroundings. The result is an expansion as the region of heated molecules pushes outward and increases the volume. Since the same amount of matter now occupies a larger volume, the overall density has been decreased (Figure 4.15).

In fluids, expansion sets the stage for convection. Warm, less dense fluid is pushed upward by the cooler, more dense fluid around it. In general, cooler air is more dense; it sinks and flows downhill. Cold air, being more dense, flows out near the bottom of an open refrigerator. You can feel the cold, dense air pouring from the bottom of a refrigerator to your toes on the floor. On the other hand, you hold your hands *over* a heater because the warm, less dense air is pushed upward. In a room, warm air is pushed upward from a heater. The warm air spreads outward along the ceiling and is slowly displaced as newly warmed air is pushed upward to the ceiling. As the air cools it sinks over another part of the room, setting up a circulation pattern known as a *convection current* (Figure 4.16). Convection currents can also be observed in a large pot of liquid that is heating on a range. You can see the warmer liquid being forced upward over the warmer parts of the range element, then sink over the cooler parts. Overall, convection currents give the pot of liquid an appearance of turning over as it warms.

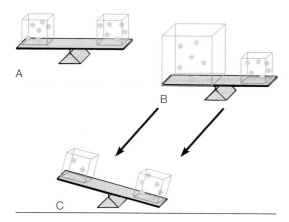

FIGURE 4.15

(*A*) Two identical volumes of air are balanced, since they have the same number of molecules and the same mass. (*B*) Increased temperature causes one volume to expand from the increased kinetic energy of the gas molecules. (*C*) The same volume of the expanded air now contains fewer gas molecules and is less dense, and it is buoyed up by the cooler, more dense air.

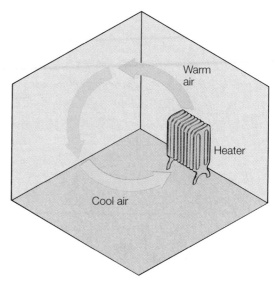

FIGURE 4.16

Convection currents move warm air throughout a room as the air over the heater becomes warmed, expands, and is moved upward by cooler air.

Concepts Applied

How Convection Works

Convection takes place in fluids where a temperature difference exists. To see why this occurs, obtain a balloon filled with very cold water and a second balloon filled with the same volume of very hot water. Carefully put the balloon with cold water in a large container of hot water. Place the balloon filled with hot water into a large container of cold water. What happened in each container? What does this tell you about the relationship between the temperature and density of a fluid and how convection works?

Radiation

The third way that heat transfer takes place because of a temperature difference is called **radiation**. Radiation involves the form of energy called *radiant energy*, energy that moves through space. As you will learn in chapter 7, radiant energy includes visible light and many other forms as well. All objects with a temperature above absolute zero give off radiant energy. The absolute temperature of the object determines the rate, intensity, and kinds of radiant energy emitted. You know that visible light is emitted if an object is heated to a certain temperature. A heating element on an electric range, for example, will glow with a reddish-orange light when at the highest setting, but it produces no visible light at lower temperatures, although you feel warmth in your hand when you hold it near the element. Your hand absorbs the nonvisible radiant energy being emitted

from the element. The radiant energy does work on the molecules of your hand, giving them more kinetic energy. You sense this as an increase in temperature, that is, warmth.

All objects above absolute zero emit radiant energy, but all objects also absorb radiant energy. A hot object, however, emits more radiant energy than a cold object. The hot object will emit more energy than it absorbs from the colder object, and the colder object will absorb more energy from the hot object than it emits. There is, therefore, a net energy transfer that will take place by radiation as long as there is a temperature difference between the two objects.

Concepts Applied

Candle Wax, Bubbles, and a Lightbulb

Here are three experiments you can do on heat flow:

Conduction: Use melted candle wax to stick thumbtack heads to a long metal rod. Heat one end of the rod with a flame. Record any evidence you observe that heat moves across the rod by conduction.

Convection: Choose a calm, warm day with plenty of strong sunlight. Make toy soap bubbles to study convection currents between a grass field and an adjacent asphalt parking lot. Find other adjacent areas where you think you might find convection currents and study them with soap bubbles, too. Record your experiments, findings, and explanations for what you observed.

Radiation: Hold your hand under an unlighted electric light-bulb, and then turn on the bulb. Describe evidence that what you feel traveled to you by radiation, not conduction or convection. Describe any experiments you can think of to prove you felt radiant energy.

Energy, Heat, and Molecular Theory

The kinetic molecular theory of matter is based on evidence from different fields of physical science, not just one subject area. Chemists and physicists developed some convincing conclusions about the structure of matter over the past 150 years, using carefully designed experiments and mathematical calculations that explained observable facts about matter. Step by step, the detailed structure of this submicroscopic, invisible world of particles became firmly established. Today, an understanding of this particle structure is basic to physics, chemistry, biology, geology, and practically every other science subject. This understanding has also resulted in present-day technology.

Phase Change

Solids, liquids, and gases are the three common phases of matter, and each phase is characterized by different molecular arrangements. The motion of the molecules in any of the three common phases can be increased by (1) adding heat through a temperature difference or (2) the absorption of one of the five forms of energy, which results in heating. In either case the temperature of the solid, liquid, or gas increases according to the specific heat of the substance, and more heating generally means higher temperatures.

More heating, however, does not always result in increased temperatures. When a solid, liquid, or gas changes from one phase to another, the transition is called a **phase change.** A phase change always absorbs or releases energy, *a quantity of heat that is not associated with a temperature change.* Since the quantity of heat associated with a phase change is not associated with a temperature change, it is called *latent heat.* Latent heat refers to the "hidden" energy of phase changes, which is energy (heat) that goes into or comes out of *internal potential energy* (Figure 4.17).

There are three kinds of major phase changes that can occur: (1) *solid-liquid,* (2) *liquid-gas,* and (3) *solid-gas.* In each case the phase change can go in either direction. For example, the solid-liquid phase change occurs when a solid melts to a liquid or when a liquid freezes to a solid. Ice melting to water and water freezing to ice are common examples of this phase change and its two directions. Both occur at a temperature called the *freezing point* or the *melting point,* depending on the direction of the phase change. In either case, however, the freezing and melting points are the same temperature.

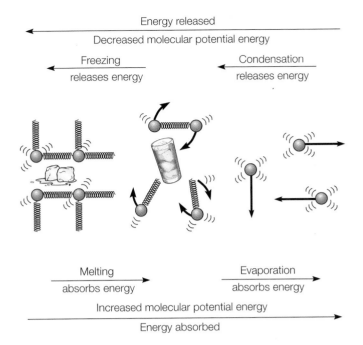

FIGURE 4.17

Each phase change absorbs or releases a quantity of latent heat, which goes into or is released from molecular potential energy.

The liquid-gas phase change also occurs in two different directions. The temperature at which a liquid boils and changes to a gas (or vapor) is called the *boiling point.* The temperature at which a gas or vapor changes back to a liquid is called the *condensation point.* The boiling and condensation points are the same temperature. There are conditions other than boiling under which liquids may undergo liquid-gas phase changes, and these conditions are discussed in the next section, "Evaporation and Condensation."

You probably are not as familiar with solid-gas phase changes, but they are common. A phase change that takes a solid directly to a gas or vapor is called *sublimation.* Mothballs and dry ice (solid CO_2) are common examples of materials that undergo sublimation, but frozen water, meaning common ice, also sublimates under certain conditions. Perhaps you have noticed ice cubes in a freezer become smaller with time as a result of sublimation. The frost that forms in a freezer, on the other hand, is an example of a solid-gas phase change that takes place in the other direction. In this case, water vapor forms the frost without going through the liquid state, a solid-gas phase change that takes place in an opposite direction to sublimation.

For a specific example, consider the changes that occur when ice is subjected to a constant source of heat (Figure 4.18). Starting at the left side of the graph, you can see that the temperature of the ice increases from the constant input of heat. The ice warms according to $Q = mc\Delta T$, where c is the specific heat of ice. When the temperature reaches the melting point (0°C), it stops increasing as the ice begins to melt. More and

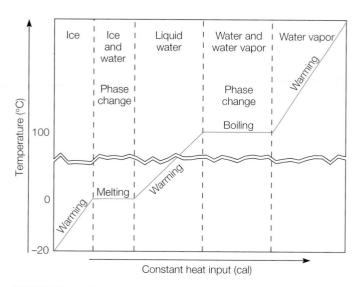

FIGURE 4.18

This graph shows three warming sequences and two phase changes with a constant input of heat. The ice warms to the melting point, then absorbs heat during the phase change as the temperature remains constant. When all the ice has melted, the now-liquid water warms to the boiling point, where the temperature again remains constant as heat is absorbed during this second phase change from liquid to gas. After all the liquid has changed to gas, continued warming increases the temperature of the water vapor.

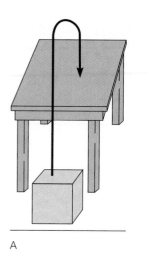
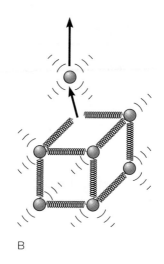

A   B

FIGURE 4.19

(A) Work is done against gravity to lift an object, giving the object more gravitational potential energy. (B) Work is done against intermolecular forces in separating a molecule from a solid, giving the molecule more potential energy.

more liquid water appears as the ice melts, but the temperature *remains* at 0°C even though heat is still being added at a constant rate. It takes a certain amount of heat to melt all of the ice. Finally, when all the ice is completely melted, the temperature again increases at a constant rate between the melting and boiling points. Then, at constant temperature the addition of heat produces another phase change, from liquid to gas. The quantity of heat involved in this phase change is used in doing the work of breaking the molecule-to-molecule bonds in the solid, making a liquid with molecules that are now free to move about and roll over one another. Since the quantity of heat (Q) is absorbed without a temperature change, it is called the **latent heat of fusion** (L_f). The latent heat of fusion is *the heat involved in a solid-liquid phase change in melting or freezing.* You learned in a previous chapter that when you do work on something you give it energy. In this case, the work done in breaking the molecular bonds in the solid gave the molecules more *potential* energy (Figure 4.19). This energy is "hidden," or latent, since heat was absorbed but a temperature increase did not take place. This same potential energy is given up when the molecules of the liquid return to the solid state. A melting solid absorbs energy and a freezing liquid releases this *same amount* of energy, warming the surroundings. Thus, you put ice in a cooler because the melting ice absorbs the latent heat of fusion from the beverage cans, cooling them. Citrus orchards are flooded with water when

freezing temperatures are expected because freezing water releases the latent heat of fusion, which warms the air around the trees. For water, the latent heat of fusion is 80.0 cal/g (144.0 Btu/lb). This means that every gram of ice that melts in your cooler *absorbs* 80.0 cal of heat. Every gram of water that freezes *releases* 80.0 cal. The total heat involved in a solid-liquid phase change depends on the mass of the substance involved, so

$$Q = mL_f$$

equation 4.5

where L_f is the latent heat of fusion for the substance involved.

Refer again to Figure 4.18. After the solid-liquid phase change is complete, the constant supply of heat increases the temperature of the water according to $Q = mc\Delta T$, where c is now the specific heat of liquid water. When the water reaches the boiling point, the temperature again remains constant even though heat is still being supplied at a constant rate. The quantity of heat involved in the liquid-gas phase change again goes into doing the work of overcoming the attractive molecular forces. This time the molecules escape from the liquid state to become single, independent molecules of gas. The quantity of heat (Q) absorbed or released during this phase change is called the **latent heat of vaporization** (L_v). The latent heat of vaporization is *the heat involved in a liquid-gas phase change where there is evaporation or condensation.* The latent heat of vaporization is the energy gained by the gas molecules as work is done in overcoming molecular forces. Thus, the escaping molecules absorb energy from the surroundings, and a condensing gas (or vapor) releases this *exact same amount of energy.* For water, the latent heat of vaporization is 540.0 cal/g (970.0 Btu/lb). This means that every gram of water vapor that condenses on your

Passive solar application is an economically justifiable use of solar energy today. Passive solar design uses a structure's construction to heat a living space with solar energy. There are few electric fans, motors, or other energy sources used. The passive solar design takes advantage of free solar energy; it stores and then distributes this energy through natural conduction, convection, and radiation.

Sunlight that reaches the earth's surface is mostly absorbed. Buildings, the ground, and objects become warmer as the radiant energy is absorbed. Nearly all materials, however, reradiate the absorbed energy at longer wavelengths, wavelengths too long to be visible to the human eye. The short wavelengths of sunlight pass readily through ordinary window glass, but the longer, reemitted wavelengths cannot. Therefore, sunlight passes through a window and warms objects inside a house. The reradiated longer wavelengths cannot pass readily back through the glass but are absorbed by certain molecules in the air. The temperature of the air is thus increased. This is called the "greenhouse effect." Perhaps you have experienced the effect when you left your car windows closed on a sunny, summer day.

In general, a passive solar home makes use of the materials from which it is constructed to capture, store, and distribute solar energy to its occupants. Sunlight enters the house through large windows facing south and warms a thick layer of concrete, brick, or stone. This energy "storage mass" then releases energy during the day, and, more important, during the night. This release of energy can be by direct radiation to occupants, by conduction to adjacent air, or by convection of air across the surface of the storage mass. The living space is thus heated without special plumbing or forced air circulation. As you can imagine, the key to a successful passive solar home is to consider every detail of natural energy flow, including the materials of which floors and walls are constructed, convective air circulation patterns, and the size and placement of windows. In addition, a passive solar home requires a different lifestyle and living patterns. Carpets, for example, would defeat the purpose of a storage-mass floor, since it would insulate the storage mass from sunlight. Glass is not a good insulator, so windows must have curtains or movable insulation panels to slow energy loss at night. This requires the daily activity of closing curtains or moving insulation panels at night and then opening curtains and moving panels in the morning. Passive solar homes, therefore, require a high level of personal involvement by the occupants.

There are three basic categories of passive solar design: (1) direct solar gain, (2) indirect solar gain, and (3) isolated solar gain.

A *direct solar gain* home is one in which solar energy is collected in the actual living space of the home (Box Figure 4.1). The advantage of this design is the large, open window space with a calculated overhang, which admits maximum solar energy in the winter but prevents solar gain in the summer. The disadvantage is that the occupants are living in the collection and storage components of the design and can place nothing (such as carpets and furniture) that would interfere with warming the storage mass in the floors and walls.

An *indirect solar gain* home uses a massive wall inside a window that serves as a storage mass. Such a wall, called a *Trombe wall*, is shown in Box Figure 4.2. The Trombe wall collects and stores solar energy, then warms the living space with radiant energy and convection currents. The disadvantage to the indirect solar gain design is that large windows are blocked by the Trombe wall. The advantage is that the occupants are not in direct contact with the solar collection and storage area, so they can place carpets and furniture as they wish. Controls to prevent energy loss at night are still necessary with this design.

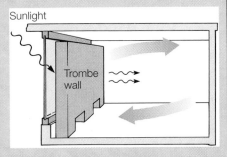

BOX FIGURE 4.1
The direct solar gain design collects and stores solar energy in the living space.

BOX FIGURE 4.2
The indirect solar gain design uses a Trombe wall to collect, store, and distribute solar energy.

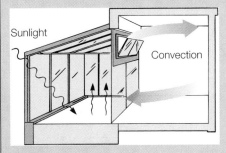

BOX FIGURE 4.3
The isolated solar gain design uses a separate structure to collect and store solar energy.

—*Continued top of next page*

An *isolated solar gain* home uses a structure that is separated from the living space to collect and store solar energy. Examples of an isolated gain design are an attached greenhouse or sun porch (Box Figure 4.3, on p. 101). Energy flow between the attached structure and the living space can be by conduction, convection, and radiation, which can be controlled by opening or closing off the attached structure. This design provides the best controls, since it can be completely isolated, opened to the living space as needed, or directly used as living space when the conditions are right. Additional insulation is needed for the glass at night, however, and for sunless winter days.

It has been estimated that building a passive solar home would cost about 10 percent more than building a traditional home of the same size. Considering the possible energy savings, you might believe that most homes would now have a passive solar design. They do not, however, as most new buildings require technology and large amounts of energy to maximize comfort. Yet, it would not require too much effort to consider where to place windows in relation to the directional and seasonal intensity of the sun and where to plant trees. Perhaps in the future you will have an opportunity to consider using the environment to your benefit through the natural processes of conduction, convection, and radiation.

---

bathroom mirror releases 540.0 cal, which warms the bathroom. The total heating depends on how much water vapor condensed, so

$$Q = mL_v$$

equation 4.6

where L_v is the latent heat of vaporization for the substance involved. The relationships between the quantity of heat absorbed during warming and phase changes are shown in Figure 4.20. Some physical constants for water and heat are summarized in Table 4.4.

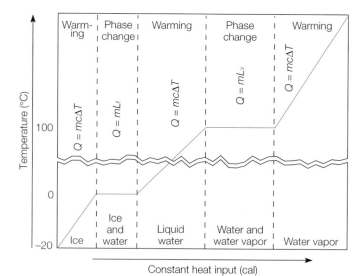

FIGURE 4.20

Compare this graph to the one in Figure 4.18. This graph shows the relationships between the quantity of heat absorbed during warming and phase changes as water is warmed from ice at −20°C to water vapor at some temperature above 100°C. Note that the specific heat for ice, liquid water, and water vapor (steam) have different values.

TABLE 4.4

Some physical constants for water and heat

Specific Heat (c)	
Water	$c = 1.00 \text{ cal/gC}°$
Ice	$c = 0.500 \text{ cal/gC}°$
Steam	$c = 0.480 \text{ cal/gC}°$
Latent Heat of Fusion	
L_f (water)	$L_f = 80.0 \text{ cal/g}$
Latent Heat of Vaporization	
L_v (water)	$L_v = 540.0 \text{ cal/g}$
Mechanical Equivalent of Heat	
1 kcal	4,184 J

EXAMPLE 4.7

How much energy does a refrigerator remove from 100.0 g of water at 20.0°C to make ice at −10.0°C?

Solution

This type of problem is best solved by subdividing it into smaller steps that consider (1) the heat added or removed and the resulting temperature changes *for each phase* of the substance and (2) the heat flow resulting from any *phase change* that occurs within the ranges of changes as identified by the problem (see Figure 4.20). The heat involved in each phase change and the heat involved in the heating or cooling of each phase are identified as Q_1, Q_2, and so forth. Temperature readings are calculated with *absolute values,* so you ignore any positive or negative signs.

1. Water in the liquid state cools from 20.0°C to 0°C (the freezing point) according to the relationship $Q = mc\Delta T$, where c is the specific heat of water, and

$$Q_1 = mc\Delta T$$
$$= (100.0 \text{ g})\left(1.00 \frac{\text{cal}}{\text{g°C}}\right)(0° - 20.0°\text{C})$$
$$= (100.0)(1.00)(20.0°) \frac{\text{g·cal·°C}}{\text{g°C}}$$
$$= 2{,}000 \text{ cal}$$
$$Q_1 = 2.00 \times 10^3 \text{ cal}$$

2. The latent heat of fusion must now be removed as water at 0°C becomes ice at 0°C through a phase change, and

$$Q_2 = mL_f$$
$$= (100.0 \text{ g})\left(80.0 \frac{\text{cal}}{\text{g}}\right)$$
$$= (100.0)(80.0) \frac{\text{g·cal}}{\text{g}}$$
$$= 8{,}000 \text{ cal}$$
$$Q_2 = 8.00 \times 10^3 \text{ cal}$$

3. The ice is now at 0°C and is cooled to −10°C as specified in the problem. The ice cools according to $Q = mc\Delta T$, where c is the specific heat of ice. The specific heat of ice is 0.500 cal/gC°, and

$$Q_3 = mc\Delta T$$
$$= (100.0 \text{ g})\left(0.500 \frac{\text{cal}}{\text{gC°}}\right)(10.0° - 0°\text{C})$$
$$= (100.0)(0.500)(10.0°) \frac{\text{g·cal·°C}}{\text{gC°}}$$
$$= 500 \text{ cal}$$
$$Q_3 = 5.00 \times 10^2 \text{ cal}$$

The total energy removed is then

$$Q_t = Q_1 + Q_2 + Q_3$$
$$= (2.00 \times 10^3 \text{ cal}) + (8.00 \times 10^3 \text{ cal}) + (5.00 \times 10^2 \text{ cal})$$
$$= 10.50 \text{ kcal}$$
$$\boxed{Q_t = 10.50 \times 10^3 \text{ kcal}}$$

Evaporation and Condensation

Liquids do not have to be at the boiling point to change to a gas and, in fact, tend to undergo a phase change at any temperature when left in the open. The phase change occurs at any temperature but does occur more rapidly at higher temperatures. The temperature of the water is associated with the *average* kinetic energy of the water molecules. The word *average* implies that some of the molecules have a greater energy and some have less (refer to Figure 4.4). If a molecule of water that has an exceptionally high energy is near the surface, and is headed in the right direction, it may overcome the attractive forces of the other water molecules and escape the liquid to become a gas. This is the process of *evaporation.* Evaporation reduces a volume of liquid water as water molecules leave the liquid state to become water vapor in the atmosphere (Figure 4.21).

Water molecules that evaporate move about in all directions, and some will return, striking the liquid surface. The same forces that they escaped from earlier capture the molecules, returning them to the liquid state. This is called the process of condensation. Condensation is the opposite of evaporation. In *evaporation,* more molecules are leaving the liquid state than are returning. In *condensation,* more molecules are returning to the liquid state than are leaving. This is a dynamic, ongoing process with molecules leaving and returning continuously. The net number leaving or returning determines whether evaporation or condensation is taking place (Figure 4.22).

When the condensation rate *equals* the evaporation rate, the air above the liquid is said to be *saturated.* The air immediately next to a surface may be saturated, but the condensation of water molecules is easily moved away with air movement. There

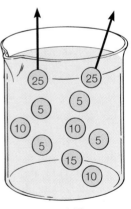
Average $= \frac{115}{10} = 11.5$

Average $= \frac{65}{8} = 8.1$

FIGURE 4.21

Temperature is associated with the average energy of the molecules of a substance. These numbered circles represent arbitrary levels of molecular kinetic energy that, in turn, represent temperature. The two molecules with the higher kinetic energy values [25 in (A)] escape, which lowers the average values from 11.5 to 8.1 (B). Thus evaporation of water molecules with more kinetic energy contributes to the cooling effect of evaporation in addition to the absorption of latent heat.

FIGURE 4.22

The inside of this closed bottle is isolated from the environment, so the space above the liquid becomes saturated. While it is saturated, the evaporation rate equals the condensation rate. When the bottle is cooled, condensation exceeds evaporation and droplets of liquid form on the inside surfaces.

thus increase the net rate of evaporation. This is why things dry more rapidly on a windy day. (4) *Pressure* is defined as *force per unit area*, which can be measured in lb/in² or N/m². Gases exert a pressure, which is interpreted in terms of the kinetic molecular theory. Atmospheric pressure is discussed in detail in chapter 23. For now, consider that the atmosphere exerts a pressure of about 10.0 N/cm² (14.7 lb/in²) at sea level. The atmospheric pressure and the intermolecular forces tend to hold water molecules in the liquid state. Thus, reducing the atmospheric pressure will reduce one of the forces holding molecules in a liquid state. Perhaps you have noticed that wet items dry more quickly at higher elevations, where the atmospheric pressure is less.

is no net energy flow when the air is saturated, since the heat carried away by evaporation is returned by condensation. This is why you fan your face when you are hot. The moving air from the fanning action pushes away water molecules from the air near your skin, preventing the adjacent air from becoming saturated, thus increasing the rate of evaporation. Think about this process the next time you see someone fanning his or her face.

There are four ways to increase the rate of evaporation. (1) An increase in the temperature of the liquid will increase the average kinetic energy of the molecules and thus increase the number of high-energy molecules able to escape from the liquid state. (2) Increasing the surface area of the liquid will also increase the likelihood of molecular escape to the air. This is why you spread out wet clothing to dry or spread out a puddle you want to evaporate. (3) Removal of water vapor from near the surface of the liquid will prevent the return of the vapor molecules to the liquid state and

Thermodynamics

The branch of physical science called *thermodynamics* is concerned with the study of heat and its relationship to mechanical energy, including the science of heat pumps, heat engines, and the transformation of energy in all its forms. The *laws of thermodynamics* describe the relationships concerning what happens as energy is transformed to work and the reverse, also serving as useful intellectual tools in meteorology, chemistry, and biology.

Mechanical energy is easily converted to heat through friction, but a special device is needed to convert heat to mechanical energy. A *heat engine* is a device that converts heat into mechanical energy. The operation of a heat engine can be explained by the kinetic molecular theory, as shown in Figure 4.23. This illustration shows a cylinder, much like a big can, with a

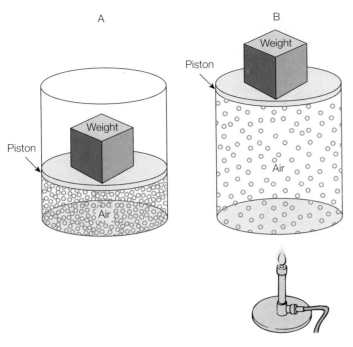

A   B

FIGURE 4.23

A very simple heat engine. The air in (B) has been heated, increasing the molecular motion and thus the pressure. Some of the heat is transferred to the increased gravitational potential energy of the weight as it is converted to mechanical energy.

closely fitting piston that traps a sample of air. The piston is like a slightly smaller cylinder and has a weight resting on it, supported by the trapped air. If the air in the large cylinder is now heated, the gas molecules will acquire more kinetic energy. This results in more gas molecule impacts with the enclosing surfaces, which results in an increased pressure. Increased pressure results in a net force, and the piston and weight move upward as shown in Figure 4.23B. Thus, some of the heat has now been transformed to the increased gravitational potential energy of the weight.

Thermodynamics is concerned with the *internal energy* (U), the total internal potential and kinetic energies of molecules making up a substance, such as the gases in the simple heat engine. The variables of temperature, gas pressure, volume, heat, and so forth characterize the total internal energy, which is called the *state* of the system. Once the system is identified, everything else is called the *surroundings*. A system can exist in a number of states since the variables that characterize a state can have any number of values and combinations of values. Any two systems that have the same values of variables that characterize internal energy are said to be in the same state.

The First Law of Thermodynamics

Any thermodynamic system has a unique set of properties that will identify the internal energy of the system. This state can be changed two ways, by (1) heat flowing into (Q_{in}) or out (Q_{out})

of the system, or (2) by the system doing work (W_{out}) or by work being done on the system (W_{in}). Thus work (W) and heat (Q) can change the internal energy of a thermodynamic system according to

$$JQ - W = U_2 - U_1$$

equation 4.7

where J is the mechanical equivalence of heat ($J = 4.184$ joule/calorie) and ($U_2 - U_1$) is the internal energy difference between two states. This equation represents the **first law of thermodynamics,** which states that the energy supplied to a thermodynamic system in the form of heat, minus the work done by the system, is equal to the change in internal energy. The first law of thermodynamics is an application of the law of conservation of energy, which applies to all energy matters. The first law of thermodynamics is concerned specifically with a thermodynamic system. As an example, consider energy conservation that is observed in the thermodynamic system of a heat engine (see Figure 4.24). As the engine cycles to the original state of internal energy ($U_2 - U_1 = 0$), all the external work accomplished must be equal to all the heat absorbed in the cycle. The heat supplied to the engine from a high temperature source

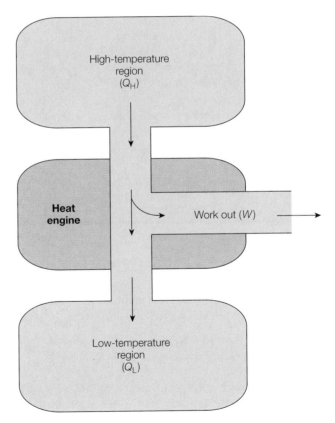

FIGURE 4.24

The heat supplied (Q_H) to a heat engine goes into the mechanical work (W), and the remainder is expelled in the exhaust (Q_L). The work accomplished is therefore the difference in the heat input and output ($Q_H - Q_L$), so the work accomplished represents the heat used, $W = J(Q_H - Q_L)$.

There is an interesting toy called a "Dippy Bird," "Happy Drinking Bird," (Box Figure 4.4) or various other names, that "drinks" water from a glass. The bird rocks back and forth, dipping its head into a glass of water, then bobbing up and rocking back and forth again before dipping down for another "drink." The bird will continue rocking and bobbing until the water in the glass is too low to reach.

Why does the Dippy Bird rock back and forth, then dip its head in the water? This question can be answered with some basic concepts of evaporation, heat, and vapor pressure, but first we need some information about the construction of the bird. The head and body of the bird are similar-sized glass bulbs connected with a short glass tube. The tube is attached directly to the upper bulb, but it extends well inside the lower bulb, almost to the bottom of the bulb. The tube and lower body are sealed, with a measured amount of a red liquid. The red liquid is probably methylene chloride, or some other volatile liquid that vaporizes at room temperature. The volume of liquid is sufficient to cover the tube in the lower bulb when the bird is upright, but not enough to cover both ends of the tube when the bird is nearly horizontal.

A metal pivot bar is clamped on the glass tube, near the middle, with ends resting on a stand sometimes made to look like bird legs. The bar is bent so the bird leans slightly in a forward direction. The upper bulb is coated with some fuzzy stuff, and has an attached beak, hat, eyes, and feathers. To start the bird on its cycles of dipping, the head and beak are wet with water. The bird is then set upright in front of a glass of water, filled and placed so the bird's beak will dip into the water when it rotates forward.

Before starting, the red liquid inside the two bulbs is at equilibrium with the same temperature, vapor pressure, and volume. Equilibrium is upset when the fuzzy material on the upper bulb is wet. Evaporation of water cools the head, and some of the vapor inside the upper bulb condenses. The greater pressure from the warmer vapor in the lower bulb is now able to push some fluid up the tube and into the upper bulb. The additional fluid in the upper bulb changes the pivot point and the bird dips forward. However, the tube opening is no longer covered by fluid, and some of the warmer vapor is now able to move to the upper bulb. This equalizes the pressure between the two bulbs, and fluid drains back into the lower bulb. This changes the balance, returning the bird to an upright position to rock back and forth, then repeat the events in the cycle.

When it is in the nearly horizontal position, the beak dips into the glass of water.

BOX FIGURE 4.4

A Dippy Bird

The material on the beak keeps the head wet, which makes it possible to continue the cycle of cooling and warming the upper bulb. Since the Dippy Bird operation depends on the cooling effects of evaporation, it works best in dry air and performs poorly in humid air.

(Q_H) is partly converted to work (W) and the rest is rejected in the lower-temperature exhaust (Q_L). The work accomplished is therefore the difference in the heat input and the heat output ($Q_H - Q_L$), so the work accomplished represents the heat used,

$$W = J(Q_H - Q_L)$$

equation 4.8

where J is the mechanical equivalence of heat ($J = 4.184$ joules/calorie). A schematic diagram of this relationship is shown in Figure 4.24. You can increase the internal energy (produce heat) as long as you supply mechanical energy (or do work). The first law of thermodynamics states that the conversion of work to heat is reversible, meaning that heat can be changed to work. There are several ways of converting heat to work, for example, the use of a steam turbine or gasoline automobile engine.

The Second Law of Thermodynamics

A heat pump is the *opposite* of a heat engine as shown schematically in Figure 4.25. The heat pump does work (W) in compressing vapors and moving heat from a region of lower temperature (Q_L) to a region of higher temperature (Q_H). That work is required to move heat this way is in accord with the observation that heat naturally flows from a region of higher temperature to a region of lower temperature. Energy is required for the opposite, moving heat from a cooler region to a warmer region. The natural direction of this process is called the **second law of thermodynamics,** which is that heat flows from objects with a higher temperature to objects with a cooler temperature. In other words, if you want heat to flow from a colder region to a warmer one you must *cause* it to do so by using energy. And if you do, such as with the use of a heat pump, you necessarily cause changes elsewhere, particularly in the energy sources used

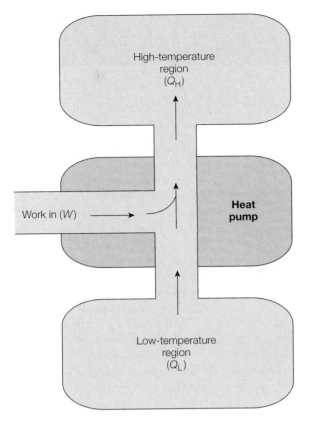

FIGURE 4.25

A heat pump uses work (W) to move heat from a low temperature region (Q_L) to a high temperature region (Q_H). The heat moved (Q_L) requires work (W), so $JQ_L = W$. A heat pump can be used to chill things at the Q_L end, or warm things at the Q_H end.

in the generation of electricity. Another statement of the second law is that it is impossible to convert heat completely into mechanical energy. This does not say that you cannot convert mechanical energy completely into heat, for example, in the brakes of a car when the brakes bring it to a stop. The law says that the reverse process is not possible, that you cannot convert 100 percent of a heat source into mechanical energy. Both of the preceding statements of the second law are concerned with a *direction* of thermodynamic processes, and the implications of this direction will be discussed next.

The Second Law and Natural Processes

Energy can be viewed from two considerations of scale, (1) the observable *external energy* of an object, and (2) the *internal energy* of the molecules, or particles that make up an object. A ball, for example, has kinetic energy after it is thrown through the air, and the entire system of particles making up the ball acts like a single massive particle as the ball moves. The motion and energy of the single system can be calculated from the laws of motion and from the equations representing the concepts of work and energy. All of the particles are moving together, in *coherent motion* when the external kinetic energy is considered.

But the particles making up the ball have another kind of kinetic energy, with the movements and vibrations of internal kinetic energy. In this case the particles are not moving uniformly together, but are vibrating with motions in many different directions. Since there is a lack of net motion and a lack of correlation, the particles have a jumbled *incoherent motion*, which is often described as chaotic. This random, chaotic motion is sometimes called *thermal motion*.

Thus there are two kinds of motion that the particles of an object can have, (1) a coherent motion where they move together, in step, and (2) an incoherent, chaotic motion of individual particles. These two types of motion are related to the two modes of energy transfer, working and heating. The relationship is that *work* on an object is associated with its *coherent motion* while *heating* an object is associated with its internal *incoherent motion*.

The second law of thermodynamics implies a direction to the relationship between work (coherent motion) and heat (incoherent motion) and this direction becomes apparent as you analyze what happens to the motions during energy conversions. Some forms of energy, such as electrical and mechanical, have a greater amount of order since they involve particles moving together in a coherent motion. The term *quality of energy* is used to identify the amount of coherent motion. Energy with high order and coherence is called a *high-quality energy*. Energy with less order and less coherence, on the other hand, is called *low-quality energy*. In general, high-quality energy can be easily converted to work, but low-quality energy is less able to do work.

High-quality electrical and mechanical energy can be used to do work, but then become dispersed as heat through energy-form conversions and friction. The resulting heat can be converted to do more work only if there is a sufficient temperature difference. The temperature differences do not last long, however, as conduction, convection, and radiation quickly disperse the energy even more. Thus the transformation of high-quality energy into lower-quality energy is a natural process. Energy tends to disperse, both from the conversion of an energy form to heat and from the heat flow processes of conduction, convection, and radiation. Both processes flow in one direction only and cannot be reversed. This is called the *degradation of energy*, which is the transformation of high-quality energy to lower-quality energy. In every known example it is a natural process of energy to degrade, becoming less and less available to do work. The process is *irreversible* even though it is possible to temporarily transform heat to mechanical energy through a heat engine or to upgrade the temperature through the use of a heat pump. Eventually the upgraded mechanical energy will degrade to heat and the increased heat will disperse through the processes of heat flow.

The apparent upgrading of energy by a heat pump or heat engine is always accompanied by a greater degrading of energy someplace else. The electrical energy used to run the heat pump, for example, was produced by the downgrading of chemical or nuclear energy at an electrical power plant. The overall result is that the *total* energy was degraded toward a more disorderly state.

Count Rumford (Benjamin Thompson) (1753–1814)

Count Rumford was a U.S.-born physicist who first demonstrated conclusively that heat is not a fluid but a form of motion.

Thompson was born into a farming family at Woburn, Massachusetts, on March 26, 1753. At the age of 19 he became a schoolmaster as a result of much self-instruction and some help from local clergy. He moved to Rumford (now Concord, New Hampshire) and almost immediately married a wealthy widow many years his senior. Thompson's first activities seem to have been political. When the War of Independence broke out, he remained loyal to the crown and acted as some sort of secret agent. Obliged to flee to London in 1776 (having separated from his wife the year before), he was rewarded with government work and the appointment as lieutenant colonel of a British regiment in New York. After the war, he retired from the army and lived permanently in exile in Europe. Thompson moved to Bavaria and spent the next few years with the civil administration there, becoming war and police minister as well as grand chamberlain to the elector.

In 1781 Thompson was made a fellow of the Royal Society on the basis of a paper on gunpowder and cannon vents. He had studied the relationship between various gunpowders and the apparent force with which the cannonballs were shot forth. This was a topic, part of a basic interest in guns and other weapons, to which he frequently returned, and in 1797 he produced a gunpowder standard.

In 1791 Thompson was made a count of the Holy Roman Empire in recognition of his work in Bavaria. He took the title from Rumford in his homeland, and it is by this name that we know him today.

Rumford was greatly concerned with the promotion of science, and in 1796 he established the Rumford medals in the Royal Society and in the Academy of Arts and Science, Boston. These were the best-endowed prizes of the time, and they still exist today. In 1799, Rumford returned to England and with Joseph Banks (1743–1820) founded the Royal Institution, choosing Humphry Davy

(1778–1829) as lecturer. Its aim was the popularization of science and technology, a tradition that still continues there. Two years later Rumford resumed his travels. He settled in Paris and married the widow of Antoine Lavoisier (1743–1794), who had produced the caloric theory of heat that Rumford overthrew. However, this was a second unsuccessful match, and after separating from his wife, Rumford lived at Auteuil near Paris until his death on August 21, 1814. In his will he endowed the Rumford chair at Harvard, still occupied by a succession of distinguished scientists.

Rumford's early work in Bavaria shows him at his most versatile and innovative. He combined social experiments with his lifelong interests concerning heat in all its aspects. When he employed beggars from the streets to manufacture military uniforms, he was faced with a feeding problem. A study of nutrition led to the recognition of the importance of water and vegetables, and Rumford decided that soups would fit the requirements. He devised many recipes and developed cheap food emphasizing the potato. Meanwhile soldiers were being employed in gardening to produce the vegetables. Rumford's interest in gardens and landscape gave Munich its huge Englischer Garten, which he planned and which remains an important feature of the city today. The uniform enterprise led to a study of insulation and to the conclusion that heat was lost mainly through convection; thus clothing should be designed to inhibit this.

No application of heat technology was too humble for Rumford's scrutiny. He devised the domestic range—the "fire in a box"—and special utensils to go with it. In the interest of fuel efficiency, he devised a calorimeter to compare the heats of combustion of various fuels. Smoky fireplaces also drew his attention, and after a study of the various air movements, he produced designs incorporating all the features now considered essential in open fires and chimneys, such as the smoke shelf and damper. His search for an alternative to alcoholic drinks led to the promotion of coffee and the design of the first percolator.

The work for which Rumford is best remembered took place in 1798. As military commander for the elector of Bavaria, he was concerned with the manufacture of cannons. These were bored from blocks of iron with drills, and it was believed that the cannons became hot because as the drills cut into the cannon, heat was escaping in the form of a fluid called caloric. However, Rumford noticed that heat production increased as the drills became blunter and cut less into the metal. If a very blunt drill was used, no metal was removed yet the heat output appeared to be limitless. Clearly heat could not be a fluid in the metal, but must be related to the work done in turning the drill. Rumford also studied the expansion of liquids of different densities and different specific heats and showed by careful weighings that the expansion was not due to caloric taking up the extra space.

Rumford's contribution to science in demolishing the caloric theory of heat was very important, because it paved the way to the realization that heat is a form of energy and that all forms of energy are interconvertible. However it took several decades to establish the view that caloric does not exist, as the caloric theory readily explained the important conclusions on heat radiation made by Pierre Prévost (1751–1839) in 1791 and on the motive power of heat made by Sadi Carnot (1796–1832) in 1824.

A *thermodynamic measure of disorder* is called **entropy.** Order means patterns and coherent arrangements. Disorder means dispersion, no patterns, and a randomized, or spread-out arrangement. Entropy is therefore a measure of chaos, and this leads to another statement about the second law of thermodynamics and the direction of natural change, that

the total entropy of the universe continually increases.

Note the use of the words *total* and *universe* in this statement of the second law. The entropy of a system can decrease (more order), for example when a heat pump cools and condenses the random, chaotically moving water vapor molecules into the more ordered state of liquid water. When the energy source for the production, transmission, and use of electrical energy is considered, however, the *total* entropy will be seen as increasing. Likewise, the total entropy increases during the growth of a plant or animal. When all the food, waste products, and products of metabolism are considered, there is again an increase in *total* entropy.

Thus the *natural process* is for a state of order to degrade into a state of disorder with a corresponding increase in entropy. This means that all the available energy of the universe is gradually diminishing, and over time, the universe should therefore approach a limit of maximum disorder called the *heat death* of the universe. The *heat death* of the universe is the theoretical limit of disorder, with all molecules spread far, far apart, vibrating slowly with a uniform low temperature.

The heat death of the universe seems to be a logical consequence of the second law of thermodynamics, but scientists are not certain if the second law should apply to the whole universe. What do you think? Will the universe with all its complexities of organization end with the simplicity of spread-out and slowly vibrating molecules? As has been said, nature is full of symmetry—so why should the universe begin with a bang and end with a whisper?

EXAMPLE **4.8**

A heat engine operates with 65.0 kcal of heat supplied, and exhausts 40.0 kcal of heat. How much work did the engine do?

Solution

Listing the known and unknown quantities:

heat input	$Q_H = 65.0$ kcal
heat rejected	$Q_L = 40.0$ kcal
mechanical equivalent of heat	1 kcal = 4,184 J

The relationship between these quantities is found in equation 4.8 $W = J(Q_H - Q_L)$. This equation states a relationship between the heat supplied to the engine from a high-temperature source (Q_H), which is partly converted to work (W), with the rest rejected in a lower-temperature exhaust (Q_L). The work accomplished is therefore the difference in the heat input and the heat output ($Q_H - Q_L$), so the work accomplished represents the heat used,

where J is the mechanical equivalence of heat (1 kcal = 4,184 J). Therefore,

$$W = J(Q_H - Q_L)$$
$$= 4{,}184\,\frac{J}{kcal}(65.0\text{ kcal} - 40.0\text{ kcal})$$
$$= 4{,}184\,\frac{J}{kcal}(25.0\text{ kcal})$$
$$= 4{,}184 \times 25.0\,\frac{J \cdot kcal}{kcal}$$
$$= 104{,}600\text{ J}$$
$$= \boxed{105\text{ kJ}}$$

Concepts Applied

Thermodynamics in Action

The laws of thermodynamics are concerned with changes in energy and heat. This application explores some of these relationships.

Obtain an electric blender and a thermometer. Fill the blender halfway, then let it remain undisturbed until the temperature is constant as shown by two consecutive temperature readings.

Run the blender at the highest setting for a short time, then stop and record the temperature of the water. Repeat this procedure several times.

Explain your observations in terms of thermodynamics. See if you can think of other experiments that show relationships between changes in energy and heat.

Summary

The kinetic theory of matter assumes that all matter is made up of tiny, ultimate particles of matter called *molecules*. A molecule is defined as the smallest particle of a compound, or a gaseous element, that can exist and still retain the characteristic properties of that substance. Molecules interact, attracting each other through a force of *cohesion*. Liquids, solids, and gases are the *phases of matter* that are explained by the molecular arrangements and forces of attraction between their molecules. A *solid* has a definite shape and volume because it has molecules that vibrate in a fixed equilibrium position with strong cohesive forces. A *liquid* has molecules that have cohesive forces strong enough to give it a definite volume but not strong enough to give it a definite shape. The molecules of a liquid can flow, rolling over each other. A *gas* is composed of molecules that are far apart, with weak cohesive forces. Gas molecules move freely in a constant, random motion.

Molecules can have *vibrational*, *rotational*, or *translational* kinetic energy. The *temperature* of an object is related to the *average kinetic energy* of the molecules making up the object. A measure of temperature tells how hot or cold an object is on two arbitrary scales, the *Fahrenheit scale* and the *Celsius scale*. The *absolute scale*, or *Kelvin scale*, has the coldest temperature possible (−273°C) as zero (0 K).

The observable potential and kinetic energy of an object is the *external energy* of that object, while the potential and kinetic energy of the molecules making up the object is the *internal energy* of the object. Heat refers to the total internal energy and is a transfer of energy that takes place (1) because of a *temperature difference* between two objects, or (2) because of an *energy-form conversion*. An energy-form conversion is actually an energy conversion involving work at the molecular level, so all energy transfers involve *heating* and *working*.

A quantity of heat can be measured in *joules* (a unit of work or energy) or *calories* (a unit of heat). A *kilocalorie* is 1,000 calories, another unit of heat. A *Btu*, or *British thermal unit*, is the English system unit of heat. The *mechanical equivalent of heat* is 4,184 J = 1 kcal.

The *specific heat* of a substance is the amount of energy (or heat) needed to increase the temperature of 1 gram of a substance 1 degree Celsius. The specific heat of various substances is not the same because the molecular structure of each substance is different.

Energy transfer that takes place because of a temperature difference does so through conduction, convection, or radiation. *Conduction* is the transfer of increased kinetic energy from molecule to molecule. Substances vary in their ability to conduct heat, and those that are poor conductors are called *insulators*. Gases, such as air, are good insulators. The best insulator is a vacuum. *Convection* is the transfer of heat by the displacement of large groups of molecules with higher kinetic energy. Convection takes place in fluids, and the fluid movement that takes place because of density differences is called a *convection current*. *Radiation* is radiant energy that moves through space. All objects with an absolute temperature above zero give off radiant energy, but all objects absorb it as well. Energy is transferred from a hot object to a cold one through radiation.

The transition from one phase of matter to another is called a *phase change*. A phase change always absorbs or releases a quantity of *latent heat* not associated with a temperature change. Latent heat is energy that goes into or comes out of *internal potential energy*. The *latent heat of fusion* is absorbed or released at a solid-liquid phase change. The latent heat of fusion for water is 80.0 cal/g (144.0 Btu/lb). The *latent heat of vaporization* is absorbed or released at a liquid-gas phase change. The latent heat of vaporization for water is 540.0 cal/g (970.0 Btu/lb).

Molecules of liquids sometimes have a high enough velocity to escape the surface through the process called *evaporation*. Evaporation is a cooling process, since the escaping molecules remove the latent heat of vaporization in addition to their high molecular energy. Vapor molecules return to the liquid state through the process called *condensation*. Condensation is the opposite of evaporation and is a warming process. When the condensation rate equals the evaporation rate, the air is said to be *saturated*. The rate of evaporation can be *increased* by (1) increased temperature, (2) increased surface area, (3) removal of evaporated molecules, and (4) reduced atmospheric pressure.

Thermodynamics is the study of heat and its relationship to mechanical energy, and the *laws of thermodynamics* describe these relationships: The *first law of thermodynamics* states that the energy supplied to a thermodynamic system in the form of heat, minus the work done by the system, is equal to the change in internal energy. The *second law of thermodynamics* states that heat flows from objects with a higher temperature to objects with a cooler temperature. The second law implies a *degradation of energy* as *high-quality* (more ordered) energy sources undergo *degradation* to *low-quality* (less ordered) sources. *Entropy* is a thermodynamic measure of disorder; it is seen as continually increasing in the universe and may result in the maximum disorder called the *heat death* of the universe.

Summary of Equations

4.1
$$T_F = \frac{9}{5}T_C + 32°$$

4.2
$$T_C = \frac{5}{9}(T_F - 32°)$$

4.3
$$T_K = T_C + 273$$

4.4
Quantity of heat = (mass)(specific heat)(temperature change)
$$Q = mc\Delta T$$

4.5
Heat absorbed or released = (mass)(latent heat of fusion)
$$Q = mL_f$$

4.6
Heat absorbed or released = (mass)(latent heat of vaporization)
$$Q = mL_v$$

4.7
(mechanical equivalence of heat) − (work) = internal energy difference between two states
$$JQ - W = U_2 - U_1$$

4.8
work = (mechanical equivalence of heat)(difference in heat input and heat output)
$$W = J(Q_H - Q_L)$$

KEY TERMS

absolute scale (p. **91**)	kilocalorie (p. **94**)
British thermal unit (p. **94**)	kinetic molecular theory (p. **86**)
calorie (p. **93**)	
Celsius scale (p. **91**)	latent heat of fusion (p. **100**)
conduction (p. **96**)	latent heat of vaporization (p. **100**)
convection (p. **97**)	
entropy (p. **109**)	molecule (p. **87**)
external energy (p. **92**)	phase change (p. **99**)
Fahrenheit scale (p. **90**)	radiation (p. **98**)
first law of thermodynamics (p. **105**)	second law of thermodynamics (p. **106**)
heat (p. **92**)	specific heat (p. **94**)
internal energy (p. **92**)	temperature (p. **88**)

1. The temperature of a gas is proportional to the
   a. average velocity of the gas molecules.
   b. internal potential energy of the gas.
   c. number of gas molecules in a sample.
   d. average kinetic energy of the gas molecules.

2. The kinetic molecular theory explains the expansion of a solid material with increases of temperature as basically the result of
   a. individual molecules expanding.
   b. increased translational kinetic energy.
   c. molecules moving a little farther apart.
   d. heat taking up the spaces between molecules.

3. Two degree intervals on the Celsius temperature scale are
   a. equivalent to 3.6 Fahrenheit degree intervals.
   b. equivalent to 35.6 Fahrenheit degree intervals.
   c. twice as hot as 1 Celsius degree.
   d. none of the above.

4. A temperature reading of 2°C is
   a. equivalent to 3.6°F.
   b. equivalent to 35.6°F.
   c. twice as hot as 1°C.
   d. none of the above.

5. The temperature known as *room temperature* is nearest to
   a. 0°C.
   b. 20°C.
   c. 60°C.
   d. 100°C.

6. Using the absolute temperature scale, the freezing point of water is correctly written as
   a. 0 K.
   b. 0°K.
   c. 273 K.
   d. 273°K.

7. The metric unit of heat called a *calorie* is
   a. equivalent to 3.6 Fahrenheit degrees.
   b. the energy needed to increase the temperature of 1 gram of water 1 degree Celsius.
   c. equivalent to a little over 8 joules of mechanical work.
   d. all of the above.

8. Which of the following is a shorthand way of stating that "the temperature change of a substance is directly proportional to the quantity of heat added"?
   a. $Q \propto m$
   b. $m \propto T_f - T_i$
   c. $Q \propto \Delta T$
   d. $Q = T_f - T_i$

9. The quantity known as *specific heat* is
   a. any temperature reported on the more specific absolute temperature scale.
   b. the energy needed to increase the temperature of 1 gram of a substance 1 degree Celsius.
   c. any temperature of a 1 kg sample reported in degrees Celsius.
   d. the heat needed to increase the temperature of any size sample by 1°C.

10. Table 4.2 lists the specific heat of soil as 0.20 kcal/kgC° and the specific heat of water as 1.00 kcal/kgC°. This means that if 1 kg of soil and 1 kg of water each receive 1 kcal of energy, ideally,
    a. the water will be warmer than the soil by 0.8°C.
    b. the soil will be 5°C.
    c. the water will be 5°C.
    d. the water will warm by 1°C, and the soil will warm by 0.2°C.

11. The heat transfer that takes place by energy moving directly from molecule to molecule is called
    a. conduction.
    b. convection.
    c. radiation.
    d. none of the above.

12. The heat transfer that does not require matter is
    a. conduction.
    b. convection.
    c. radiation.
    d. impossible, for matter is always required.

13. Styrofoam is a good insulating material because
    a. it is a plastic material that conducts heat poorly.
    b. it contains many tiny pockets of air.
    c. of the structure of the molecules making up the Styrofoam.
    d. it is not very dense.

14. The transfer of heat that takes place because of density difference in fluids is
    a. conduction.
    b. convection.
    c. radiation.
    d. none of the above.

15. When a solid, liquid, or a gas changes from one physical state to another, the change is called
    a. melting.
    b. entropy.
    c. a phase change.
    d. sublimation.

16. Latent heat is "hidden" because it
    a. goes into or comes out of internal potential energy.
    b. is a fluid (caloric) that cannot be sensed.
    c. does not actually exist.
    d. is a form of internal kinetic energy.

17. As a solid undergoes a phase change to a liquid state, it
   a. releases heat while remaining at a constant temperature.
   b. absorbs heat while remaining at a constant temperature.
   c. releases heat as the temperature decreases.
   d. absorbs heat as the temperature increases.

18. The condensation of water vapor actually
   a. warms the surroundings.
   b. cools the surroundings.
   c. sometimes warms and sometimes cools the surroundings, depending on the relative humidity at the time.
   d. neither warms nor cools the surroundings.

19. Water molecules move back and forth between the liquid and the gaseous state
   a. only when the air is saturated.
   b. at all times, with evaporation, condensation, and saturation defined by the net movement.
   c. only when the outward movement of vapor molecules produces a pressure equal to the atmospheric pressure.
   d. only at the boiling point.

20. A heat engine is designed to
   a. drive heat from a cool source to a warmer location.
   b. drive heat from a warm source to a cooler location.
   c. convert mechanical energy into heat.
   d. convert heat into mechanical energy.

21. The work that a heat engine is able to accomplish is ideally equivalent to the
   a. difference between the heat supplied and the heat rejected.
   b. heat that was produced in the cycle.
   c. heat that appears in the exhaust gases.
   d. sum total of the heat input and the heat output.

Answers

1. d 2. c 3. a 4. b 5. b 6. c 7. b 8. c 9. b 10. b 11. a 12. c 13. b 14. b 15. c
16. a 17. b 18. a 19. b 20. d 21. a

1. What is temperature? What is heat?
2. Explain why most materials become less dense as their temperature is increased.
3. Would the tight packing of more insulation, such as glass wool, in an enclosed space increase or decrease the insulation value? Explain.
4. A true vacuum bottle has a double-walled, silvered bottle with the air removed from the space between the walls. Describe how this design keeps food hot or cold by dealing with conduction, convection, and radiation.
5. Why is cooler air found in low valleys on calm nights?
6. Why is air a good insulator?
7. Explain the meaning of the mechanical equivalent of heat.
8. What do people really mean when they say that a certain food "has a lot of Calories"?
9. A piece of metal feels cooler than a piece of wood at the same temperature. Explain why.
10. Explain how the latent heat of fusion and the latent heat of vaporization are "hidden."
11. What is condensation? Explain, on a molecular level, how the condensation of water vapor on a bathroom mirror warms the bathroom.
12. Which provides more cooling for a Styrofoam cooler, one with 10 lb of ice at 0°C or one with 10 lb of ice water at 0°C? Explain your reasoning.
13. Explain why a glass filled with a cold beverage seems to "sweat." Would you expect more sweating inside a house during the summer or during the winter? Explain.
14. Explain why a burn from 100°C steam is more severe than a burn from water at 100°C.
15. Briefly describe, using sketches as needed, how a heat pump is able to move heat from a cooler region to a warmer region.
16. Which has more entropy—ice, liquid water, or water vapor? Explain your reasoning.
17. Suppose you use a heat engine to do the work to drive a heat pump. Could the heat pump be used to provide the temperature difference to run the heat engine? Explain.

The exercises in groups A and B cover the same concepts. Solutions to group A exercises are located in appendix D.

Note: *Neglect all frictional forces in all exercises.*

Group A

1. The average human body temperature is 98.6°F. What is the equivalent temperature on the Celsius scale?
2. An electric current heats a 221 g copper wire from 20.0°C to 38.0°C. How much heat was generated by the current? (c_{copper} = 0.093 kcal/kgC°)
3. A bicycle and rider have a combined mass of 100.0 kg. How many calories of heat are generated in the brakes when the bicycle comes to a stop from a speed of 36.0 km/h?

Group B

1. The Fahrenheit temperature reading is 98° on a hot summer day. What is this reading on the Kelvin scale?
2. A 0.25 kg length of aluminum wire is warmed 10.0°C by an electric current. How much heat was generated by the current? ($c_{aluminum}$ = 0.22 kcal/kgC°)
3. A 1,000.0 kg car with a speed of 90.0 km/h brakes to a stop. How many cal of heat are generated by the brakes as a result?

4. A 15.53 kg loose bag of soil falls 5.50 m at a construction site. If all the energy is retained by the soil in the bag, how much will its temperature increase? (c_{soil} = 0.200 kcal/kgC°)

5. A 75.0 kg person consumes a small order of french fries (250.0 Cal) and wishes to "work off" the energy by climbing a 10.0 m stairway. How many vertical climbs are needed to use all the energy?

6. A 0.5 kg glass bowl (c_{glass} = 0.2 kcal/kgC°) and a 0.5 kg iron pan (c_{iron} = 0.11 kcal/kgC°) have a temperature of 68°F when placed in a freezer. How much heat will the freezer have to remove from each to cool them to 32°F?

7. A sample of silver at 20.0°C is warmed to 100.0°C when 896 cal is added. What is the mass of the silver? (c_{silver} = 0.056 kcal/kgC°)

8. A 300.0 W immersion heater is used to heat 250.0 g of water from 10.0°C to 70.0°C. About how many minutes did this take?

9. A 100.0 g sample of metal is warmed 20.0°C when 60.0 cal is added. What is the specific heat of this metal?

10. How much heat is needed to change 250.0 g of ice at 0°C to water at 0°C?

11. How much heat is needed to change 250.0 g of water at 80.0°C to steam at 100.0°C?

12. A 100.0 g sample of water at 20.0°C is heated to steam at 125.0°C. How much heat was absorbed?

13. In an electric freezer, 400.0 g of water at 18.0°C is cooled, frozen, and the ice is chilled to −5.00°C. (a) How much total heat was removed from the water? (b) If the latent heat of vaporization of the Freon refrigerant is 40.0 cal/g, how many grams of Freon must be evaporated to absorb this heat?

14. A heat engine is supplied with 300.0 cal and rejects 200.0 cal in the exhaust. How many joules of mechanical work was done?

15. A refrigerator removes 40.0 kcal of heat from the freezer and releases 55.0 kcal through the condenser on the back. How much work was done by the compressor?

4. A 1.0 kg metal head of a geology hammer strikes a solid rock with a velocity of 5.0 m/s. Assuming all the energy is retained by the hammer head, how much will its temperature increase? (c_{head} = 0.11 kcal/kgC°)

5. A 60.0 kg person will need to climb a 10.0 m stairway how many times to "work off" each excess Cal (kcal) consumed?

6. A 50.0 g silver spoon at 20.0°C is placed in a cup of coffee at 90.0°C. How much heat does the spoon absorb from the coffee to reach a temperature of 89.0°C?

7. If the silver spoon placed in the coffee in problem 6 causes it to cool 0.75°C, what is the mass of the coffee? (Assume c_{coffee} = 1.0 cal/gC°)

8. How many minutes would be required for a 300.0 W immersion heater to heat 250.0 g of water from 20.0°C to 100.0°C?

9. A 200.0 g china serving bowl is warmed 65.0°C when it absorbs 2.6 kcal of heat from a serving of hot food. What is the specific heat of the china dish?

10. A 1.00 kg block of ice at 0°C is added to a picnic cooler. How much heat will the ice remove as it melts to water at 0°C?

11. A 500.0 g pot of water at room temperature (20.0°C) is placed on a stove. How much heat is required to change this water to steam at 100.0°C?

12. Spent steam from an electric generating plant leaves the turbines at 120.0°C and is cooled to 90.0°C liquid water by water from a cooling tower in a heat exchanger. How much heat is removed by the cooling tower water for each kg of spent steam?

13. Lead is a soft, dense metal with a specific heat of 0.028 kcal/kgC°, a melting point of 328.0°C, and a heat of fusion of 5.5 kcal/kg. How much heat must be provided to melt a 250.0 kg sample of lead with a temperature of 20.0°C?

14. A heat engine converts 100.0 cal from a supply of 400.0 cal into work. How much mechanical work was done?

15. A heat pump releases 60.0 kcal as it removes 40.0 kcal at the evaporator coils. How much work does this heat pump ideally accomplish?

Compared to the sounds you hear on a calm day in the woods, the sounds from a waterfall can carry up to a million times more energy.

CHAPTER

5

Wave Motions
and Sound

S ometimes you can feel the floor of a building shake for a moment when something heavy is dropped. You can also feel prolonged vibrations in the ground when a nearby train moves by. The floor of a building and the ground are solids that transmit vibrations from a disturbance. Vibrations are common in most solids because the solids are elastic, having a tendency to rebound, or snap back, after a force or an impact deforms them. Usually you cannot see the vibrations in a floor or the ground, but you sense they are there because you can feel them.

There are many examples of vibrations that you can see. You can see the rapid blur of a vibrating guitar string (Figure 5.1). You can see the vibrating up-and-down movement of a bounced-upon diving board. Both the vibrating guitar string and the diving board set up a vibrating motion of air that you identify as a sound. You cannot see the vibrating motion of the air, but you sense it is there because you hear sounds.

There are many kinds of vibrations that you cannot see but can sense. Heat, as you have learned, is associated with molecular vibrations that are too rapid and too tiny for your senses to detect other than as an increase in temperature. Other invisible vibrations include electrons that vibrate, generating spreading electromagnetic radio waves or visible light. Thus vibrations are not only observable motions of objects but are also characteristics of sound, heat, electricity, and light. The vibrations involved in all these phenomena are alike in many ways, and all involve energy. Therefore, many topics of physical science are concerned with vibrational motion. In this chapter you will learn about the nature of vibrations and how they produce waves in general. These concepts will be applied to sound in this chapter and to electricity, light, and radio waves in later chapters.

Forces and Elastic Materials

If you drop a rubber ball, it bounces because it is capable of recovering its shape when it hits the floor. A ball of clay, on the other hand, does not recover its shape and remains a flattened blob on the floor. An *elastic* material is one that is capable of recovering its shape after a force deforms it. A rubber ball is elastic and a ball of clay is not elastic. You know a metal spring is elastic because you can stretch it or compress it and it always recovers its shape.

There is a direct relationship between the extent of stretching or compression of a spring and the amount of force applied to it. A large force stretches a spring a lot; a small force stretches it a little. As long as the applied force does not exceed the elastic limit of the spring, it will always return to its original shape when you remove the applied force. There are three important considerations about the applied force and the response of the spring:

1. The greater the applied force, the greater the compression or stretch of the spring from its original shape.
2. The spring appears to have an *internal restoring force,* which returns it to its original shape.
3. The farther the spring is pushed or pulled, the *stronger* the restoring force that returns the spring to its original shape.

Forces and Vibrations

A **vibration** is a back-and-forth motion that repeats itself. Such a motion is not restricted to any particular direction, and it can be in many different directions at the same time. Almost any

FIGURE 5.1

Vibrations are common in many elastic materials, and you can see and hear the results of many in your surroundings. Other vibrations in your surroundings, such as those involved in heat, electricity, and light, are invisible to the senses.

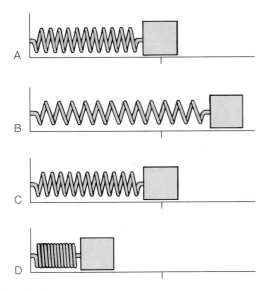

FIGURE 5.2

A mass on a frictionless surface is at rest at an equilibrium position (A) when undisturbed. When the spring is stretched (B) or compressed (D), then released (C), the mass vibrates back and forth because restoring forces pull opposite to and proportional to the displacement.

solid can be made to vibrate if it is elastic. To see how forces are involved in vibrations, consider the spring and mass in Figure 5.2. The spring and mass are arranged so that the mass can freely move back and forth on a frictionless surface. When the mass has not been disturbed, it is at rest at an *equilibrium position* (Figure 5.2A). At the equilibrium position the spring is not compressed or stretched, so it applies no force on the mass. If, however, the mass is pulled to the right (Figure 5.2B) the spring is stretched and applies a restoring force on the mass toward the left. The farther the mass is displaced, the greater the stretch of the spring and thus the greater the restoring force. The restoring force is proportional to the displacement and is in the opposite direction of the applied force.

If the mass is now released, the restoring force is the only force acting (horizontally) on the mass, so it accelerates back toward the equilibrium position. This force will continuously decrease until the moving mass arrives back at the equilibrium position, where the force is zero (Figure 5.2C). The mass will have a maximum velocity when it arrives, however, so it overshoots the equilibrium position and continues moving to the left (Figure 5.2D). As it moves to the left of the equilibrium position, it compresses the spring, which exerts an increasing force on the mass. The moving mass comes to a temporary halt, but now the restoring force again starts it moving back toward the equilibrium position. The whole process repeats itself again and again as the mass moves back and forth over the same path.

The periodic vibration, or oscillation, of the mass is similar to many vibrational motions found in nature called *simple harmonic motion*. Simple harmonic motion is defined as the vibratory motion that occurs when there is a restoring force opposite to and proportional to a displacement.

The vibrating mass and spring system will continue to vibrate for a while, slowly decreasing with time until the vibrations stop completely. The slowing and stopping is due to air resistance and internal friction. If these could be eliminated or compensated for with additional energy, the mass would continue to vibrate with a repeating, or *periodic*, motion.

Describing Vibrations

A vibrating mass is described by measuring several variables (Figure 5.3). The extent of displacement from the equilibrium position is called the **amplitude.** A vibration that has a mass displaced a greater distance from equilibrium thus has a greater amplitude than a vibration with less displacement.

A complete vibration is called a **cycle.** A cycle is the movement from some point, say the far left, all the way to the far right, and back to the same point again, the far left in this example. The **period** (T) is the number of seconds per cycle. For example, suppose 0.1 s is required for an object to move through one complete cycle, to complete the motion from one point, then back to that point. The period of this vibration is 0.1 s.

Sometimes it is useful to know how frequently a vibration completes a cycle every second. The number of cycles per second is called the **frequency** (f). For example, a vibrating object moves through 10 cycles in 1 s. The frequency of this vibration is 10 cycles per second. Frequency is measured in a unit called a **hertz** (Hz). The unit for a hertz is 1/s since a cycle does not have dimensions. Thus a frequency of 10 cycles per second is referred to as 10 hertz or 10 1/s.

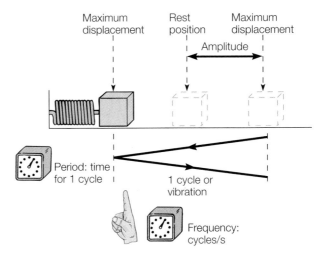

FIGURE 5.3

A vibrating mass attached to a spring is displaced from the rest or equilibrium position, and then released. The maximum displacement is called the *amplitude* of the vibration. A cycle is one complete vibration. The period is the number of seconds per cycle. The frequency is a count of how many cycles are completed in 1 s.

The period and frequency are two ways of describing the time involved in a vibration. Since the period (T) is the number of seconds per cycle and the frequency (f) is the number of cycles per second, the relationship is

$$T = \frac{1}{f}$$

equation 5.1

or

$$f = \frac{1}{T}$$

equation 5.2

EXAMPLE 5.1

A vibrating system has a period of 0.5 s. What is the frequency in Hz?

Solution

$$T = 0.5 \text{ s}$$
$$f = ?$$

$$f = \frac{1}{T}$$

$$= \frac{1}{0.5 \text{ s}}$$

$$= \frac{1}{0.5} \frac{1}{\text{s}}$$

$$= 2 \frac{1}{\text{s}}$$

$$= \boxed{2 \text{ Hz}}$$

You can obtain a graph of a vibrating object, which makes it easier to measure the amplitude, period, and frequency. If a pen is fixed to a vibrating mass and a paper is moved beneath it at a steady rate, it will draw a curve, as shown in Figure 5.4. The greater the amplitude of the vibrating mass, the greater the height of this curve. The greater the frequency, the closer together the peaks and valleys. Note the shape of this curve. This shape is characteristic of simple harmonic motion and is called a *sinusoidal*, or sine, graph. It is so named because it is the same shape as a graph of the sine function in trigonometry.

Waves

A vibration can be a repeating, or *periodic,* type of motion that disturbs the surroundings. A *pulse* is a disturbance of a single event of short duration. Both pulses and periodic vibrations can create a physical **wave** in the surroundings. A wave is a disturbance that moves through a medium such as a solid or the air. A

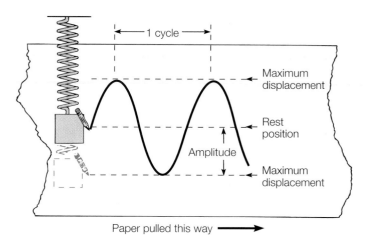

FIGURE 5.4
A graph of simple harmonic motion is described by a sinusoidal curve.

heavy object dropped on the floor, for example, makes a pulse that sends a mechanical wave that you feel. It might also make a sound wave in the air that you hear. In either case, the medium that transported a wave (solid floor or air) returns to its normal state after the wave has passed. The medium does not travel from place to place, the wave does. Two major considerations about a wave are that (1) a wave is a traveling disturbance, and (2) a wave transports energy.

You can observe waves when you drop a rock into a still pool of water. The rock pushes the water into a circular mound as it enters the water. Since it is forcing the water through a distance, it is doing work to make the mound. The mound starts to move out in all directions, in a circle, leaving a depression behind. Water moves into the depression, and a circular wave—mound and depression—moves from the place of disturbance outward. Any floating object in the path of the wave, such as a leaf, exhibits an up-and-down motion as the mound and depression of the wave pass. But the leaf merely bobs up and down and after the wave has passed, it is much in the same place as before the wave. Thus, it was the disturbance that traveled across the water, not the water itself. If the wave reaches a leaf floating near the edge of the water, it may push the leaf up and out of the water, doing work on the leaf. Thus, the wave is a moving disturbance that transfers energy from one place to another.

Kinds of Waves

If you could see the motion of an individual water molecule near the surface as a water wave passed, you would see it trace out a circular path as it moves up and over, down and back. This circular motion is characteristic of the motion of a particle reacting to a water wave disturbance. There are other kinds of waves, and each involves particles in a characteristic motion.

A **longitudinal wave** is a disturbance that causes particles to move closer together or farther apart in the same direction that the wave is moving. If you attach one end of a coiled spring to a wall and pull it tight, you will make longitudinal

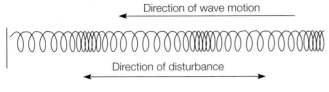

A Longitudinal wave

Direction of wave motion

Direction of disturbance

B Transverse wave

Direction of wave motion

Direction of disturbance

FIGURE 5.5

(*A*) Longitudinal waves are created in a spring when the free end is moved back and forth parallel to the spring. (*B*) Transverse waves are created in a spring when the free end is moved up and down.

waves in the spring if you grasp the spring and then move your hand back and forth parallel to the spring. Each time you move your hand toward the length of the spring, a pulse of closer-together coils will move across the spring (Figure 5.5A). Each time you pull your hand back, a pulse of farther-apart coils will move across the spring. The coils move back and forth in the same direction that the wave is moving, which is the characteristic movement in reaction to a longitudinal wave.

You will make a different kind of wave in the stretched spring if you now move your hand up and down perpendicular to the length of the spring. This creates a **transverse wave.** A transverse wave is a disturbance that causes motion perpendicular to the direction that the wave is moving. Particles responding to a transverse wave do not move closer together or farther apart in response to the disturbance; rather, they vibrate back and forth or up and down in a direction perpendicular to the direction of the wave motion (see Figure 5.5B).

Concepts Applied

Making Waves

Obtain a Slinky or a long, coiled spring and stretch it out on the floor. Have another person hold the opposite end stationary while you make waves move along the spring. Make longitudinal and transverse waves, observing how the disturbance moves in each case. If the spring is long enough, measure the distance, then time the movement of each type of wave. How fast were your waves?

Whether you make mechanical longitudinal or transverse waves depends not only on the nature of the disturbance creating the waves, but also on the nature of the medium. Mechanical transverse waves can move through a material only if there is some interaction, or attachment, between the molecules making up the medium. In a gas, for example, the molecules move about freely without attachments to one another. A pulse can cause these molecules to move closer together or farther apart, so a gas can carry a longitudinal wave. But if a gas molecule is caused to move up or down, there is no reason for other molecules to do the same, since they are not attached. Thus, a gas will carry mechanical longitudinal waves but not mechanical transverse waves. Likewise a liquid will carry mechanical longitudinal waves but not mechanical transverse waves because the liquid molecules simply slide past one another. The surface of a liquid, however, is another story because of surface tension. A surface water wave is, in fact, a combination of longitudinal and transverse wave patterns that produce the circular motion of a disturbed particle. Solids can and do carry both longitudinal and transverse waves because of the strong attachments between the molecules.

Waves in Air

Because air is fluid, waves in air can only be longitudinal; therefore, sound waves in air must be longitudinal waves. A familiar situation will be used to describe the nature of a longitudinal wave moving through air before we consider sound specifically. The situation involves a small room with no open windows and two doors that open into the room. When you open one door into the room the other door closes. Why does this happen? According to the kinetic molecular theory, the room contains many tiny, randomly moving gas molecules that make up the air. As you opened the door, it pushed on these gas molecules, creating a jammed-together zone of molecules immediately adjacent to the door. This jammed-together zone of air now has a greater density and pressure, which immediately spreads outward from the door as a pulse. The disturbance is rapidly passed from molecule to molecule, and the pulse of compression spreads through the room. In the example of the closing door, the pulse of greater density and increased pressure of air reached the door at the other side of the room, and the composite effect of the molecules hitting the door, that is, the increased pressure, caused it to close.

If the door at the other side of the room does not latch, you can probably cause it to open again by pulling on the first door quickly. By so doing, you send a pulse of thinned-out molecules of lowered density and pressure. The door you pulled quickly pushed some of the molecules out of the room. Other molecules quickly move into the region of less pressure, then back to their normal positions. The overall effect is the movement of a thinned-out pulse that travels through the room. When the pulse of slightly reduced pressure reaches the other door, molecules exerting their normal pressure on the other side of the door cause it to move. After a pulse has passed a particular place, the molecules are very soon homogeneously distributed again due to their rapid, random movement.

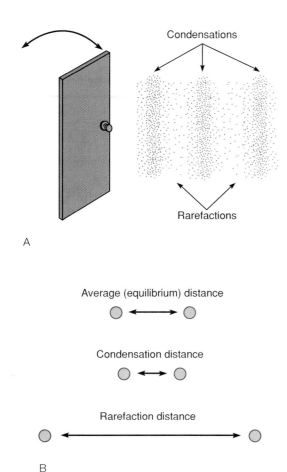

Condensations

Rarefactions

A

Average (equilibrium) distance

Condensation distance

Rarefaction distance

B

FIGURE 5.6

(A) Swinging the door inward produces pulses of increased density and pressure called *condensations*. Pulling the door outward produces pulses of decreased density and pressure called *rarefactions*. (B) In a condensation, the average distance between gas molecules is momentarily decreased as the pulse passes. In a rarefaction, the average distance is momentarily increased.

If you were to swing a door back and forth, it would be a vibrating object. As it vibrates back and forth it would have a certain frequency in terms of the number of vibrations per second. As the vibrating door moves toward the room, it creates a pulse of jammed-together molecules called a *condensation* (or compression) that quickly moves throughout the room. As the vibrating door moves away from the room, a pulse of thinned-out molecules called a *rarefaction* quickly moves throughout the room. The vibrating door sends repeating pulses of condensation (increased density and pressure) and rarefaction (decreased density and pressure) through the room as it moves back and forth (Figure 5.6). You know that the pulses transmit energy because they produce movement, or do work on, the other door. Individual molecules execute a harmonic motion about their equilibrium position and can do work on a movable object. Energy is thus transferred by this example of longitudinal waves.

Hearing Waves in Air

You cannot hear a vibrating door because the human ear normally hears sounds originating from vibrating objects with a frequency between 20 and 20,000 Hz. Longitudinal waves with frequencies less than 20 Hz are called **infrasonic.** You usually *feel* sounds below 20 Hz rather than hearing them, particularly if you are listening to a good sound system. Longitudinal waves above 20,000 Hz are called **ultrasonic.** Although 20,000 Hz is usually considered the upper limit of hearing, the actual limit varies from person to person and becomes lower and lower with increasing age. Humans do not hear infrasonic or ultrasonic sounds, but various animals have different limits. Dogs, cats, rats, and bats can hear higher frequencies than humans. Dogs can hear an ultrasonic whistle when a human hears nothing, for example. Some bats make and hear sounds of frequencies up to 100,000 Hz as they navigate and search for flying insects in total darkness. Scientists discovered recently that elephants communicate with extremely low-frequency sounds over distances of several kilometers. Humans cannot detect such low-frequency sounds. This raises the possibility of infrasonic waves that other animals can detect that we cannot.

A tuning fork that vibrates at 260 Hz makes longitudinal waves much like the swinging door, but these longitudinal waves are called *sound waves* because they are within the frequency range of human hearing. The prongs of a struck tuning fork vibrate, moving back and forth. This is more readily observed if the prongs of the fork are struck, then held against a sheet of paper or plunged into a beaker of water. In air, the vibrating prongs first move toward you, pushing the air molecules into a condensation of increased density and pressure. As the prongs then move back, a rarefaction of decreased density and pressure is produced. The alternation of increased and decreased pressure pulses moves from the vibrating tuning fork and spreads outward equally in all directions, much like the surface of a rapidly expanding balloon (Figure 5.7). When the pulses reach your eardrum, it is forced in and out by the pulses. It now vibrates with the same frequency as the tuning fork. The vibrations of the eardrum are transferred by three tiny bones to a fluid in a coiled chamber (Figure 5.8). Here, tiny hairs respond

to the frequency and size of the disturbance, activating nerves that transmit the information to the brain. The brain interprets a frequency as a sound with a certain **pitch.** High-frequency sounds are interpreted as high-pitched musical notes, for example, and low-frequency sounds are interpreted as low-pitched musical notes. The brain then selects certain sounds from all you hear, and you "tune" to certain ones, enabling you to listen to whatever sounds you want while ignoring the background noise, which is made up of all the other sounds.

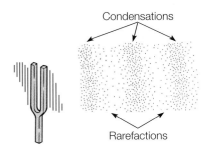

FIGURE 5.7

A vibrating tuning fork produces a series of condensations and rarefactions that move away from the tuning fork. The pulses of increased and decreased pressure reach your ear, vibrating the eardrum. The ear sends nerve signals to the brain about the vibrations, and the brain interprets the signals as sounds.

Describing Waves

A tuning fork vibrates with a certain frequency and amplitude, producing a longitudinal wave of alternating pulses of increased-pressure condensations and reduced-pressure rarefactions. A graph of the frequency and amplitude of the vibrations is shown in Figure 5.9A, and a representation of the condensations and rarefactions is shown in Figure 5.9B. The wave pattern can also be represented by a graph of the changing air pressure of the traveling sound wave, as shown in Figure 5.9C. This graph can be used to define some interesting concepts associated with sound waves. Note the correspondence between (1) the amplitude, or displacement, of the vibrating prong, (2) the pulses of condensations and rarefactions, and (3) the changing air pressure. Note also the correspondence between the frequency of the vibrating prong and the frequency of the wave cycles.

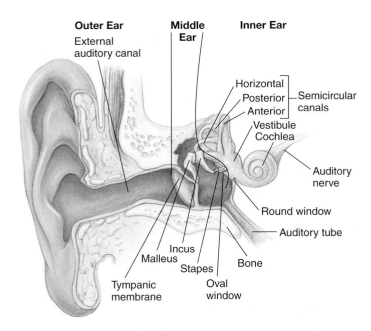

FIGURE 5.8

Anatomy of the ear. Sound enters the outer ear and, upon reaching the middle ear, impinges upon the tympanic membrane, which vibrates three bones (malleus, incus, and stapes). The vibrating stapes hits the oval window, and hair cells in the cochlea convert the vibrations into action potentials, which follow the auditory nerve to the brain. Hair cells in the semicircular canals and in the vestibule sense balance. The auditory tube connects the middle ear to the throat, equalizing air pressure.

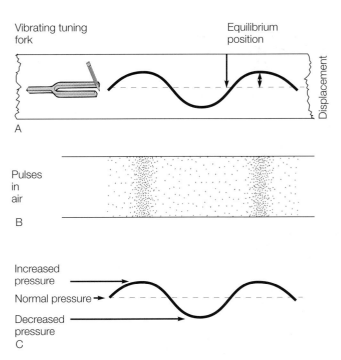

FIGURE 5.9

Compare the (A) back-and-forth vibrations of a tuning fork with (B) the resulting condensations and rarefactions that move through the air and (C) the resulting increases and decreases of air pressure on a surface that intercepts the condensations and rarefactions.

Ultrasonic waves are mechanical waves that have frequencies above the normal limit of hearing of the human ear. The arbitrary upper limit is about 20,000 Hz, so an ultrasonic wave has a frequency of 20,000 Hz or greater. Intense ultrasonic waves are used in many ways in industry and medicine.

Industrial and commercial applications of ultrasound utilize lower-frequency ultrasonic waves in the 20,000 to 60,000 Hz range. Commercial devices that send ultrasound through the air include burglar alarms and rodent repellers. An ultrasonic burglar alarm sends ultrasonic spherical waves through the air of a room. The device is adjusted to ignore echoes from the contents of the room. The presence of a person provides a new source of echoes, which activates the alarm. Rodents emit ultrasonic frequencies up to 150,000 Hz that are used in rodent communication when they are disturbed or during aggressive behavior. The ultrasonic rodent repeller generates ultrasonic waves of similar frequency. Other commercial applications of ultrasound include sonar and depth measurements, cleaning and drilling, welding plastics and metals, and material flaw detection. Ultrasonic cleaning baths are used to remove dirt and foreign matter from solid surfaces, usually within a liquid solvent. The ultrasonic waves create vapor bubbles in the liquid, which vibrate and emit audible and ultrasonic sound waves. The audible frequencies are often heard as a hissing or frying sound.

Medical applications of ultrasound use frequencies in the 1,000,000 to 20,000,000 Hz range. Ultrasound in this frequency range cannot move through the air because the required displacement amplitudes of the gas molecules in the air are less than the average distance between the molecules. Thus, a gas molecule that is set into motion in this frequency range cannot collide with other gas molecules to transmit the energy of the wave. Intense ultrasound is used for cleaning teeth and disrupting kidney stones. Less intense ultrasound is used for therapy (heating and reduction of pain). The least intense ultrasounds are used for diagnostic imaging. The largest source of exposure of humans to ultrasound is for the purpose of ultrasonic diagnostic imaging, particularly in fertility and pregnancy cases (Box Figure 5.1).

Originally developed in the late 1950s, ultrasonic medical machines have been improved and today make images of outstanding detail and clarity. One type of ultrasonic medical machine uses a transducer probe, which emits an ultrasonic pulse that passes into the body. Echoes from the internal tissues and organs are reflected back to the transducer, which sends the signals to a computer. Another pulse of ultrasound is then sent out after the echoes from the first pulse have returned. The strength of and number of pulses per second vary with the application, ranging from hundreds to thousands of pulses per second. The computer constructs a picture from the returning echoes, showing an internal view without the use of more dangerous X rays.

Ultrasonic scanners have been refined to the point that the surface of the ovaries can now be viewed, showing the number and placement of developing eggs. The ovary scan is typically used in conjunction with fertility-stimulating drugs, where multiple births are possible, and to identify the exact time of ovulation. As early as four weeks after conception, the ultrasonic scanner is used to identify and monitor the fetus. By the thirteenth week, an ultrasonic scan can show the fetal heart movement, bone and skull size, and internal organs. The ultrasonic scanner is often used to show the position of the fetus and placenta for the purpose of amniocentesis, which involves withdrawing a sample of amniotic fluid from the uterus for testing.

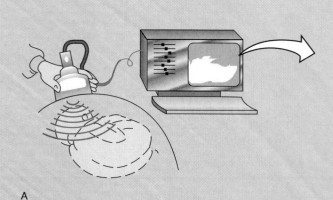

A

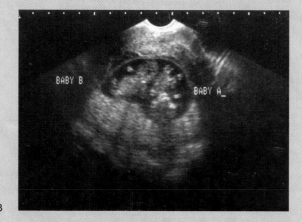

B

BOX FIGURE 5.1

(A) Physicians can "see" a baby inside its mother's body by using the handheld external probe of an ultrasonic imaging system. (B) An ultrasonic image of 10-week old human twins inside the mother's body.

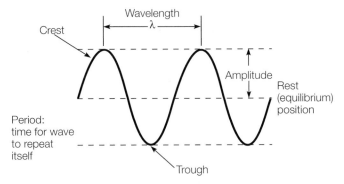

FIGURE 5.10

Here are some terms associated with periodic waves. The *wavelength* is the distance from a part of one wave to the same part in the next wave, such as from one crest to the next. The *amplitude* is the displacement from the rest position. The *period* is the time required for a wave to repeat itself, that is, the time for one complete wavelength to move past a given location.

Figure 5.10 shows the terms commonly associated with waves from a continuously vibrating source. The wave *crest* is the maximum disturbance from the undisturbed (rest) position. For a sound wave, this would represent the maximum increase of air pressure. The wave *trough* is the maximum disturbance in the opposite direction from the rest position. For a sound wave, this would represent the maximum decrease of air pressure. The *amplitude* of a wave is the displacement from rest to the crest *or* from rest to the trough. The time required for a wave to repeat itself is the *period* (*T*). To repeat itself means the time required to move through one full wave, such as from the crest of one wave to the crest of the next wave. This length in which the wave repeats itself is called the **wavelength** (the symbol is λ, which is the Greek letter lambda). Wavelength is measured in centimeters or meters just like any other length.

There is a relationship between the wavelength, period, and speed of a wave. Recall that speed is

$$v = \frac{\text{distance}}{\text{time}}$$

Since it takes one period (*T*) for a wave to move one wavelength (λ), then the speed of a wave can be measured from

$$v = \frac{\text{one wavelength}}{\text{one period}} = \frac{\lambda}{T}$$

The *frequency*, however, is more convenient than the period for dealing with waves that repeat themselves rapidly. Recall the relationship between frequency (*f*) and the period (*T*) is

$$f = \frac{1}{T}$$

Substituting *f* for 1/*T* yields

$$v = \lambda f$$

equation 5.3

which we will call the *wave equation*. This equation tells you that the velocity of a wave can be obtained from the product of the wavelength and the frequency. Note that it also tells you that the wavelength and frequency are inversely proportional at a given velocity.

EXAMPLE 5.2

A sound wave with a frequency of 260 Hz has a wavelength of 1.27 m. With what speed would you expect this sound wave to move?

Solution

$$f = 260 \text{ Hz} \qquad v = \lambda f$$
$$\lambda = 1.27 \text{ m}$$
$$v = ? \qquad\qquad = (1.27 \text{ m})\left(260\frac{1}{s}\right)$$
$$= 1.27 \times 260 \ \text{m} \times \frac{1}{s}$$
$$= \boxed{330 \frac{\text{m}}{\text{s}}}$$

EXAMPLE 5.3

In general, the human ear is most sensitive to sounds at 2,500 Hz. Assuming that sound moves at 330 m/s, what is the wavelength of sounds to which people are most sensitive? (Answer: 13 cm)

Sound Waves

The transmission of a sound wave requires a medium, that is, a solid, liquid, or gas to carry the disturbance. Therefore, sound does not travel through the vacuum of outer space, since there is nothing to carry the vibrations from a source. The nature of the molecules making up a solid, liquid, or gas determines how well or how rapidly the substance will carry sound waves. The two variables are (1) the inertia of the molecules and (2) the strength of the interaction, if the molecules are attached to one another. Thus, hydrogen gas, with the least massive molecules with no interaction or attachments, will carry a sound wave at 1,284 m/s (4,213 ft/s) when the temperature is 0°C. More massive helium gas molecules have more inertia and carry a sound wave at only 965 m/s (3,166 ft/s) at the same temperature. A solid, however, has molecules that are strongly attached, so vibrations are passed rapidly from molecule to molecule. Steel, for example, is highly elastic, and sound will move through a steel rail at 5,940 m/s (19,488 ft/s). Thus there is a reason for the old saying, "Keep your ear to the ground," because sounds move through solids more rapidly than through a gas (Table 5.1).

TABLE 5.1
Speed of sound in various materials

Medium	m/s	ft/s
Carbon dioxide (0°C)	259	850
Dry air (0°C)	331	1,087
Helium (0°C)	965	3,166
Hydrogen (0°C)	1,284	4,213
Water (25°C)	1,497	4,911
Seawater (25°C)	1,530	5,023
Lead	1,960	6,430
Glass	5,100	16,732
Steel	5,940	19,488

Velocity of Sound in Air

Most people have observed that sound takes some period of time to move through the air. If you watch a person hammering on a roof a block away, the sounds of the hammering are not in sync with what you see. Light travels so rapidly that you can consider what you see to be simultaneous with what is actually happening for all practical purposes. Sound, however, travels much more slowly, and the sounds arrive late in comparison to what you are seeing. This is dramatically illustrated by seeing a flash of lightning, then hearing thunder seconds later. Perhaps you know of a way to estimate the distance to a lightning flash by timing the interval between the flash and boom. If not, you will learn a precise way to measure this distance shortly.

The air temperature influences how rapidly sound moves through the air. The gas molecules in warmer air have a greater kinetic energy than those of cooler air. The molecules of warmer air therefore transmit an impulse from molecule to molecule more rapidly. More precisely, the speed of a sound wave increases 0.600 m/s (2.00 ft/s) for *each* Celsius degree increase in temperature above 0°C. So, it will be easier for your car or airplane to break the sound barrier on a cold day than on a warm day. How much easier? In *dry* air at sea-level density (normal pressure) at 0°C, the speed of sound is about 331 m/s (1,087 ft/s). If the air temperature is 30°C, sound will travel at 0.600 m/s faster for each degree above 0°C, or (0.600 m/s per 0°C)(30°C) = 18 m/s. Adding this to the speed of sound at 0°C, you have 331 m/s + 18 m/s = 349 m/s. You would need to move at 349 m/s to travel at the speed of sound when the air temperature is 30°C, but you could also travel at the speed of sound at 331 m/s when the air temperature is 0°C.

The simple relationship of the speed of sound at 0°C plus the fractional increase per degree above 0°C can be combined as in the following equations:

$$v_{Tp}(\text{m/s}) = v_0 + \left(\frac{0.600 \text{ m/s}}{°C}\right)(T_p)$$

equation 5.4

where v_{Tp} is the velocity of sound at the present temperature, v_0 is the velocity of sound at 0°C, and T_p is the present temperature.

This equation tells you that the velocity of a sound wave increases 0.6 m/s for each degree C above 0°C. For units of ft/s,

$$v_{Tp}(\text{ft/s}) = v_0 + \left(\frac{2.00 \text{ ft/s}}{°C}\right)(T_p)$$

equation 5.5

Equation 5.5 tells you that the velocity of a sound wave increases 2.0 ft/s for each degree Celsius above 0°C.

EXAMPLE 5.4

What is the velocity of sound in m/s at room temperature (20.0°C)?

Solution

$v_0 = 331$ m/s
$T_p = 20.0°C$
$v_{Tp} = ?$

$$v_{Tp} = v_0 + \left(\frac{0.600 \text{ m/s}}{°C}\right)(T_p)$$

$$= 331 \text{ m/s} + \left(\frac{0.600 \text{ m/s}}{°C}\right)(20.0°C)$$

$$= 331 + (0.600 \times 20.0) \text{ m/s} + \frac{\text{m/s}}{°C} \times °C$$

$$= 331 + 12.0 \text{ m/s} + \frac{\text{m/s/}°C}{°C}$$

$$= \boxed{343 \text{ m/s}}$$

EXAMPLE 5.5

The air temperature is 86.0°F. What is the velocity of sound in ft/s? (Note that °F must be converted to °C for equation 5.5.) (Answer: 1.15×10^3 ft/s)

Refraction and Reflection

When you drop a rock into a still pool of water, circular patterns of waves move out from the disturbance. These water waves are on a flat, two-dimensional surface. Sound waves, however, move in three-dimensional space like a rapidly expanding balloon. Sound waves are *spherical waves* that move outward from the source. Spherical waves of sound move as condensations and rarefactions from a continuously vibrating source at the center. If you identify the same part of each wave in the spherical waves, you have identified a *wave front*. For example, the crest of each

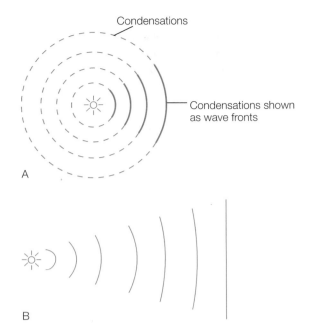

FIGURE 5.11

(*A*) Spherical waves move outward from a sounding source much like a rapidly expanding balloon. This two-dimensional sketch shows the repeating condensations as spherical wave fronts. (*B*) Some distance from the source, a spherical wave front is considered a linear, or plane, wave front.

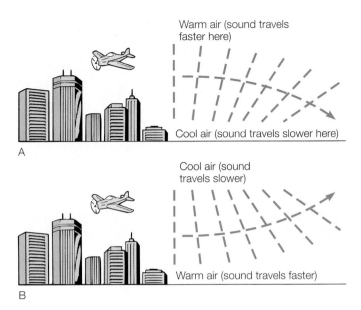

FIGURE 5.12

(*A*) Since sound travels faster in warmer air, a wave front becomes bent, or refracted, toward the earth's surface when the air is cooler near the surface. (*B*) When the air is warmer near the surface, a wave front is refracted upward, away from the surface.

condensation could be considered a wave front. From one wave front to the next, therefore, identifies one complete wave or wavelength. At some distance from the source a small part of a spherical wave front can be considered a *linear wave front* (Figure 5.11).

Waves move within a homogeneous medium such as a gas or a solid at a fairly constant rate but gradually lose energy to friction. When a wave encounters a different condition (temperature, humidity, or nature of material) however, drastic changes may occur rapidly. The division between two physical conditions is called a *boundary.* Boundaries are usually encountered (1) between different materials or (2) between the same materials with different conditions. An example of a wave moving between different materials is a sound made in the next room that moves through the air to the wall and through the wall to the air in the room where you are. The boundaries are air-wall and wall-air. If you have ever been in a room with "thin walls," it is obvious that sound moved through the wall and air boundaries.

Refraction

An example of sound waves moving through the same material with different conditions is found when a wave front moves through air of different temperatures. Since sound travels faster in warm air than in cold air, the wave front becomes bent. The bending of a wave front between boundaries is called **refraction.** Refraction changes the direction of travel of a wave

front. Consider, for example, that on calm, clear nights the air near the earth's surface is cooler than air farther above the surface. Air at rooftop height above the surface might be four or five degrees warmer under such ideal conditions. Sound will travel faster in the higher, warmer air than it will in the lower, cooler air close to the surface. A wave front will therefore become bent, or refracted, toward the ground on a cool night and you will be able to hear sounds from farther away than on warm nights (Figure 5.12A). The opposite process occurs during the day as the earth's surface becomes warmer from sunlight (Figure 5.12B). Wave fronts are refracted upward because part of the wave front travels faster in the warmer air near the surface. Thus, sound does not seem to carry as far in the summer as it does in the winter. What is actually happening is that during the summer the wave fronts are refracted away from the ground before they travel very far.

Reflection

When a wave front strikes a boundary that is parallel to the front the wave may be absorbed, transmitted, or undergo **reflection,** depending on the nature of the boundary medium, or the wave may be partly absorbed, partly transmitted, partly reflected, or any combination thereof. Some materials, such as hard, smooth surfaces, reflect sound waves more than they absorb them. Other materials, such as soft, ruffly curtains, absorb sound waves more than they reflect them. If you have ever been in a room with smooth, hard walls and with no curtains, carpets, or furniture, you know that sound waves may be reflected several times before they are finally absorbed.

FIGURE 5.13
This closed-circuit TV control room is acoustically treated by covering the walls with sound-absorbing baffles. Reverberation and echoes cannot occur in this treated room because absorbed sounds are not reflected.

Do you sing in the shower? Many people do because the tone is more pleasing than singing elsewhere. The walls of a shower are usually hard and smooth, reflecting sounds back and forth several times before they are absorbed. The continuation of many reflections causes a tone to gain in volume. Such mixing of reflected sounds with the original is called **reverberation**. Reverberation adds to the volume of a tone, and it is one of the factors that determine the acoustical qualities of a room, lecture hall, or auditorium. An open-air concert sounds flat without the reverberation of an auditorium and is usually enhanced electronically to make up for the lack of reflected sounds. Too much reverberation in a room or classroom is not good because the spoken word is not as sharp. Sound-absorbing materials are therefore used on the walls and floors where clear, distinct speech is important (Figure 5.13). The carpet and drapes you see in a movie theater are not decorator items but are there to absorb sounds.

If a reflected sound arrives after 0.10 s, the human ear can distinguish the reflected sound from the original sound. A reflected sound that can be distinguished from the original is called an **echo.** Thus, a reflected sound that arrives before 0.10 s is perceived as an increase in volume and is called a reverberation, but a sound that arrives after 0.10 s is perceived as an echo.

EXAMPLE 5.6

The human ear can distinguish a reflected sound pulse from the original sound pulse if 0.10 s or more elapses between the two sounds. What is the minimum distance to a reflecting surface from which we can hear an echo (see Figure 5.14A) if the speed of sound is 343 m/s?

Solution

$$t = 0.10 \text{ s} \text{ (minimum)}$$
$$v = 343 \text{ m/s}$$
$$d = ?$$

$$v = \frac{d}{t} \quad \therefore \quad d = vt$$

$$= \left(343 \frac{\text{m}}{\text{s}}\right)(0.10 \text{ s})$$

$$= 343 \times 0.10 \frac{\text{m}}{\text{s}} \times \text{s}$$

$$= 34.3 \frac{\text{m} \cdot \text{s}}{\text{s}}$$

$$= 34 \text{ m}$$

Since the sound pulse must travel from the source to the reflecting surface, then back to the source,

$$34 \text{ m} \times 1/2 = \boxed{17 \text{ m}}$$

The minimum distance to a reflecting surface from which we hear an echo when the air is at room temperature is therefore 17 m (about 56 ft).

EXAMPLE 5.7

An echo is heard exactly 1.00 s after a sound when the speed of sound is 1,147 ft/s. How many feet away is the reflecting surface? (Answer: 574 ft)

Sound wave echoes are measured to determine the depth of water or to locate underwater objects by a *sonar* device. The word *sonar* is taken from *so*und *na*vigation *r*anging. The device generates an underwater ultrasonic sound pulse, then measures the elapsed time for the returning echo. Sound waves travel at about 1,531 m/s (5,023 ft/s) in seawater at 25°C (77°F). A 1 s lapse between the ping of the generated sound and the echo return would mean that the sound traveled 5,023 ft for the round trip. The bottom would be half this distance below the surface (Figure 5.14B).

Interference

Waves interact with a boundary much as a particle would, reflecting or refracting because of the boundary. A moving ball, for example, will bounce from a surface at the same angle it strikes the surface, just as a wave does. A particle or a ball, however, can be in only one place at a time, but waves can be spread over a distance at the same time. You know this since many different people in different places can hear the same sound at the same time.

Constructive and Destructive

Another difference between waves and particles is that two or more waves can exist in the same place at the same time. When two patterns of waves meet, they pass through each other

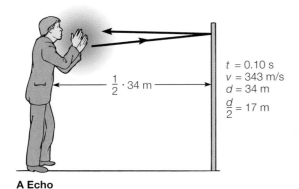

$t = 0.10$ s
$v = 343$ m/s
$d = 34$ m
$\frac{d}{2} = 17$ m

A Echo

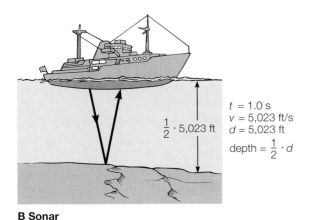

$t = 1.0$ s
$v = 5{,}023$ ft/s
$d = 5{,}023$ ft

depth $= \frac{1}{2} \cdot d$

B Sonar

FIGURE 5.14

(A) At room temperature, sound travels at 343 m/s. In 0.10 s, sound would travel 34 m. Since the sound must travel to a surface and back in order for you to hear an echo, the distance to the surface is one-half the total distance. (B) Sonar measures a depth by measuring the elapsed time between an ultrasonic sound pulse and the echo. The depth is one-half the round trip.

without refracting or reflecting. However, at the place where they meet the waves interfere with each other, producing a *new* disturbance. This new disturbance has a different amplitude, which is the algebraic sum of the amplitudes of the two separate wave patterns. If the wave crests or wave troughs arrive at the same place at the same time, the two waves are said to be *in phase*. The result of two waves arriving in phase is a new disturbance with a crest and trough that has greater displacement than either of the two separate waves. This is called *constructive interference* (Figure 5.15A). If the trough of one wave arrives at the same place and time as the crest of another wave, the waves are completely *out of phase*. When two waves are completely out of phase, the crest of one wave (positive displacement) will cancel the trough of the other wave (negative displacement), and the result is zero total disturbance, or no wave. This is called *destructive interference* (Figure 5.15B). If the two sets of wave patterns do not have the same amplitudes or wavelengths, they will be neither completely in phase nor completely out of phase. The result will be partly constructive or destructive interference, depending on the exact nature of the two wave patterns.

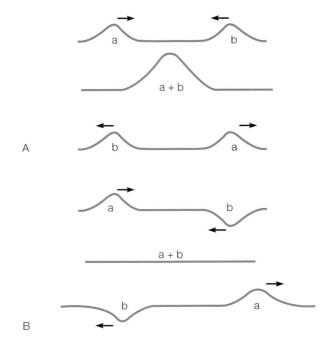

FIGURE 5.15

(A) Constructive interference occurs when two equal, in-phase waves meet. (B) Destructive interference occurs when two equal, out-of-phase waves meet. In both cases, the wave displacements are superimposed when they meet, but they then pass through one another and return to their original amplitudes.

Beats

Suppose that two vibrating sources produce sounds that are in phase, equal in amplitude, and equal in frequency. The resulting sound will be increased in volume because of constructive interference. But suppose the two sources are slightly different in frequency, for example, 350 and 352 Hz. You will hear a regularly spaced increase and decrease of sound known as **beats**. Beats occur because the two sound waves experience alternating constructive and destructive interferences (Figure 5.16). The phase relationship changes because of the difference in frequency, as you can see in Figure 5.16. These alternating constructive and destructive interference zones are moving from the source to the receiver, and the receiver hears the results as a rapidly rising and falling sound level. The beat frequency is the difference between the frequencies of the two sources. A 352 Hz source and 350 Hz source sounded together would result in a beat frequency of 2 Hz. Thus, the frequencies are closer and closer together, and fewer beats will be heard per second. You may be familiar with the phenomenon of beats if you have ever flown in an airplane with two engines. If one engine is running slightly faster than the other, you hear a slow beat. The beat frequency (f_b) is equal to the absolute difference in frequency of two interfering waves with slightly different frequencies, or

$$f_b = f_2 - f_1$$

equation 5.6

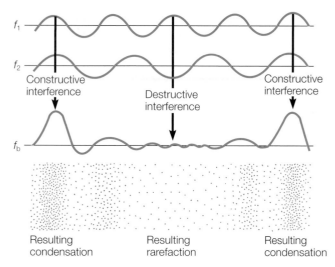

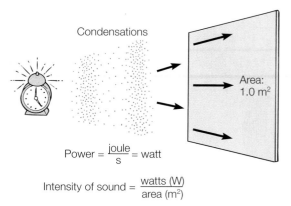

$$Power = \frac{joule}{s} = watt$$

$$Intensity\ of\ sound = \frac{watts\ (W)}{area\ (m^2)}$$

FIGURE 5.17

The intensity of a sound wave is the rate of energy transferred to an area perpendicular to the waves. Intensity is measured in watts per square meter, W/m^2.

FIGURE 5.16

Two waves of equal amplitude but slightly different frequencies interfere destructively and constructively. The result is an alternation of loudness called a *beat*.

Energy and Sound

All waves involve the transportation of energy, including sound waves. The vibrating mass and spring in Figure 5.2 vibrate with an amplitude that depends on how much work you did on the mass in moving it from its equilibrium position. More work on the mass results in a greater displacement and a greater amplitude of vibration. A vibrating object that is producing sound waves will produce more intense condensations and rarefactions if it has a greater amplitude. The intensity of a sound wave is a measure of the energy the sound wave is carrying (Figure 5.17). *Intensity* is defined as the power (in watts) transmitted by a wave to a unit area (in square meters) that is perpendicular to the waves. Intensity is therefore measured in watts per square meter (W/m^2) or

$$Intensity = \frac{Power}{Area}$$

$$I = \frac{P}{A}$$

equation 5.7

Loudness

The *loudness* of a sound is a subjective interpretation that varies from person to person. Loudness is also related to (1) the energy of a vibrating object, (2) the condition of the air the sound wave travels through, and (3) the distance between you and the vibrating source. Furthermore, doubling the amplitude of the vibrating source will quadruple the *intensity* of the resulting sound wave, but the sound will not be perceived as four times as loud. The relationship between perceived loudness and the intensity of a sound wave is not a linear relationship. In fact, a sound that is perceived as twice as loud requires ten times the intensity, and quadrupling the loudness requires a one-hundred-fold increase in intensity.

The human ear is very sensitive. It is capable of hearing sounds with intensities as low as 10^{-12} W/m^2 and is not made uncomfortable by sound until the intensity reaches about 1 W/m^2. The second intensity is a million million (10^{12}) times greater than the first. Within this range, the subjective interpretation of intensity seems to vary by powers of ten. This observation led to the development of the **decibel scale** to measure the intensity level. The scale is a ratio of the intensity level of a given sound to the threshold of hearing, which is defined as 10^{-12} W/m^2 at 1,000 Hz. In keeping with the power-of-ten subjective interpretations of intensity, a logarithmic scale is used rather than a linear scale. Originally the scale was the logarithm of the ratio of the intensity level of a sound to the threshold of hearing. This definition set the zero point at the threshold of human hearing. The unit was named the *bel* in honor of Alexander Graham Bell. This unit was too large to be practical, so it was reduced by one-tenth and called a *decibel*. The intensity level of a sound is therefore measured in decibels (Table 5.2). Compare the decibel noise level of familiar sounds listed in Table 5.2, and note that each increase of 10 on the decibel scale is matched by a *multiple* of 10 on the intensity level. For example, moving from a decibel level of 10 to a decibel level of 20 requires *ten times* more intensity. Likewise, moving from a decibel level of 20 to 40 requires a one-hundred-fold increase in the intensity level. As you can see, the decibel scale is not a simple linear scale.

Resonance

You know that sound waves transmit energy when you hear a thunderclap rattle the windows. In fact, the sharp sounds from an explosion have been known not only to rattle but also break windows. The source of the energy is obvious when

TABLE 5.2
Comparison of noise levels in decibels with intensity

Example	Response	Decibels	Intensity W/m^2
Least needed for hearing	Barely perceived	0	1×10^{-12}
Calm day in woods	Very, very quiet	10	1×10^{-11}
Whisper (15 ft)	Very quiet	20	1×10^{-10}
Library	Quiet	40	1×10^{-8}
Talking	Easy to hear	65	3×10^{-6}
Heavy street traffic	Conversation difficult	70	1×10^{-5}
Pneumatic drill (50 ft)	Very loud	95	3×10^{-3}
Jet plane (200 ft)	Discomfort	120	1

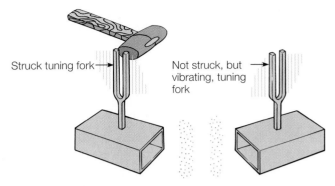

FIGURE 5.18

When the frequency of an applied force, including the force of a sound wave, matches the natural frequency of an object, energy is transferred very efficiently. The condition is called *resonance*.

thunderclaps or explosions are involved. But sometimes energy transfer occurs through sound waves when it is not clear what is happening. A truck drives down the street, for example, and one window rattles but the others do not. A singer shatters a crystal water glass by singing a single note, but other objects remain undisturbed. A closer look at the nature of vibrating objects and the transfer of energy will explain these phenomena.

Almost any elastic object can be made to vibrate and will vibrate freely at a constant frequency after being sufficiently disturbed. Entertainers sometimes discover this fact and appear on late-night talk shows playing saws, wrenches, and other odd objects as musical instruments. All material objects have a *natural frequency* of vibration determined by the materials and shape of the objects. The natural frequencies of different wrenches enable an entertainer to use the suspended tools as if they were the bars of a xylophone.

If you have ever pumped a swing, you know that small forces can be applied at any frequency. If the frequency of the applied forces matches the natural frequency of the moving swing, there is a dramatic increase in amplitude. When the two frequencies match, energy is transferred very efficiently. This condition, when the frequency of an external force matches the natural frequency, is called **resonance.** The natural frequency of an object is thus referred to as the *resonant frequency,* that is, the frequency at which resonance occurs.

A silent tuning fork will resonate if a second tuning fork with the same frequency is struck and vibrates nearby (Figure 5.18). You will hear the previously silent tuning fork sounding if you stop the vibrations of the struck fork by touching it. The waves of condensations and rarefactions produced by the struck tuning fork produce a regular series of impulses that match the natural frequency of the silent tuning fork. This illustrates that at resonance, relatively little energy is required to start vibrations.

A truck causing vibrations as it is driven past a building may cause one window to rattle while others do not. Vibrations caused by the truck have matched the natural frequency of this window but not the others. The window is undergoing resonance from the sound wave impulses that matched its natural frequency. It is also resonance that enables a singer to break a water glass. If the tone is at the resonant frequency of the glass, the resulting vibrations may be large enough to shatter it.

Concepts Applied

A Singing Glass

Did you ever hear a glass "sing" when the rim was rubbed? The trick to make the glass sing is to remove as much oil from your finger as possible. Then you lightly rub around and on the top of the glass rim at the correct speed. Without oil, your wet finger will imperceptively catch on the glass as you rub the rim. With the appropriate pressure and speed your catching finger might match the natural frequency of the glass. The resonant vibration will cause the glass to "sing" with a high-pitched note.

Resonance considerations are important in engineering. A large water pump, for example, was designed for a nuclear power plant. Vibrations from the electric motor matched the resonant frequency of the impeller blades, and they shattered after a short period of time. When the problem was discovered, the blades were redesigned to have a different natural frequency. Resonance vibrations are particularly important in the design of buildings.

Matchmaking

Set up two identical pendulums as shown in Figure 5.19A. The bobs should be identical and suspended from identical strings of the same length attached to a tight horizontal string. Start one pendulum vibrating by pulling it back, then releasing it. Observe the vibrations and energy exchange between the two pendulums for the next several minutes. Now change the frequency of vibrations of *one* of the pendulums by shortening the string (Figure 5.19B). Again start one pendulum vibrating and observe for several minutes. Compare what you observe when the frequencies are matched and when they are not. Explain what happens in terms of resonance.

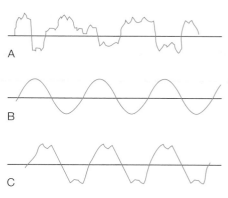

FIGURE 5.20
Different sounds that you hear include (A) noise, (B) pure tones, and (C) musical notes.

Sources of Sounds

All sounds have a vibrating object as their source. The vibrations of the object send pulses or waves of condensations and rarefactions through the air. These sound waves have physical properties that can be measured, such as frequency and intensity.

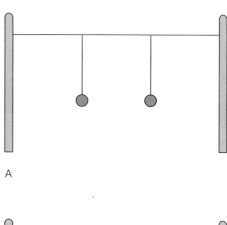

A

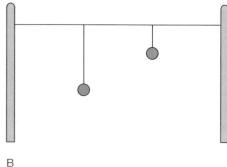

B

FIGURE 5.19
The "Concepts Applied" section on this page demonstrates that one of these pendulum arrangements will show resonance, and the other will not. Can you predict which one will show resonance?

Subjectively, your response to frequency is to identify a certain pitch. A high-frequency sound is interpreted as a high-pitched sound, and a low-frequency sound is interpreted as a low-pitched sound. Likewise, a greater intensity is interpreted as increased loudness, but there is not a direct relationship between intensity and loudness as there is between frequency and pitch.

There are other subjective interpretations about sounds. Some sounds are bothersome and irritating to some people but go unnoticed by others. In general, sounds made by brief, irregular vibrations such as those made by a slamming door, dropped book, or sliding chair are called *noise*. Noise is characterized by sound waves with mixed frequencies and jumbled intensities (Figure 5.20). On the other hand, there are sounds made by very regular, repeating vibrations such as those made by a tuning fork. A tuning fork produces a *pure tone* with a sinusoidal curved pressure variation and regular frequency. Yet a tuning fork produces a tone that most people interpret as bland. You would not call a tuning fork sound a musical note! Musical sounds from instruments have a certain frequency and loudness, as do noise and pure tones, but you can readily identify the source of the very same musical note made by two different instruments. You recognize it as a musical note, not noise and not a pure tone. You also recognize if the note was produced by a violin or a guitar. The difference is in the wave form of the sounds made by the two instruments, and the difference is called the *sound quality*. How does a musical instrument produce a sound of a characteristic quality? The answer may be found by looking at the instruments that make use of vibrating strings.

Vibrating Strings

A stringed musical instrument, such as a guitar, has strings that are stretched between two fixed ends. When a string is plucked, waves of many different frequencies travel back and forth on the string, reflecting from the fixed ends. Many of these waves quickly fade away but certain frequencies resonate, setting up patterns of waves. Before considering these resonant patterns in detail, keep in mind that (1) two or more waves can be in the same place at the same time, traveling through one another

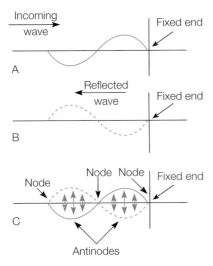

FIGURE 5.21

An incoming wave on a cord with a fixed end (*A*) meets a reflected wave (*B*) with the same amplitude and frequency, producing a standing wave (*C*). Note that a standing wave of one wavelength has three nodes and two antinodes.

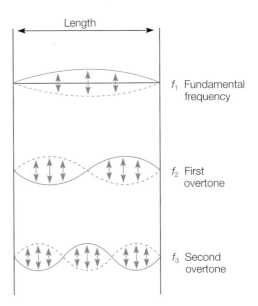

FIGURE 5.22

A stretched string of a given length has a number of possible resonant frequencies. The lowest frequency is the fundamental, f_1; the next higher frequencies, or overtones, shown are f_2 and f_3.

from opposite directions; (2) a confined wave will be reflected at a boundary, and the reflected wave will be inverted (a crest becomes a trough); and (3) reflected waves interfere with incoming waves of the same frequency to produce **standing waves.** Figure 5.21 is a graphic "snapshot" of what happens when reflected wave patterns meet incoming wave patterns. The incoming wave is shown as a solid line, and the reflected wave is shown as a dotted line. The result is (1) places of destructive interference, called *nodes*, which show no disturbance, and (2) loops of constructive interference, called *antinodes*, which take place where the crests and troughs of the two wave patterns produce a disturbance that rapidly alternates upward and downward. This pattern of alternating nodes and antinodes does not move along the string and is thus called a *standing wave*. Note that the standing wave for *one wavelength* will have a node at both ends and in the center and also two antinodes. Standing waves occur at the natural, or resonant, frequencies of the string, which are a consequence of the nature of the string, the string length, and the tension in the string. Since the standing waves are resonant vibrations, they continue as all other waves quickly fade away.

Since the two ends of the string are not free to move, the ends of the string will have nodes. The *longest* wave that can make a standing wave on such a string has a wavelength (λ) that is twice the length (L) of the string. Since frequency (f) is inversely proportional to wavelength ($f = v/\lambda$ from equation 5.3), this longest wavelength has the lowest frequency possible, called the **fundamental frequency.** The fundamental frequency has one antinode, which means that the length of the string has one-half a wavelength. The fundamental frequency (f_1) determines the pitch of the *basic* musical note being sounded and is called the first harmonic. Other resonant frequencies occur at the same time, however, since other standing waves can also fit

onto the string. A higher frequency of vibration (f_2) could fit two half-wavelengths between the two fixed nodes. An even higher frequency (f_3) could fit three half-wavelengths between the two fixed nodes (Figure 5.22). Any whole number of halves of the wavelength will permit a standing wave to form. The frequencies (f_2, f_3, etc.) of these wavelengths are called the *overtones*, or harmonics. It is the presence and strength of various overtones that give a musical note from a certain instrument its characteristic quality. The fundamental and the overtones add together to produce the characteristic *sound quality*, which is different for the same-pitched note produced by a violin and by a guitar (Figure 5.23).

Since nodes must be located at the ends, only half-wavelengths ($1/2\ \lambda$) can fit on a string of a given length (L), so the fundamental frequency of a string is $1/2\ \lambda = L,\ or\ \lambda = 2L$.

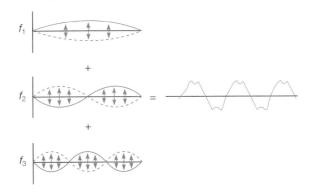

FIGURE 5.23

A combination of the fundamental and overtone frequencies produces a composite waveform with a characteristic sound quality.

Substituting this value in the wave equation (solved for frequency, f) will give the relationship for finding the fundamental frequency and the overtones when the string length and velocity of waves on the string are known. The relationship is

$$f_n = \frac{nv}{2L}$$

equation 5.8

where $n = 1, 2, 3, 4 \ldots$, and $n = 1$ is the fundamental frequency and $n = 2$, $n = 3$, and so forth are the overtones.

EXAMPLE 5.8

What is the fundamental frequency of a 0.5 m string if wave speed on the string is 400 m/s?

Solution

The length (L) and the velocity (v) are given. The relationship between these quantities and the fundamental frequency ($n = 1$) is given in equation 5.8, and

$$L = 0.5 \text{ m}$$
$$v = 400 \text{ m/s}$$
$$f_1 = ?$$

$$f_n = \frac{nv}{2L} \text{ where } n = 1 \text{ for the fundamental frequency}$$

$$f_1 = \frac{1 \times 400 \text{ m/s}}{2 \times 0.5 \text{ m}}$$

$$= \frac{400}{1} \frac{\text{m}}{\text{s}} \times \frac{1}{\text{m}}$$

$$= 400 \frac{\text{m}}{\text{s·m}}$$

$$= 400 \frac{1}{\text{s}}$$

$$= \boxed{400 \text{ Hz}}$$

EXAMPLE 5.9

What is the frequency of the first overtone in a 0.5 m string when the wave speed is 400 m/s? (Answer: 800 Hz)

The vibrating string produces a waveform with overtones, so instruments that have vibrating strings are called *harmonic instruments.* Instruments that use an air column as a sound maker are also harmonic instruments. These include all the wind instruments such as the clarinet, flute, trombone, trumpet, pipe organ, and many others. The various wind instruments have different ways of making a column of air vibrate. In the flute, air vibrates as it moves over a sharp edge, while in the clarinet, saxophone, and other reed instruments, it vibrates through fluttering thin reeds. The air column in brass instruments, on the other hand, is vibrated by the tightly fluttering lips of the player.

The length of the air column determines the frequency, and woodwind instruments have holes in the side of a tube that are opened or closed to change the length of the air column. The resulting tone depends on the length of the air column and the resonant overtones.

Sounds from Moving Sources

When the source of a sound is stationary, equally spaced sound waves expand from a source in all directions. But, if the sounding source starts moving, then successive sound waves become displaced in the direction of movement, and this changes the pitch. For example, the siren of an approaching ambulance seems to change pitch when the ambulance passes you. The sound wave is "squashed" as the ambulance approaches you, and you hear a higher-frequency siren than the people inside the ambulance. When the ambulance passes you, the sound waves are "stretched" and you hear a lower-frequency siren (Figure 5.24). The overall effect of a higher pitch as a source approaches, and then a lower pitch as it moves away, is called the **Doppler effect.** The Doppler effect is evident if you stand by a street and an approaching car sounds its horn as it drives by you. You will hear a higher-pitched horn

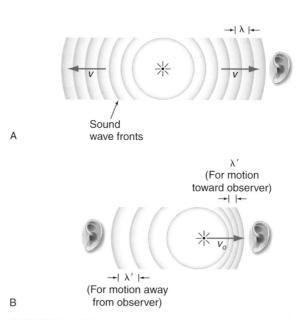

A Sound wave fronts

B (For motion away from observer)

FIGURE 5.24

(A) Sound waves emitted by a stationary source and observed by a stationary observer. (B) Sound waves emitted by a source in motion toward the right. An observer on the right receives wavelengths that are shortened; an observer on the left receives wavelengths that are lengthened.

Johann Christian Doppler (1803–1853)

Johann Doppler was an Austrian physicist who discovered the Doppler effect, which relates the observed frequency of a wave to the relative motion of the source and the observer. The Doppler effect is readily observed in moving sound sources, producing a fall in pitch as the source passes the observer, but it is of most use in astronomy, where it is used to estimate the velocities and distances of distant bodies.

Doppler was born in Salzburg, Austria, on November 29, 1803, the son of a stonemason. He showed early promise in mathematics, and attended the Polytechnic Institute in Vienna from 1822 to 1825. He then returned to Salzburg and continued his studies privately while tutoring in physics and mathematics. From 1829 to 1833, Doppler went back to Vienna to work as a mathematical assistant and produced his first papers on mathematics and electricity. Despairing of ever obtaining an academic post, he decided in 1835 to emigrate to the United States. Then, on the point of departure, he was offered a professorship of mathematics at the State Secondary School in Prague and changed his mind. He subsequently obtained professorships in mathematics at the State Technical Academy in Prague in 1841, and at the Mining Academy in Schemnitz in 1847. Doppler returned to Vienna the following year and, in 1850, became director of the new Physical Institute and Professor of Experimental Physics at the Royal

Imperial University of Vienna. He died from a lung disease in Venice on March 17, 1853.

Doppler explained the effect that bears his name by pointing out that sound waves from a source moving toward an observer will reach the observer at a greater frequency than if the source is stationary, thus increasing the observed frequency and raising the pitch of the sound. Similarly, sound waves from a source moving away from the observer reach the observer more slowly, resulting in a decreased frequency and a lowering of pitch. In 1842, Doppler put forward this explanation and derived the observed frequency mathematically in Doppler's principle.

The first experimental test of Doppler's principle was made in 1845 at Utrecht in Holland. A locomotive was used to carry a group of trumpeters in an open carriage to and fro past some musicians able to sense the pitch of the notes being played. The variation of pitch produced by the motion of the trumpeters verified Doppler's equations.

Doppler correctly suggested that his principle would apply to any wave motion and cited light as an example as well as sound. He believed that all stars emit white light and that differences in color are observed on earth because the motion of stars affects the observed frequency of the light and hence its color. This idea was not universally true, as stars vary in their basic color. However, Armand Fizeau (1819–1896) pointed out in 1848 that shifts in the spectral lines of stars could be observed and ascribed to the Doppler effect and hence enable their motion to be determined. This idea was first applied in 1868 by William Huggins (1824–1910), who found that Sirius is moving away from the solar system by detecting a small redshift in its spectrum. With the linking of the velocity of a galaxy to its distance by Edwin Hubble (1889–1953) in 1929, it became possible to use the redshift to determine the distances of galaxies. Thus the principle that Doppler discovered to explain an everyday and inconsequential effect in sound turned out to be of truly cosmological importance.

From the Hutchinson *Dictionary of Scientific Biography*. © Research Machines plc [2003] All Rights Reserved. Helicon Publishing is a division of Research Machines.

as the car approaches, which shifts to a lower-pitched horn as the waves go by you. The driver of the car, however, will hear the continual, true pitch of the horn because the driver is moving with the source.

A Doppler shift is also noted if the observer is moving and the source of sound is stationary. When the observer moves toward the source, the wave fronts are encountered more frequently than if the observer were standing still. As the observer moves away from the source, the wave fronts are encountered less frequently than if the observer were not moving. An observer on a moving train approaching a crossing with a sounding bell thus hears a high-pitched bell that shifts to a lower-pitched bell as the train whizzes by the crossing.

When an object moves through the air at the speed of sound, it keeps up with its own sound waves. All the successive wave fronts pile up on one another, creating a large wave disturbance

The Doppler effect was named after the Austrian scientist Johann Doppler, who first demonstrated the effect using sound waves back in 1842. The same principle applies to electromagnetic radiation as well as sound, but now the shifts are in the frequency of the radiation. A lower frequency is observed when a source of light is moving away, and this is called a "redshift." Also, a "blueshift" toward a higher frequency occurs when a source of light is moving toward an observer. Radio waves will also experience such shifts of frequency, and weather radar that measures frequency changes as a result of motion is called *Doppler radar.*

Weather radar broadcasts short radio waves from an antenna. When directed at a storm, the waves are reflected back to the antenna by rain, snow, and hail. Reflected radar waves are electronically converted and displayed on a monitor, showing the location and intensity of precipitation. A Doppler radar also measures frequency shifts in the reflected radio waves. Waves from objects moving toward the antenna show a higher frequency, and waves from objects moving away from the antenna show a lower frequency. These shifts of frequency are measured, then displayed as the speed and direction of winds that move raindrops and other objects in the storm.

Weather forecasters can direct a Doppler radar machine to measure different elevations of a storm system. This shows highly accurate information that can be used to identify, for example, where and when a tornado might form, the intensity of storm winds in a given area, and even how much precipitation fell from the storm.

called a **shock wave** (Figure 5.25). The shock wave from a supersonic airplane is a cone-shaped shock wave of intense condensations trailing backward at an angle dependent on the speed of the aircraft. Wherever this cone of superimposed crests passes, a **sonic boom** occurs. The many crests have been added together, each contributing to the pressure increase. The human ear cannot differentiate between such a pressure wave created by a supersonic aircraft and a pressure wave created by an explosion.

Does a sonic boom occur just when an airplane breaks the sound barrier? The answer is no; an airplane traveling at or faster than the speed of sound produces the shock wave continuously, and a sonic boom will be heard everywhere the plane drags its cone-shaped shock wave. In addition, high-speed airplanes often produce two or more shock waves, which are associated with the nose, tail, and other projections on the aircraft. Can you find evidence of shock waves associated with projections on the airplane pictured in Figure 5.26?

The Austrian physicist Ernst Mach published a paper in 1877 laying out the principles of supersonics. He also came up with the idea of using a ratio of the velocity of an object to the velocity of sound. Today, this ratio is called the *Mach number.* A plane traveling at the speed of sound has a Mach number of 1, a plane traveling at twice the speed of sound has a Mach number of 2, and so on. Ernst Mach was also the first to describe what is

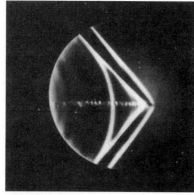

Shock waves are produced when a projectile travels faster than the speed of sound in air.

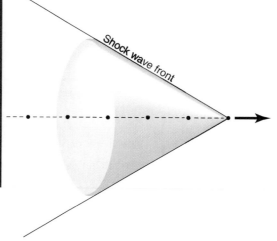

FIGURE 5.25

A sound source moves with velocity *greater* than the speed of sound in the medium. The envelope of spherical wavefront forms the conical shock wave.

FIGURE 5.26

A cloud sometimes forms just as a plane accelerates to break the sound barrier. Moist air is believed to form the cloud water droplets as air pressure drops behind the shock wave.

happening to produce a sonic boom, and he observed the existence of a conical shock wave formed by a projectile as it approached the speed of sound.

Summary

Elastic objects *vibrate,* or move back and forth, in a repeating motion when disturbed by some external force. They are able to do this because they have an *internal restoring force* that returns them to their original positions after they are deformed by some external force. If the internal restoring force is opposite to and proportional to the deforming displacement, the vibration is called a *simple harmonic motion.* The extent of displacement is called the *amplitude,* and one complete back-and-forth motion is one *cycle.* The time required to complete one cycle is a *period.* The *frequency* is the number of cycles per second, and the unit of frequency is the *hertz.* A *graph* of the displacement as a function of time for a simple harmonic motion produces a *sinusoidal* graph.

Periodic, or repeating, vibrations or the *pulse* of a single disturbance can create *waves,* disturbances that carry energy through a medium. A wave that disturbs particles in a back-and-forth motion in the direction of the wave travel is called a *longitudinal wave.* A wave that disturbs particles in a motion perpendicular to the direction of wave travel is called a *transverse wave.* The nature of the medium and the nature of the disturbance determine the type of wave created.

Waves that move through the air are longitudinal and cause a back-and-forth motion of the molecules making up the air. A zone of molecules forced closer together produces a *condensation,* a pulse of increased density and pressure. A zone of reduced density and pressure is a *rarefaction.* A vibrating object produces condensations and rarefactions that expand outward from the source. If the frequency is between 20 and 20,000 Hz, the human ear perceives the waves as *sound* of a certain *pitch.* High frequency is interpreted as high-pitched sound, and low frequency as low-pitched sound.

A graph of pressure changes produced by condensations and rarefactions can be used to describe sound waves. The condensations produce crests, and the rarefactions produce *troughs.* The *amplitude* is the maximum change of pressure from the normal. The *wavelength* is the distance between any two successive places on a wave train, such as the distance from one crest to the next crest. The *period* is the time required for a wave to repeat itself. The *velocity* of a wave is how quickly a wavelength passes. The frequency can be calculated from the *wave equation,* $v = \lambda f$.

Sound waves can move through any medium but not a vacuum. The velocity of sound in a medium depends on the molecular inertia and strength of interactions. Sound, therefore, travels most rapidly through a solid, then a liquid, then a gas. In air, sound has a greater velocity in warmer air than in cooler air because the molecules of air are moving about more rapidly, therefore transmitting a pulse more rapidly.

Sound waves are *reflected* or *refracted* from a *boundary,* which means a change in the transmitting medium. Reflected waves that are *in phase* with incoming waves undergo *constructive interference* and waves that are *out of phase* undergo *destructive interference.* Two waves that are otherwise alike but with slightly different frequencies produce an alternating increasing and decreasing of loudness called *beats.*

The *energy* of a sound wave is called the wave *intensity,* which is measured in watts per square meter. The intensity of sound is expressed on the *decibel scale,* which relates it to changes in loudness as perceived by the human ear.

All elastic objects have *natural frequencies* of vibration that are determined by the materials they are made of and their shapes. When energy is transferred at the natural frequencies, there is a dramatic increase of amplitude called *resonance.* The natural frequencies are also called *resonant frequencies.*

Sounds are compared by pitch, loudness, and quality. The *quality* is determined by the instrument sounding the note. Each instrument has its own characteristic quality because of the resonant frequencies that it produces. The basic, or *fundamental,* frequency is the longest standing wave that it can make. The fundamental frequency determines the basic note being sounded, and other resonant frequencies, or standing waves called *overtones* or *harmonics,* combine with the fundamental to give the instrument its characteristic quality.

A moving source of sound or a moving observer experiences an apparent shift of frequency called the *Doppler effect.* If the source is moving as fast or faster than the speed of sound, the sound waves pile up into a *shock wave* called a *sonic boom.* A sonic boom sounds very much like the pressure wave from an explosion.

Summary of Equations

5.1

$$\text{period} = \frac{1}{\text{frequency}}$$

$$T = \frac{1}{f}$$

5.2

$$\text{frequency} = \frac{1}{\text{period}}$$

$$f = \frac{1}{T}$$

5.3

$$\text{velocity} = (\text{wavelength})(\text{frequency})$$

$$v = \lambda f$$

5.4

velocity of sound (m/s) at present temperature	=	velocity of sound at 0°C	+	0.600 m/s increase per degree Celsius	×	present temperature in °C

$$v_{Tp}(\text{m/s}) = v_0 + \left(\frac{0.600 \text{ m/s}}{\text{°C}}\right)(T_p)$$

5.5

velocity of sound (ft/s) at present temperature	=	velocity of sound at 0°C	+	2.00 ft/s increase per degree Celsius	×	present temperature in °C

$$v_{Tp}(\text{ft/s}) = v_0 + \left(\frac{2.00 \text{ ft/s}}{\text{°C}}\right)(T_p)$$

5.6

$$\text{beat frequency} = \text{one frequency} - \text{other frequency}$$

$$f_b = f_2 - f_1$$

5.7

$$\text{Intensity} = \frac{\text{Power}}{\text{Area}}$$

$$I = \frac{P}{A}$$

5.8

$$\text{resonant frequency} = \frac{\text{number} \times \text{velocity on string}}{2 \times \text{length of string}}$$

where number 1 = fundamental frequency, and numbers 2, 3, 4, and so on = overtones.

$$f_n = \frac{nv}{2L}$$

KEY TERMS

amplitude (p. **117**)
beats (p. **127**)
cycle (p. **117**)
decibel scale (p. **128**)
Doppler effect (p. **132**)
echo (p. **126**)
frequency (p. **117**)
fundamental frequency (p. **131**)
hertz (p. **117**)
infrasonic (p. **120**)
longitudinal wave (p. **118**)
period (p. **117**)

pitch (p. **121**)
reflection (p. **125**)
refraction (p. **125**)
resonance (p. **129**)
reverberation (p. **126**)
shock wave (p. **134**)
sonic boom (p. **134**)
standing waves (p. **131**)
transverse wave (p. **119**)
ultrasonic (p. **120**)
vibration (p. **116**)
wave (p. **118**)
wavelength (p. **123**)

APPLYING THE CONCEPTS

1. Simple harmonic motion is
   a. any motion that results in a pure tone.
   b. a combination of overtones.
   c. a vibration with a restoring force proportional and opposite to a displacement.
   d. all of the above.

2. The displacement of a vibrating object is measured by
   a. a cycle.
   b. amplitude.
   c. the period.
   d. frequency.

3. The time required for a vibrating object to complete one full cycle is the
   a. frequency.
   b. amplitude.
   c. period.
   d. hertz.

4. The number of cycles that a vibrating object moves through during a time interval of 1 second is the
   a. amplitude.
   b. period.
   c. full cycle.
   d. frequency.

5. The unit of cycles per second is called a
   a. hertz.
   b. lambda.
   c. wave.
   d. watt.

6. The period of a vibrating object is related to the frequency, since they are
   a. directly proportional.
   b. inversely proportional.
   c. frequently proportional.
   d. not proportional.

7. A wave is
   a. the movement of material from one place to another place.
   b. a traveling disturbance that carries energy.
   c. a wavy line that moves through materials.
   d. all of the above.

8. A longitudinal mechanical wave causes particles of a material to move
   a. back and forth in the same direction the wave is moving.
   b. perpendicular to the direction the wave is moving.
   c. in a circular motion in the direction the wave is moving.
   d. in a circular motion opposite the direction the wave is moving.

9. A transverse mechanical wave causes particles of a material to move
   a. back and forth in the same direction the wave is moving.
   b. perpendicular to the direction the wave is moving.
   c. in a circular motion in the direction the wave is moving.
   d. in a circular motion opposite the direction the wave is moving.

10. Transverse mechanical waves will move only through
    a. solids.
    b. liquids.
    c. gases.
    d. all of the above.
11. Longitudinal mechanical waves will move only through
    a. solids.
    b. liquids.
    c. gases.
    d. all of the above.
12. A pulse of jammed-together molecules that quickly moves away from a vibrating object
    a. is called a condensation.
    b. causes an increased air pressure when it reaches an object.
    c. has a greater density than the surrounding air.
    d. includes all of the above.
13. The characteristic of a wave that is responsible for what you interpret as pitch is the wave
    a. amplitude.
    b. shape.
    c. frequency.
    d. height.
14. The extent of displacement of a vibrating tuning fork is related to the resulting sound wave characteristic of
    a. frequency.
    b. amplitude.
    c. wavelength.
    d. period.
15. The number of cycles that a vibrating tuning fork experiences each second is related to the resulting sound wave characteristic of
    a. frequency.
    b. amplitude.
    c. wave height.
    d. quality.
16. From the wave equation of $v = \lambda f$, you know that the wavelength and frequency at a given velocity are
    a. directly proportional.
    b. inversely proportional.
    c. directly or inversely proportional, depending on the pitch.
    d. not related.
17. Since $v = \lambda f$, then f must equal
    a. $v\lambda$.
    b. v/λ.
    c. λ/v.
    d. $v\lambda/\lambda$.
18. Sound waves travel faster in
    a. solids as compared to liquids.
    b. liquids as compared to gases.
    c. warm air as compared to cooler air.
    d. all of the above.
19. The difference between an echo and a reverberation is
    a. an echo is a reflected sound; reverberation is not.
    b. the time interval between the original sound and the reflected sound.
    c. the amplitude of an echo is much greater.
    d. reverberation comes from acoustical speakers; echoes come from cliffs and walls.
20. Sound interference is necessary to produce the phenomenon known as
    a. resonance.
    b. decibels.
    c. beats.
    d. reverberation.
21. The efficient transfer of energy that takes place at a natural frequency is known as
    a. resonance.
    b. beats.
    c. the Doppler effect.
    d. reverberation.
22. The fundamental frequency of a standing wave on a string has
    a. one node and one antinode.
    b. one node and two antinodes.
    c. two nodes and one antinode.
    d. two nodes and two antinodes.
23. An observer on the ground will hear a sonic boom from an airplane traveling faster than the speed of sound
    a. only when the plane breaks the sound barrier.
    b. as the plane is approaching.
    c. when the plane is directly overhead.
    d. after the plane has passed by.

Answers

1. c 2. b 3. c 4. d 5. a 6. b 7. b 8. a 9. b 10. a 11. d 12. d 13. c 14. b 15. a
16. b 17. b 18. d 19. b 20. c 21. a 22. c 23. d

QUESTIONS FOR THOUGHT

1. What is a wave?
2. Is it possible for a transverse wave to move through air? Explain.
3. A piano tuner hears three beats per second when a tuning fork and a note are sounded together and six beats per second after the string is tightened. What should the tuner do next, tighten or loosen the string? Explain.
4. Why do astronauts on the moon have to communicate by radio even when close to one another?
5. What is resonance?
6. Explain why sounds travel faster in warm air than in cool air.
7. Do all frequencies of sound travel with the same velocity? Explain your answer by using one or more equations.
8. What eventually happens to a sound wave traveling through the air?
9. What gives a musical note its characteristic quality?

10. Does a supersonic aircraft make a sonic boom only when it cracks the sound barrier? Explain.

11. What is an echo?

12. Why are fundamental frequencies and overtones also called resonant frequencies?

PARALLEL EXERCISES

The exercises in groups A and B cover the same concepts. Solutions to group A exercises are located in appendix D.

Group A

1. A grasshopper floating in water generates waves at a rate of three per second with a wavelength of two centimeters. (a) What is the period of these waves? (b) What is the wave velocity?

2. The upper limit for human hearing is usually considered to be 20,000 Hz. What is the corresponding wavelength if the air temperature is 20.0°C?

3. A tone with a frequency of 440 Hz is sounded at the same time as a 446 Hz tone. How many beats per second are produced?

4. Medical applications of ultrasound use frequencies up to 2.00×10^7 Hz. What is the wavelength of this frequency in air?

5. A baseball fan is 150.0 m from the home plate. How much time elapses between the instant the fan sees the batter hit the ball and the moment the fan hears the sound?

6. An echo is heard from a building 0.500 s after you shout "Hello." How many feet away is the building if the air temperature is 20.0°C?

7. A sonar signal is sent from an oceangoing ship and the signal returns from the bottom 1.75 s later. How deep is the ocean beneath the ship if the speed of sound in seawater is 1,530 m/s?

8. A sound wave in a steel rail of a railroad track has a frequency of 660 Hz and a wavelength of 9.0 m. What is the speed of sound in this rail?

9. According to the condensed steam released, a factory whistle blows 2.5 s before you hear the sound. If the air temperature is 20.0°C, how many meters are you from the whistle?

10. Compare the distance traveled in 8.00 s as a given sound moves through (a) air at 0°C, and (b) a steel rail.

11. A vibrating object produces periodic waves with a wavelength of 50 cm and a frequency of 10 Hz. How fast do these waves move away from the object?

12. The distance between the center of a condensation and the center of an adjacent rarefaction is 1.50 m. If the frequency is 112.0 Hz, what is the speed of the wave front?

13. Water waves are observed to pass under a bridge at a rate of one complete wave every 4.0 s. (a) What is the period of these waves? (b) What is the frequency?

14. A sound wave with a frequency of 260 Hz moves with a velocity of 330 m/s. What is the distance from one condensation to the next?

15. The following sound waves have what velocity?
    (a) Middle C, or 256 Hz and 1.34 m λ.
    (b) Note A, or 440.0 Hz and 78.0 cm λ.
    (c) A siren at 750.0 Hz and λ of 45.7 cm.
    (d) Note from a stereo at 2,500.0 Hz and λ of 13.72 cm.

Group B

1. A water wave has a frequency of 6 Hz and a wavelength of 3 m. (a) What is the period of these waves? (b) What is the wave velocity?

2. The lower frequency limit for human hearing is usually considered to be 20.0 Hz. What is the corresponding wavelength for this frequency if the air temperature is 20.0 C?

3. A 520 Hz tone is sounded at the same time as a 516 Hz tone. How many beats per second will you hear?

4. The low range of frequencies used for medical applications is about 1,000,000 Hz. What is the wavelength of this frequency in air?

5. How much time will elapse between seeing and hearing an event that happens 400.0 meters from you?

6. An echo bounces from a building exactly 1.00 s after you honk your horn. How many feet away is the building if the air temperature is 20.0°C?

7. A submarine sends a sonar signal, which returns from another ship 2.250 s later. How far away is the other ship if the speed of sound in seawater is 1,530.0 m/s?

8. A student under water clicks two rocks together and makes a sound with a frequency of 600.0 Hz and a wavelength of 2.5 m. What is the speed of this underwater sound?

9. You see condensed steam expelled from a ship's whistle 2.50 s before you hear the sound. If the air temperature is 20.0°C, how many meters are you from the ship?

10. Compare the distance traveled in 6.00 s as a given sound moves through (a) water at 25.0°C, and (b) through seawater at 25.0°C.

11. A tuning fork vibrates 440.0 times a second, producing sound waves with a wavelength of 78.0 cm. What is the velocity of these waves?

12. The distance between the center of a condensation and the center of an adjacent rarefaction is 65.23 cm. If the frequency is 256.0 Hz, how fast are these waves moving?

13. A warning buoy is observed to rise every 5.0 s as crests of waves pass by it. (a) What is the period of these waves? (b) What is the frequency?

14. Sound from the siren of an emergency vehicle has a frequency of 750.0 Hz and moves with a velocity of 343.0 m/s. What is the distance from one condensation to the next?

15. The following sound waves have what velocity?
    (a) 20.0 Hz, λ of 17.15 m
    (b) 200.0 Hz, λ of 1.72 m
    (c) 2,000.0 Hz, λ of 17.15 cm
    (d) 20,000.0 Hz, λ of 1.72 cm

16. What is the speed of sound, in ft/s, if the air temperature is:
    (a) 0.0°C
    (b) 20.0°C
    (c) 40.0°C
    (d) 80.0°C

17. An echo is heard from a cliff 4.80 s after a rifle is fired. How many feet away is the cliff if the air temperature is 43.7°F?

18. The air temperature is 80.00°F during a thunderstorm, and thunder was timed 4.63 s after lightning was seen. How many feet away was the lightning strike?

19. If the velocity of a 440 Hz sound is 1,125 ft/s in the air and 5,020 ft/s in seawater, find the wavelength of this sound (a) in air, (b) in seawater.

16. How much time is required for a sound to travel 1 mile (5,280.0 ft) if the air temperature is:
    (a) 0.0°C
    (b) 20.0°C
    (c) 40.0°C
    (d) 80.0°C

17. A ship at sea sounds a whistle blast, and an echo returns from the coastal land 10.0 s later. How many km is it to the coastal land if the air temperature is 10.0°C?

18. How many seconds will elapse between seeing lightning and hearing the thunder if the lightning strikes 1 mile (5,280 ft) away and the air temperature is 90.0°F?

19. A 600.0 Hz sound has a velocity of 1,087.0 ft/s in the air and a velocity of 4,920.0 ft/s in water. Find the wavelength of this sound (a) in the air, (b) in the water.

A thunderstorm produces an interesting display of electrical discharge. Each bolt can carry over 150,000 amperes of current with a voltage of 100 million volts.

Electricity

The previous chapters have been concerned with *mechanical* concepts, explanations of the motion of objects that exert forces on one another. These concepts were used to explain straight-line motion, the motion of free fall, and the circular motion of objects on the earth as well as the circular motion of planets and satellites. The mechanical concepts were based on Newton's laws of motion and are sometimes referred to as Newtonian physics. The mechanical explanations were then extended into the submicroscopic world of matter through the kinetic molecular theory. The objects of motion were now particles, molecules that exert force on one another, and concepts associated with heat were interpreted as the motion of these particles. In a further extension of Newtonian concepts, mechanical explanations were given for concepts associated with sound, a mechanical disturbance that follows the laws of motion as it moves through the molecules of matter.

You might wonder, as did the scientists of the 1800s, if mechanical interpretations would also explain other natural phenomena such as electricity, chemical reactions, and light. A mechanical model would be very attractive because it already explained so many other facts of nature, and scientists have always looked for basic, unifying theories. Mechanical interpretations were tried, as electricity was considered a moving fluid, and light was considered a mechanical wave moving through a material fluid. There were many unsolved puzzles with such a model, and gradually it was recognized that electricity, light, and chemical reactions could not be explained by mechanical interpretations. Gradually, the point of view changed from a study of particles to a study of the properties of the space around the particles. In this chapter you will learn about electric charge in terms of the space around particles. This model of electric charge, called the *field model*, will be used to develop concepts about electric current, the electric circuit, and electrical work and power. A relationship between electricity and the fascinating topic of magnetism is discussed next, including what magnetism is and how it is produced. The relationship is then used to explain the mechanical production of electricity (Figure 6.1), how electricity is measured, and how electricity is used in everyday technological applications.

Electric Charge

You are familiar with the use of electricity in many electrical devices such as lights, toasters, radios, and calculators. You are also aware that electricity is used for transportation and for heating and cooling places where you work and live. Many people accept electrical devices as part of their surroundings, with only a hazy notion of how they work. To many people electricity seems to be magical. Electricity is not magical, and it can be understood, just as we understand any other natural phenomenon. There are theories that explain observations, quantities that can be measured, and relationships between these quantities, or laws, that lead to understanding. All of the observations, measurements, and laws begin with an understanding of *electric charge.*

Electron Theory of Charge

It was a big mystery for thousands of years. No one could figure out why a rubbed piece of amber, which is fossilized tree resin, would attract small pieces of paper, thread, and hair. This unexplained attraction was called the "amber effect." Then about one hundred years ago, Joseph J. Thomson found the answer while experimenting with electric currents. From these experiments, Thomson was able to conclude that negatively charged particles were present in all matter, and in fact might be the stuff of which matter is made.

The amber effect was traced to the movement of these particles, so they were called *electrons* after the Greek word for amber. The word *electricity* is also based on the Greek word for amber.

Today, we understand that the basic unit of matter is the *atom*, which is made up of electrons and other particles such as *protons* and *neutrons*. The atom is considered to have a dense center part called a *nucleus* that contains the closely situated protons and neutrons. The electrons move around the nucleus at some relatively greater distance (Figure 6.2). Details on the nature of protons, neutrons, electrons, and models of how the atom is constructed will be considered in chapter 8. For understanding electricity, you need only consider the protons in the nucleus, the electrons that move around the nucleus, and the fact that electrons can be moved from an atom and caused to move to or from one object to another. Basically, the electrical, light, and chemical phenomena involve the *electrons* and not the more massive nucleus. The massive nuclei remain in a relatively fixed position in a solid, but some of the electrons can move about from atom to atom.

Electric Charge and Electrical Forces

Electrons and protons have a property called electric charge. Electrons have a *negative electric charge* and protons have a *positive electric charge*. The negative or positive description simply means that these two properties are opposite; it does not mean that one is better than the other. Charge is as fundamental to

FIGURE 6.1

The importance of electrical power seems obvious in a modern industrial society. What is not so obvious is the role of electricity in magnetism, light, chemical change, and as the very basis for the structure of matter. All matter, in fact, is electrical in nature, as you will see.

these subatomic particles as gravity is to masses. This means that you cannot separate gravity from a mass, and you cannot separate charge from an electron or a proton.

Electric charges interact to produce what is called the *electrical force*. Like charges produce a repulsive electrical force as positive repels positive and negative repels negative. Unlike charges produce an attractive electrical force as positive and negative charges attract each other. You can remember how this happens with the simple rule of, "*like charges repel and unlike charges attract.*"

Ordinary atoms are usually neutral because there is a balance between the number of positively charged protons and the

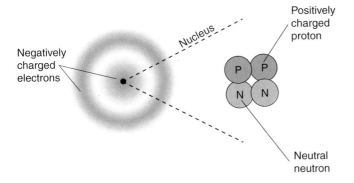

FIGURE 6.2

A very highly simplified model of an atom has most of the mass in a small, dense center called the *nucleus*. The nucleus has positively charged protons and neutral neutrons. Negatively charged electrons move around the nucleus at a much greater distance than is suggested by this simplified model. Ordinary atoms are neutral because there is balance between the number of positively charged protons and negatively charged electrons.

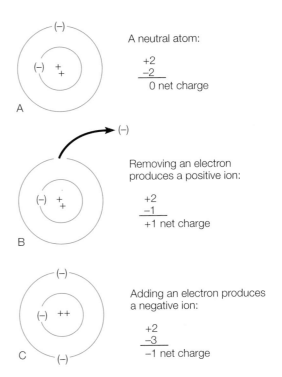

A neutral atom:

$$\begin{array}{r} +2 \\ -2 \\ \hline 0 \text{ net charge} \end{array}$$

Removing an electron produces a positive ion:

$$\begin{array}{r} +2 \\ -1 \\ \hline +1 \text{ net charge} \end{array}$$

Adding an electron produces a negative ion:

$$\begin{array}{r} +2 \\ -3 \\ \hline -1 \text{ net charge} \end{array}$$

FIGURE 6.3

(A) A neutral atom has no net charge because the numbers of electrons and protons are balanced. (B) Removing an electron produces a net positive charge; the charged atom is called a positive ion. (C) The addition of an electron produces a net negative charge and a negative ion.

number of negatively charged electrons. A number of different physical and chemical interactions can result in an atom gaining or losing electrons. In either case, the atom is said to be *ionized*, and *ions* are produced as a result. An atom that is ionized by losing electrons results in a *positive ion* beacause it has a net positive charge. An atom that is ionized by gaining electrons results in a *negative ion* because it has a net negative charge (Figure 6.3).

Electrostatic Charge

Electrons can be moved from atom to atom to create ions. They can also be moved from one object to another by friction and by other means that will be discussed soon. Since electrons are negatively charged, an object that acquires an excess of electrons becomes a negatively charged body. The loss of electrons by another body results in a deficiency of electrons, which results in a positively charged object. Thus, *electric charges on objects result from the gain or loss of electrons*. Because the electric charge is confined to an object and is not moving, it is called an **electrostatic charge.** You probably call this charge *static electricity*. Static electricity is an accumulated electric charge at rest, that is, one that is not moving. When you comb your hair with a hard rubber comb, the comb becomes negatively charged because electrons are transferred *from* your hair to the comb. Your hair becomes positively charged with a charge equal in magnitude to the charge gained by the comb (Figure 6.4). Both the negative charge on the

+3
−3
0 net charge

+3
−3
0 net charge

A

+3
−1
+2 net charge

+3
−5
−2 net charge

B

FIGURE 6.4

Arbitrary numbers of protons (+) and electrons (−) on a comb and in hair (A) before and (B) after combing. Combing transfers electrons from the hair to the comb by friction, resulting in a negative charge on the comb and a positive charge on the hair.

comb from an excess of electrons and the positive charge on your hair from a deficiency of electrons are charges that are momentarily at rest, so they are electrostatic charges.

Once charged by friction, objects such as the rubber comb soon return to a neutral, or balanced, state by the movement of electrons. This happens more quickly on a humid day because water vapor assists with the movement of electrons to or from charged objects. Thus, static electricity is more noticeable on dry days than on humid ones.

An object can become electrostatically charged (1) by *friction*, which transfers electrons from one object to another, (2) by *contact* with another charged body, which results in the transfer of electrons, or (3) by *induction*. Induction produces a charge by a redistribution of charges in a material. When you comb your hair, for example, the comb removes electrons from your hair and acquires a negative charge. When the negatively charged comb is held near small pieces of paper, it repels some electrons in the paper to the opposite side of the paper. This leaves the side of the paper closest to the comb with a positive charge, and there is an attraction between the pieces of paper and the comb, since unlike charges attract. Note that no transfer of electrons takes place in induction; the attraction results from a reorientation of the charges in the paper (Figure 6.5). Note also that charge is transferred in all three examples; it is not created or destroyed.

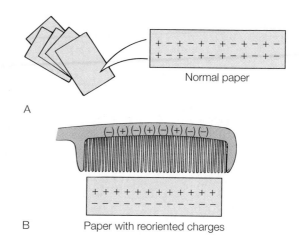

Normal paper

A

(−) (+) (−) (+) (−) (+) (−) (−)

+ + + + + + + + + + +
− − − − − − − − − − −

B Paper with reoriented charges

FIGURE 6.5

Charging by induction. The comb has become charged by friction, acquiring an excess of electrons. The paper (A) normally has a random distribution of (+) and (−) charges. (B) When the charged comb is held close to the paper, there is a reorientation of charges because of the repulsion of like charges. This leaves a net positive charge on the side close to the comb, and since unlike charges attract, the paper is attracted to the comb.

Concepts Applied

Static Charge

1. This application works best when the humidity is low. Obtain a plastic drinking straw and several sheets of light tissue paper. Tear one of the sheets into small pieces and place them on your desk. Wrap a sheet of the tissue around the straw and push and pull the straw back and forth about ten times while holding it in the tissue. Touch the end of the straw to one of the pieces of tissue on your desk to see if it is attracted. If it is, touch the attracted piece to another to see if it, too, is attracted. Depending on the humidity, you might be able to attract a long chain of pieces of tissue.

2. Suspend the straw from the edge of your desk with a length of cellophane tape. Try rubbing a plastic ballpoint pen, a black plastic comb, and other objects with a cotton cloth, flannel, fur, and other materials. Hold each rubbed object close to the straw and observe what happens.

3. Make a list of materials that seem to acquire a static charge and those that do not. See how many generalizations you can make about static electricity and materials. Describe any evidence you observed that two kinds of electric charge exist.

TABLE 6.1
Electrical conductors and insulators

Conductors	Insulators
Silver	Rubber
Copper	Glass
Gold	Carbon (diamond)
Aluminum	Plastics
Carbon (graphite)	Wood
Tungsten	
Iron	
Lead	
Nichrome	

Electrical Conductors and Insulators

When you slide across a car seat or scuff your shoes across a carpet, you are rubbing some electrons from the materials and acquiring an excess of negative charges. Because the electric charge is confined to you and is not moving, it is an electrostatic charge. The electrostatic charge is produced by friction between two surfaces and will remain until the electrons can move away because of their mutual repulsion. This usually happens when you reach for a metal doorknob, and you know when it happens because the electron movement makes a spark. Materials like the metal of a doorknob are good **electrical conductors** because they have electrons that are free to move throughout the metal. If you touch plastic or wood, however, you will not feel a shock. Materials like plastic and wood do not have electrons that are free to move throughout the material, and they are called *electrical nonconductors*. Nonconductors are also called *electrical insulators* (Table 6.1). Electrons do not move easily through an insulator, but electrons can be added or removed, and the charge tends to remain. In fact, your body is a poor conductor, which is why you become charged by friction in the first place.

Materials vary in their ability to conduct charges, and this ability is determined by how tightly or loosely the electrons are held to the nucleus. Metals have millions of free electrons that can take part in the conduction of an electric charge. Materials such as rubber, glass, and plastics hold tightly to their electrons and are good insulators. Thus, metal wires are used to conduct an electric current from one place to another, and rubber, glass, and plastics are used as insulators to keep the current from going elsewhere.

There is a third class of materials, such as silicon and germanium, that sometimes conduct and sometimes insulate, depending on the conditions and how pure they are. These materials are called *semiconductors,* and their special properties make possible a number of technological devices such as the electrostatic copying machine, solar cells, and so forth.

Measuring Electrical Charges

As you might have experienced, sometimes you receive a slight shock after walking across a carpet, and sometimes you are really zapped. You receive a greater shock when you have accumulated a greater electric charge. Since there is less electric charge at one time and more at another, it should be evident that charge occurs in different amounts, and these amounts can be measured. The size of an electric charge is identified with the number of electrons that have been transferred onto or away from an object. The quantity of such a charge (q) is measured in a unit called a **coulomb** (C). A coulomb unit is equivalent to the charge resulting from the transfer of 6.24×10^{18} of the charge carried by particles such as the electron. The coulomb is a fundamental metric unit of measure like the meter, kilogram, and second.

The coulomb is a *unit* of electric charge that is used with other metric units such as meters for distance and newtons for force. Thus, a quantity of charge (q) is described in units of coulomb (C). This is just like the process of a quantity of mass (m) being described in units of kilogram (kg). The concepts of charge and coulomb may seem less understandable than the concepts of mass and kilogram, since you cannot see charge or how it is measured. But charge does exist and it can be measured, so you can understand both the concept and the unit by working with them. Consider, for example, that an object has a net electric charge (q) because it has an unbalanced number (n) of electrons (e^-) and protons (p^+). The net charge on you after walking across a carpet depends on how many electrons you rubbed from the carpet. The net charge in this case would be the excess of electrons, or

quantity of charge = (number of electrons)(electron charge)

or

$$q = ne$$

equation 6.1

Since 1.00 coulomb is equivalent to the transfer of 6.24×10^{18} particles such as the electron, the charge on one electron must be

$$e = \frac{q}{n}$$

where q is 1.00 C, and n is 6.24×10^{18} electrons,

$$e = \frac{1.00 \text{ coulomb}}{6.24 \times 10^{18} \text{ electron}}$$

$$= 1.60 \times 10^{-19} \frac{\text{coulomb}}{\text{electron}}$$

This charge, 1.60×10^{-19} coulomb, is the *smallest* common charge known (more exactly $1.6021892 \times 10^{-19}$ C). It is the **fundamental charge** of the electron. Every electron has a charge of -1.60×10^{-19} C, and every proton has a charge of $+1.60 \times 10^{-19}$ C. To accumulate a negative charge of 1 C you would need to accumulate more than 6 billion billion electrons. All charged objects have multiples of the fundamental charge, so charge is said to be quantized.

EXAMPLE 6.1

Combing your hair on a day with low humidity results in a comb with a negative charge on the order of 1.00×10^{-8} coulomb. How many electrons were transferred from your hair to the comb?

Solution

The relationship between the quantity of charge on an object (q), the number of electrons (n), and the fundamental charge on an electron (e^-) is found in equation 6.1, $q = ne$.

$$q = 1.00 \times 10^{-8} \, C$$

$$e = 1.60 \times 10^{-19} \frac{C}{e}$$

$$n = ?$$

$$q = ne \quad \therefore \quad n = \frac{q}{e}$$

$$n = \frac{1.00 \times 10^{-8} \, C}{1.60 \times 10^{-19} \frac{C}{e}}$$

$$= \frac{1.00 \times 10^{-8}}{1.60 \times 10^{-19}} \, \cancel{C} \times \frac{e}{\cancel{C}}$$

$$= \boxed{6.25 \times 10^{10} \, e}$$

Thus, the comb acquired an excess of approximately 62.5 billion electrons. (Note that the convention in scientific notation is to express an answer with one digit to the left of the decimal. See appendix A for further information on scientific notation.)

Measuring Electrical Forces

Recall that two objects with like charges, ($-$) and ($-$) or ($+$) and ($+$), produce a repulsive force, and two objects with unlike charges, ($-$) and ($+$), produce an attractive force. The size of either force depends on the amount of charge of each object and on the distance between the objects. The relationship is know as **Coulomb's law,** which is,

$$F = k\frac{q_1 q_2}{d^2}$$

equation 6.2

where k has the value of 9.00×10^9 newton-meters^2/coulomb^2 ($9.00 \times 10^9 \, N \cdot m^2/C^2$).

The force between the two charged objects is repulsive if q_1 and q_2 are the same charge and attractive if they are different (like charges repel, unlike charges attract). Whether the force is attractive or repulsive, you know that both objects feel the same force, as described by Newton's third law of motion. In addition, the strength of this force decreases if the distance between the objects is increased (a doubling of the distance reduces the force to ¼ the original value).

EXAMPLE **6.2**

Electrons carry a negative electric charge and revolve about the nucleus of the atom, which carries a positive electric charge from the proton. The electron is held in orbit by the force of electrical attraction at a typical distance of 1.00×10^{-10} m. What is the force of electrical attraction between an electron and proton?

Solution

The fundamental charge of an electron (e^-) is 1.60×10^{-19} C, and the fundamental charge of the proton (p^+) is 1.60×10^{-19} C. The distance is given, and the force of electrical attraction can be found from equation 6.2:

$$q_1 = 1.60 \times 10^{-10} \, C$$

$$q_2 = 1.60 \times 10^{-19} \, C$$

$$d = 1.00 \times 10^{-10} \, m$$

$$k = 9.00 \times 10^9 \, N \cdot m^2/C^2$$

$$F = ?$$

$$F = k\frac{q_1 q_2}{d^2}$$

$$= \frac{\left(9.00 \times 10^9 \frac{N \cdot m^2}{C^2}\right)(1.60 \times 10^{-19} C)(1.60 \times 10^{-19} C)}{(1.00 \times 10^{-10} m)^2}$$

$$= \frac{(9.00 \times 10^9)(1.60 \times 10^{-19})(1.60 \times 10^{-19})}{1.00 \times 10^{-20}} \frac{\left(\frac{N \cdot m^2}{C^2}\right)(C^2)}{m^2}$$

$$= \frac{2.30 \times 10^{-28}}{1.00 \times 10^{-20}} \frac{N \cdot \cancel{m^2}}{\cancel{C^2}} \times \frac{\cancel{C^2}}{1} \times \frac{1}{\cancel{m^2}}$$

$$= \boxed{2.30 \times 10^{-8} \, N}$$

The electrical force of attraction between the electron and proton is 2.30×10^{-8} newton.

Force Fields

Does it seem odd to you that gravitational forces and electrical forces can act on objects that are not touching? How can gravitational forces act through the vast empty space between the earth and the sun? How can electrical forces act through a distance to pull pieces of paper to your charged comb? Such questions have bothered people since the early discovery of small, light objects being attracted to rubbed amber. There was no mental model of how such a force could act through a distance without touching. The idea of "invisible fluids" was an early attempt to develop a mental model that would help people visualize how a force could act over a distance without physical contact. Then Newton developed the law of universal gravitation, which correctly predicted the magnitude of gravitational forces acting through space. Coulomb's law of electrical forces had similar success in describing and predicting electrostatic forces acting through space. "Invisible fluids" were no longer needed to explain what was happening, because the two laws seemed to explain the results of such actions. But it was still difficult to visualize what was happening physically when forces acted through a distance, and there were a few problems with the concept of action at a distance. Not all observations were explained by the model.

The work of Michael Faraday and James Maxwell in the early 1800s finally provided a new mental model for interaction at a distance. This new model did *not* consider the force that one object exerts on another one through a distance. Instead, it considered *the condition of space* around an object. The condition of space around an electric charge is considered to be changed by the presence of the charge. The charge produces a **force field** in the space around it. Since this force field is produced by an electrical charge, it is called an **electric field.** Imagine a second electric charge, called a *test charge,* that is far enough away from the electric charge that no forces are experienced. As you move the test charge closer and closer, it will experience an increasing force as it enters the electric field. The test charge is assumed not to change the field that it is entering and can be used to identify the electric field that spreads out and around the space of an electric charge.

All electric charges are considered to be surrounded by an electric field. All *masses* are considered to be surrounded by a *gravitational field.* The earth, for example, is considered to change the condition of space around it because of its mass. A spaceship far, far from the earth does not experience a measurable force. But as it approaches the earth, it moves farther into the earth's gravitational field and eventually experiences a measurable force. Likewise, a magnet creates a *magnetic field* in the space around it. You can visualize a magnetic field by moving a magnetic compass needle around a bar magnet. Far from the bar magnet the compass needle does not respond. Moving it closer to the bar magnet, you can see where the magnetic field begins. Another way to visualize a magnetic field is to place a sheet of paper over a bar magnet, then sprinkle iron filings on the paper. The filings will clearly identify the presence of the magnetic field.

Another way to visualize a field is to make a map of the field. Consider a small positive test charge that is brought into an electric field. A *positive* test charge is always used by convention. As shown in Figure 6.6, a positive test charge is brought near a negative charge and a positive charge. The arrow points in the direction of the force that the *test charge experiences.* Thus, when brought near a negative charge, the test charge is attracted toward the unlike charge, and the arrow points that way. When brought near a positive charge, the test charge is repelled, so the arrow points away from the positive charge.

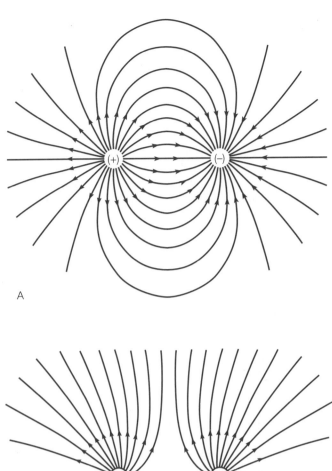

A

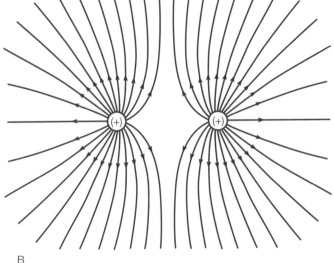

B

FIGURE 6.7

Lines of force diagrams for (*A*) a negative charge and (*B*) a positive charge when the charges are the same size as the test charge.

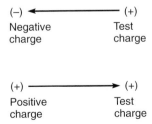

FIGURE 6.6

A positive test charge is used by convention to identify the properties of an electric field. The arrow points in the direction of the force that the test charge would experience.

An electric field is represented by drawing *lines of force* or *electric field lines* that show the direction of the field. The arrows in Figure 6.7 show field lines that could extend outward forever from isolated charges, since there is always some force on a distant test charge (review Coulomb's law; the force ideally never reaches zero). The field lines between pairs of charges in Figure 6.7 show curved field lines that originate on positive charges and end on negative charges. By convention, the field lines are closer together where the field is stronger and farther apart where the field is weaker.

The field concept explains some observations that were not explained with the Newtonian concept of action at a distance.

Suppose, for example, that a charge produces an electric field. This field is not instantaneously created all around the charge, but it is seen to build up and spread into space. If the charge is suddenly neutralized, the field that it created continues to spread outward, and then collapses back at some speed, even though the source of the field no longer exists. Consider an example with the gravitational field of the sun. If the mass of the sun were to instantaneously disappear, would the earth notice this instantaneously? Or would the gravitational field of the sun collapse at some speed, say the speed of light, to be noticed by the earth some eight minutes later? The Newtonian concept of action at a distance did not consider any properties of space, so according to this concept the gravitational force from the sun would disappear instantly. The field concept, however, explains that the disappearance would be noticed after some period of time, about eight minutes. This time delay agrees with similar observations of objects interacting with fields, so the field concept is more useful than a mysterious action-at-a-distance concept, as you will see.

Actually there are three models for explaining how gravitational, electrical, and magnetic forces operate at a distance. (1) The *action-at-a-distance model* recognizes that masses are attracted gravitationally and that electric charges and magnetic poles attract and repel each other through space, but it gives no further explanation; (2) the *field model* considers a field to be a condition of space around a mass, electric charge, or magnet, and the properties of fields are described by field lines; and (3) the *field-particle model* is a complex and highly mathematical explanation of attractive and repulsive forces as the rapid emission and absorption of subatomic particles. This model explains electrical and magnetic forces as the exchange of *virtual photons,* gravitational forces as the exchange of *gravitons,* and strong nuclear forces as the exchange of *gluons.*

Electric Potential

Recall from chapter 3 that work is accomplished as you move an object to a higher location on the earth, say by moving a book from the first shelf of a bookcase to a higher shelf. By virtue of its position, the book now has gravitational potential energy that can be measured by *mgh* (the force of the book's weight × distance), joules of gravitational potential energy. Using the field model, you could say that this work was accomplished against the gravitational field of the earth. Likewise, an electric charge has an electric field surrounding it, and work must be done to move a second charge into or out of this field. Bringing a like charged particle *into* the field of another charged particle will require work, since like charges repel, and separating two unlike charges will also require work, since unlike charges attract. In either case, the *electric potential energy* is changed, just as the gravitational potential energy is changed by moving a mass in the earth's gravitational field.

One useful way to measure electric potential energy is to consider the *potential difference* that occurs when a certain amount of work is used to move a certain quantity of charge. For example, suppose there is a firmly anchored and insulated metal sphere that

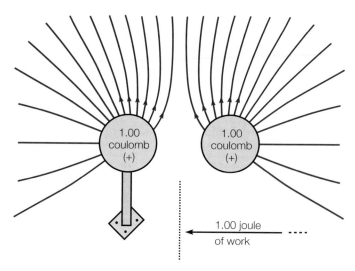

FIGURE 6.8

Electric potential results from moving a positive coulomb of charge into the electric field of a second positive coulomb of charge. When 1.00 joule of work is done in moving 1.00 coulomb of charge, 1.00 volt of potential results. A volt is a joule/coulomb.

has a positive charge (Figure 6.8). The sphere will have a positive electric field in the space around it. Suppose also that you have a second sphere that has exactly 1.00 coulomb of positive charge. You begin moving the coulomb of positive charge toward the anchored sphere. As you enter the electric field you will have to push harder and harder to overcome the increasing repulsion. If you stop moving when you have done exactly 1.00 joule of work, the repulsion will *do* one joule of work if you now release the sphere. The sphere has potential energy in the same way that a compressed spring has potential energy. In electrical matters, the *potential difference that is created by doing 1.00 joule of work in moving 1.00 coulomb of charge is defined to be 1.00 volt.* The **volt** (V) is a measure of potential difference between two points, or

$$\text{electric potential difference} = \frac{\text{work to create potential}}{\text{charge moved}}$$

$$V = \frac{W}{q}$$

equation 6.3

In units,

$$1.00 \text{ volt (V)} = \frac{1.00 \text{ joule (J)}}{1.00 \text{ coulomb (C)}}$$

The voltage of any electric charge, either static or moving, is the energy transfer per coulomb. The energy transfer can be measured by the *work that is done to move the charge* or by the *work that the charge can do* because of its position in the field. This is perfectly analogous to the work that must be done to give an object gravitational potential energy or to the work that the object can potentially do because of its new position. Thus, when a 12 volt battery is charged, 12.0 joules of work are done to transfer 1.00 coulomb of charge from an outside source against the electric field of the battery terminal. When the 12 volt battery is used, it does 12.0 joules of work for each coulomb of charge transferred from one terminal of the battery through the electrical system and back to the other terminal.

Electric Current

So far, we have considered electric charges that have been instantaneously moved by friction but then generally stayed in one place. Experiments with static electricity played a major role in the development of the understanding of electricity by identifying charge, the attractive and repulsive forces between charges, and the field concept. The work of Franklin, Coulomb, Faraday, and Maxwell thus increased studies of electrical phenomena and eventually led to insights into connections between electricity and magnetism, light, and chemistry. These connections will be discussed later, but for now, consider the sustained flowing or moving of charge, an *electric current* (I). **Electric current** means a flow of charge in the same way that "water current" means a flow of water. Since the word "current" *means* flow, you are being redundant if you speak of "flow of current." It is the *charge* that flows, and the current is defined as the flow of charge.

The Electric Circuit

When you slide across a car seat, you are acquiring electrons on your body by friction. Through friction, you did *work* on the electrons as you removed them from the seat covering. You now have a net negative charge from the imbalance of electrons, which tend to remain on you because you are a poor conductor. But the electrons are now closer than they want to be, within a repulsive electric field, and there is an electrical potential difference between you and some uncharged object, say a metal door handle. When you touch the handle, the electrons will flow, creating a momentary current in the form of a spark, which lasts only until the charge on you is neutralized.

In order to keep an electric current going, you must maintain the separation of charges and therefore maintain the electric field (or potential difference), which can push the charges through a conductor. This might be possible if you could somehow continuously slide across the car seat, but this would be a hit-and-miss way of maintaining a separation of charges and would probably result in a series of sparks rather than a continuous current. This is how electrostatic machines work.

A useful analogy for understanding the requirements for a sustained electric current is the decorative waterwheel device (Figure 6.9). Water in the upper reservoir has a greater gravitational potential energy than water in the lower reservoir. As water flows from the upper reservoir, it can do work in turning the waterwheel, but it can continue to do this only as long as the pump does the work to maintain the potential difference between the two reservoirs. This "water circuit" will do work in turning the waterwheel as long as the pump returns the water to a higher potential continuously as the water flows back to the lower potential.

So, by a water circuit analogy, a steady electric current is maintained by pumping charges to a higher potential, and the

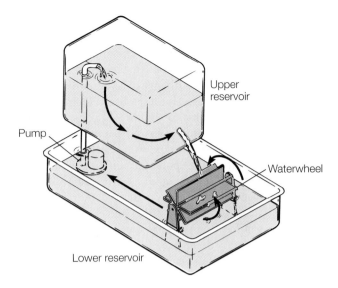

FIGURE 6.9

The falling water can do work in turning the waterwheel only as long as the pump maintains the potential difference between the upper and lower reservoirs.

charges do work as they move back to a lower potential. The higher electric potential energy is analogous to the gravitational potential energy in the waterwheel example (Figure 6.9). An **electric circuit** contains some device, such as a battery or electric generator, that acts as a source of energy as it gives charges a higher potential against an electric field. The charges do work in another part of the circuit as they light bulbs, run motors, or provide heat. The charges flow through connecting wires to make a continuous path. An electric switch is a means of interrupting or completing this continuous path.

The electrical potential difference between the two connecting wires shown in Figure 6.10 is one factor in the work done *by* the device that creates a higher electrical potential (battery, for example) and the work done *in* some device (lamp, for example). Disregarding any losses, the work done in both places

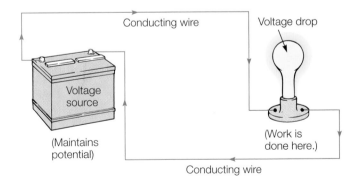

FIGURE 6.10

A simple electric circuit has a voltage source (such as a generator or battery) that maintains the electrical potential, some device (such as a lamp or motor) where work is done by the potential, and continuous pathways for the current to follow.

would be the same. Recall that work done per unit of charge is joules/coulomb, or volts (equation 6.3). The source of the electrical potential difference is therefore referred to as a *voltage source*. The device where the charges do their work causes a *voltage drop*. Electrical potential difference is measured in volts, so the term *voltage* is often used for it. Household circuits usually have a difference of potential of 120 or 240 volts. A voltage of 120 means that each coulomb of charge that moves through the circuit can do 120 joules of work in some electrical device.

Voltage describes the potential difference, in joules/coulomb, between two places in an electric circuit. By way of analogy to pressure on water in a circuit of water pipes, this potential difference is sometimes called an "electrical force" or "electromotive force" (emf). Note that in electrical matters, however, the potential difference is the *source* of a force rather than being a force such as water under pressure. Nonetheless, just as you can have a small water pipe and a large water pipe under the same pressure, the two pipes would have a different rate of water flow in gallons per minute. Electric current (I) is the rate at which charge (q) flows through a cross section of a conductor in a unit of time (t), or

$$\text{electric current} = \frac{\text{quantity of charge}}{\text{time}}$$

$$I = \frac{q}{t}$$

equation 6.4

The units of current are thus coulombs/second. A coulomb/second is called an **ampere** (A), or **amp** for short. In units, current is therefore

$$1.00 \text{ amp (A)} = \frac{1.00 \text{ coulomb (C)}}{1.00 \text{ second (s)}}$$

A 1.00 amp current is 1.00 coulomb of charge moving through a conductor each second, a 2.00 amp current is 2.00 coulombs per second, and so forth (Figure 6.11).

Using the water circuit analogy, you would expect a greater rate of water flow (gallons/minute) when the water pressure is produced by a greater gravitational potential difference. The rate of water flow is thus directly proportional to the difference in gravitational potential energy. In an electric circuit, the rate of current (coulombs/second, or amps) is directly proportional to the difference of electrical potential (joules/coulombs, or volts) between two parts of the circuit, $I \propto V$.

The Nature of Current

There are two ways to describe the current that flows outside the power source in a circuit, (1) a historically based description called *conventional current* and (2) a description based on a flow of charges called *electron current*. The *conventional* current describes current as positive charges moving from the positive to the negative terminal of a battery. This description has been used by convention ever since Ben Franklin first misnamed the charge of an object based on an accumulation, or a positive amount, of "electrical fluid." Conventional current is still used in circuit diagrams. The *electron* current description is in an opposite direction to the conventional current. The electron current describes current as the drift of negative charges that flow from the negative to the positive terminal of a battery. Today, scientists understand the role of electrons in a current, something that was unknown to Franklin. But conventional current is still used by tradition. It actually does not make any difference which description is used, since positive charges moving from the positive terminal are mathematically equivalent to negative charges moving from the negative terminal (Figure 6.12).

The description of an electron current also retains historical traces of the earlier fluid theories of electricity. Today, people understand that electricity is not a fluid but still speak of current, rate of flow, and resistance to flow (Figure 6.13). Fluid analogies

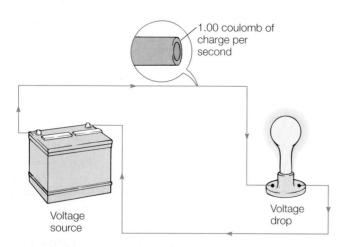

FIGURE 6.11

A simple electric circuit carrying a current of 1.00 coulomb per second through a cross section of a conductor has a current of 1.00 amp.

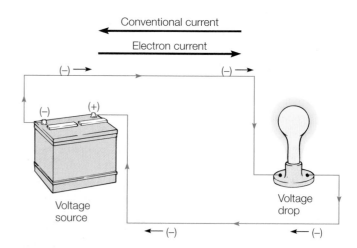

FIGURE 6.12

A conventional current describes positive charges moving from the positive terminal (+) to the negative terminal (−). An electron current describes negative charges (−) moving from the negative terminal (−) to the positive terminal (+).

FIGURE 6.13

What is the nature of the electric current carried by these conducting lines? It is an electric field that moves at near the speed of light. The field causes a net motion of electrons that constitutes a flow of charge, a current.

can be helpful because they describe the overall electrical effects. But they can also lead to incorrect concepts such as the following: (1) an electric current is the movement of electrons through a wire just as water flows through a pipe; (2) electrons are pushed out one end of the wire as more electrons are pushed in the other end; and (3) electrons must move through a wire at the speed of light since a power plant failure hundreds of miles away results in an instantaneous loss of power. Perhaps you have held one or more of these misconceptions from fluid analogies.

What is the nature of an electric current? First, consider the nature of a metal conductor without a current. The atoms making up the metal have unattached electrons that are free to move about, much as the molecules of a gas in a container. They randomly move at high speed in all directions, often colliding with each other and with stationary positive ions of the metal. This motion is chaotic, and there is no net movement in any one direction, but the motion does increase with increases in the absolute temperature of the conductor.

When a potential difference is applied to the wire in a circuit, an electric field is established everywhere in the circuit. The *electric field* travels through the conductor at nearly the speed of

light as it is established. A force is exerted on each electron by the field, which accelerates the free electrons in the direction of the force. The resulting increased velocity of the electrons is superimposed on their existing random, chaotic movement. This added motion is called the *drift velocity* of the electrons. The drift velocity of the electrons is a result of the imposed electric field. The electrons do not drift straight through the conductor, however, because they undergo countless collisions with other electrons and stationary positive ions. This results in a random zigzag motion with a net motion in one direction. *This net motion constitutes a current,* a flow of charge (Figure 6.14).

When the voltage across a conductor is zero the drift velocity is zero, and there is no current. The current that occurs when there is a voltage depends on (1) the number of free electrons per unit volume of the conducting material, (2) the charge on each electron (the fundamental charge), (3) the drift velocity, which depends on the electronic structure of the conducting material and the temperature, and (4) the cross-sectional area of the conducting wire.

The relationship between the number of free electrons, charge, drift velocity, area, and current can be used to determine the drift velocity when a certain current flows in a certain size

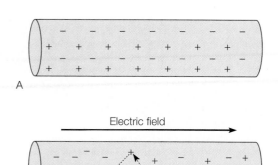

A

Electric field

B

Electric field

FIGURE 6.14

(A) A metal conductor without a current has immovable positive ions surrounded by a swarm of chaotically moving electrons. (B) An electric field causes the electrons to shift positions, creating a separation charge as the electrons move with a zigzag motion from collisions with stationary positive ions and other electrons.

wire made of copper. A 1.0 amp current in copper bell wire (#18), for example, has an average drift velocity on the order of 0.01 cm/s. At that rate, it would take over 5 h for an electron to travel the 200 cm from your car battery to the brake light of your car (Figure 6.15). Thus, it seems clear that it is the *electric field*, not electrons, that causes your brake light to come on almost instantaneously when you apply the brake. The electric field accelerates the electrons already in the filament of the brake lightbulb. Collisions between the electrons in the filament cause the bulb to glow.

Conclusions about the nature of an electric current are that (1) an electric potential difference establishes, at near the speed

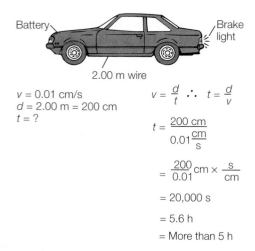

$v = 0.01$ cm/s
$d = 2.00$ m $= 200$ cm
$t = ?$

$v = \dfrac{d}{t} \quad \therefore \quad t = \dfrac{d}{v}$

$t = \dfrac{200 \text{ cm}}{0.01 \dfrac{\text{cm}}{\text{s}}}$

$= \dfrac{200}{0.01} \text{ cm} \times \dfrac{\text{s}}{\text{cm}}$

$= 20{,}000 \text{ s}$

$= 5.6 \text{ h}$

$= \text{More than 5 h}$

FIGURE 6.15

Electrons move very slowly in a direct current circuit. With a drift velocity of 0.01 cm/s, more than 5 h would be required for an electron to travel 200 cm from a car battery to the brake light. It is the electric field, not the electrons, that moves at near the speed of light in an electric circuit.

of light, an electric field throughout a circuit, (2) the field causes a net motion that constitutes a flow of charge, or current, and (3) the average velocity of the electrons moving as a current is very slow, even though the electric field that moves them travels with a speed close to the speed of light.

Another aspect of the nature of an electric current is the direction the charge is flowing. A circuit like the one described with your car battery has a current that always moves in one direction, a **direct current** (dc). Chemical batteries, fuel cells, and solar cells produce a direct current, and direct currents are utilized in electronic devices. Electric utilities and most of the electrical industry, on the other hand, use an **alternating current** (ac). An alternating current, as the name implies, moves the electrons alternately one way, then the other way. Since the electrons are simply moving back and forth, there is no electron drift along a conductor in an alternating current. Since household electric circuits use alternating current, there is no flow of electrons from the electrical outlets through the circuits. Instead, an electric field moves back and forth through the circuit at nearly the speed of light, causing electrons to move back and forth. This movement constitutes a current that flows one way, then the other with the changing field. The current changes like this 120 times a second in a 60 hertz alternating current.

Electrical Resistance

Recall the natural random and chaotic motion of electrons in a conductor and their frequent collisions with each other and with the stationary positive ions. When these collisions occur, electrons lose energy that they gained from the electric field. The stationary positive ions gain this energy, and their increased energy of vibration results in a temperature increase. Thus, there is a resistance to the movement of electrons being accelerated by an electric field and a resulting energy loss. Materials have a property of opposing or reducing a current, and this property is called **electrical resistance** (R).

Recall that the current (I) through a conductor is directly proportional to the potential difference (V) between two points in a circuit. If a conductor offers a small resistance, less voltage would be required to push an amp of current through the circuit. If a conductor offers more resistance, then more voltage will be required to push the same amp of current through the circuit. Resistance (R) is therefore a *ratio* between the potential difference (V) between two points and the resulting current (I). This ratio is

$$\text{resistance} = \frac{\text{electrical potential difference}}{\text{current}}$$

$$R = \frac{V}{I}$$

In units, this ratio is

$$1.00 \text{ ohm } (\Omega) = \frac{1.00 \text{ volt (V)}}{1.00 \text{ amp (A)}}$$

The ratio of volts/amps is the unit of resistance called an **ohm** (Ω) after the German physicist who discovered the relationship. The resistance of a conductor is therefore 1.00 ohm if 1.00 volt

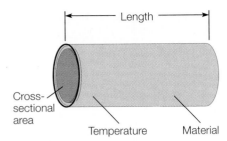

FIGURE 6.16
The four factors that influence the resistance of an electrical conductor are the length of the conductor, the cross-sectional area of the conductor, the material the conductor is made of, and the temperature of the conductor.

is required to maintain a 1.00 amp current. The ratio of volt/amp is *defined as* an ohm. Therefore,

$$\text{ohm} = \frac{\text{volt}}{\text{amp}}$$

Another way to show the relationship between the voltage, current, and resistance is

$$V = IR$$

equation 6.5

which is known as **Ohm's law.** This is one of three ways to show the relationship, but this way (solved for V) is convenient for easily solving the equation for other unknowns.

The electrical resistance of a dc electrical conductor depends on four variables (Figure 6.16):

1. *Material.* Different materials have different resistances, as shown by the list of conductors in Table 6.1. Silver, for example, is at the top of the list because it offers the least resistance, followed by copper, gold, then aluminum. Of the materials listed in the table, nichrome is the conductor with the greatest resistance. By definition, conductors have less electrical resistance than insulators, which have a very large electrical resistance.
2. *Length.* The resistance of a conductor varies directly with the length; that is, a longer wire has more resistance and a shorter wire has less resistance. The longer the wire is, the greater the resistance.
3. *Diameter.* The resistance varies inversely with the cross-sectional area of a conductor. A thick wire has a greater cross-sectional area and therefore has less resistance than a thin wire. The thinner the wire is, the greater the resistance.
4. *Temperature.* For most materials, the resistance increases with increases in temperature. This is a consequence of the increased motion of electrons and ions at higher temperatures, which increases the number of collisions. At very low temperatures, the resistance of some materials approaches zero, and the materials are said to be *superconductors.*

EXAMPLE 6.3

A lightbulb in a 120 V circuit is switched on, and a current of 0.50 A flows through the filament. What is the resistance of the bulb?

Solution

The current (I) of 0.50 A is given with a potential difference (V) of 120 V. The relationship to resistance (R) is given by Ohm's law (equation 6.5)

$$I = 0.50 \text{ A} \qquad V = IR \quad \therefore \quad R = \frac{V}{I}$$
$$V = 120 \text{ V}$$
$$R = ? \qquad\qquad = \frac{120}{0.50} \frac{\text{V}}{\text{A}}$$

$$= 240 \frac{\text{V}}{\text{A}}$$

$$= 240 \text{ ohm}$$

$$= \boxed{240 \ \Omega}$$

EXAMPLE 6.4

What current would flow through an electrical device in a circuit with a potential difference of 120 V and a resistance of 30 Ω? (Answer: 4 A)

Electrical Power and Electrical Work

All electric circuits have three parts in common: (1) a *voltage source,* such as a battery or electric generator that uses some nonelectric source of energy to do work on electrons, moving them *against* an electric field to a higher potential; (2) an *electric device,* such as a lightbulb or electric motor where work is done *by* the electric field; and, (3) *conducting wires* that maintain the potential difference across the electrical device. In a direct current circuit, the electric field moves from one terminal of a battery to the electric device through one wire. The second wire from the device carries the now low-potential field back to the other terminal, maintaining the potential difference. In an alternating current circuit, such as a household circuit, one wire supplies the alternating electric field from the electric generator of a utility company. The second wire from the device is connected to a pipe in the ground and is at the same potential as the earth. The observation that a bird can perch on a current-carrying wire without harm is explained by the fact that there is no potential difference across the bird's body. If the bird were to come into contact with the earth through a second, grounded wire, a potential difference would be established and there would be a current through it.

The work done by a voltage source (battery, electric generator) is equal to the work done by the electric field in an electric device (lightbulb, electric motor) *plus* the energy lost to resistance. Resistance is analogous to friction in a mechanical device, so low-resistance conducting wires are used to reduce this loss. Disregarding losses to resistance, electrical work can therefore be measured where the voltage source

The example of a household circuit in Box Figure 6.1 shows a light and four wall outlets in a parallel circuit. The use of the term *parallel* means that a current can flow through any of the separate branches without having to first go through any of the other devices. It does not imply that the branches are necessarily lined up with each other. Lights and outlets in this circuit all have at least two wires, one that carries the electrical load and one that maintains a potential difference by serving as a system ground. The load-carrying wire is usually black (or red) and the system ground is usually white. A third wire, usually bare or green, is used as a ground for an appliance.

Too many appliances running in a circuit—or a short circuit—can result in a very large current, great enough to cause strong heating and a fire. In a parallel circuit the total current for all the devices in the circuit flows through a fuse or circuit breaker. Circuit breakers prevent fires by disconnecting the circuit when it reaches a preset value, usually 15 or 20 amps. A fuse contains a short piece of metal that melts when heated by too large a current, creating a gap when the current through the circuit reaches the preset rating. The gap disconnects the circuit, just like cutting the wire or using a switch. The circuit breaker has the same purpose, but uses the proportional relationship between the magnitude of a current and the strength of the magnetic field that forms around the conductor. When the current reaches a preset level the magnetic field is strong enough to open a spring-loaded switch. The circuit breaker is reset by flipping the switch back to its original position.

BOX FIGURE 6.1

(*A*) One circuit breaker in this photo of a circuit breaker panel has tripped, indicating an overload or short circuit. (*B*) This could be one of the circuits protected by a simple circuit breaker.

A

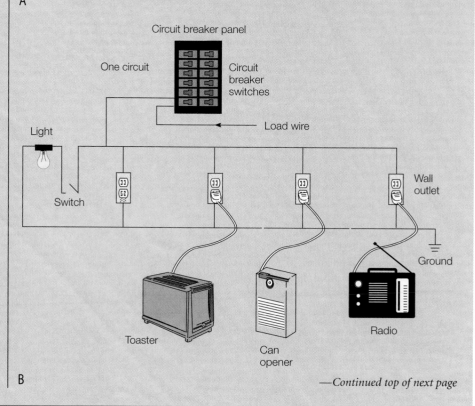

B

—*Continued top of next page*

Besides a fuse or circuit breaker, a modern household electric circuit has three-pronged plugs, polarized plugs, and ground fault interrupters (GFI) to help protect people and property from electrical damage. A *three-pronged plug* provides a grounding wire through a (usually round) prong on the plug. The grounding wire connects the housing of an appliance directly to the ground. If there is a short circuit, the current will take the path of least resistance—through the grounding wire—rather than through a person.

A *polarized plug* has one of the two flat prongs larger than the other. An alternating current moves back and forth with a frequency of 60 Hz, and polarized in this case has nothing to do with positive or negative. A polarized plug in an ac circuit means that one prong always carries the load. The smaller plug is connected to the load-carrying wire and the larger one is connected to the neutral wire. An ordinary, nonpolarized plug can fit into an outlet either way, which means there is a 50-50 chance that one of the wires will be the one that carries the load. The polarized plug always has the load-carrying wire on the same side of the circuit, so the switch can be wired in so it is always on the load-carrying side. The switch will function the same on either wire, but when it is on the ground wire the appliance contains a load-carrying wire, just waiting for a short circuit. When the switch is on the load-carrying side the appliance does not have this potential safety hazard.

Yet another safety device called a *ground-fault interrupter* (GFI) offers a different kind of protection. Normally, the current in the load-carrying and system ground wire is the same. If a short circuit occurs, some of the current might be diverted directly to the ground or to the appliance ground. A GFI device monitors the load-carrying and system ground wires and if any difference is detected, it trips, opening the circuit within a fraction of a second. This is much quicker than the regular fuse or general circuit breaker can react, and the difference might be enough to prevent a fatal shock. The GFI device is usually placed in an outside or bathroom circuit, places where currents might be diverted through people with wet feet. Note the GFI can also be tripped by a line surge that might occur during an electrical thunderstorm.

---

creates a potential difference by doing work (W) to move charges (q) to a higher potential (V). From equation 6.3, this relationship is

$$\text{work} = (\text{potential})(\text{charge})$$

or

$$W = (V)(q)$$

In units, the electrical potential is measured in joules/coulomb, and a quantity of charge is measured in coulombs. Therefore the unit of electrical work is the *joule*,

$$W = (V)(q)$$

$$\text{joule} = \frac{\text{joules}}{\text{coulomb}} \times \text{coulomb}$$

Recall that a joule is a unit of work in mechanics (a newton-meter). In electricity, a joule is also a unit of work, but it is derived from moving a quantity of charge (coulomb) to higher potential difference (joules/coulomb). In mechanics the work put into a simple machine equals the work output when you disregard *friction*. In electricity the work put into an electric circuit equals the work output when you disregard *resistance*. Thus, the work done by a voltage source is ideally equal to the work done by electrical devices in the circuit.

Recall also that mechanical power (P) was defined as work (W) per unit time (t), or

$$P = \frac{W}{t}$$

Since electrical work is $W = Vq$, then electrical power must be

$$P = \frac{Vq}{t}$$

Equation 6.4 defined a quantity of charge (q) per unit time (t) as a current (I), or $I = q/t$. Therefore electrical power is

$$P = \left(\frac{q}{t}\right)(V)$$

In units, you can see that multiplying the current A = C/s by the potential (V = J/C) yields

$$\frac{\text{coulombs}}{\text{second}} \times \frac{\text{joules}}{\text{coulombs}} = \frac{\text{joules}}{\text{second}}$$

A joule/second is a unit of power called the **watt** (W). Therefore, electrical power is measured in units of watts, and

$$P = A \cdot V$$

$$\text{power (in watts)} = \text{current (in amps)} \times \text{potential (in volts)}$$

$$\text{watts} = \text{volts} \times \text{amps}$$

This relationship is:

$$P = IV$$

equation 6.6

Household electrical devices are designed to operate on a particular voltage, usually 120 or 240 volts (Figure 6.17). They therefore draw a certain current to produce the designed power. Information about these requirements is usually found somewhere on the device. A lightbulb, for example, is usually stamped with the designed power, such as 100 watts. Other

A

B

C

FIGURE 6.17

What do you suppose it would cost to run each of these appliances for one hour? (A) This lightbulb is designed to operate on a potential difference of 120 volts and will do work at the rate of 100 W. (B) The finishing sander does work at the rate of 1.6 amp × 120 volts, or 192 W. (C) The garden shredder does work at the rate of 8 amps × 120 volts, or 960 W.

electrical devices may be stamped with amp and volt requirements. You can determine the power produced in these devices by using equation 6.6, that is, amps × volts = watts. Another handy conversion factor to remember is that 746 watts are equivalent to 1.00 horsepower.

EXAMPLE **6.5**

A 1,100 W hair dryer is designed to operate on 120 V. How much current does the dryer require?

Solution

The power (P) produced is given in watts with a potential difference of 120 V across the dryer. The relationship between the units of amps, volts, and watts is found in equation 6.6, $P = IV$

$P = 1,100$ W
$V = 120$ V
$I = ?$ A

$$P = IV \quad \therefore \quad I = \frac{P}{V}$$

$$= \frac{1,100 \, \frac{joule}{second}}{120 \, \frac{joule}{coulomb}}$$

$$= \frac{1,100}{120} \, \frac{J}{s} \times \frac{C}{J}$$

$$= 9.2 \, \frac{J \cdot C}{s \cdot J}$$

$$= 9.2 \, \frac{C}{s}$$

$$= \boxed{9.2 \text{ A}}$$

EXAMPLE **6.6**

An electric fan is designed to draw 0.5 A in a 120 V circuit. What is the power rating of the fan? (Answer: 60 W)

An electric utility charges you for the electrical power used at a rate of cents per kilowatt-hour. The rate varies from place to place across the country, depending on the cost of producing the power. You can predict the cost of running a particular electric appliance with the following equation,

$$\text{cost} = \frac{(\text{watts})(\text{time})(\text{rate})}{1,000 \, \frac{\text{watt}}{\text{kilowatt}}}$$

equation 6.7

TABLE 6.2
Summary of electrical quantities and units

Quantity	Definition*	Units
Charge	$q = ne$	1.00 coulomb (C) = charge equivalent to 6.24×10^{18} particles such as the electron
Electric potential difference	$V = \dfrac{W}{q}$	$1.00 \text{ volt (V)} = \dfrac{1.00 \text{ joule (J)}}{1.00 \text{ coulomb (C)}}$
Electric current	$I = \dfrac{q}{t}$	$1.00 \text{ amp (A)} = \dfrac{1.00 \text{ coulomb (C)}}{1.00 \text{ second (s)}}$
Electrical resistance	$R = \dfrac{V}{I}$	$1.00 \text{ ohm } (\Omega) = \dfrac{1.00 \text{ volt (V)}}{1.00 \text{ amp (A)}}$
Electrical power	$P = IV$	$1.00 \text{ watt (W)} = \dfrac{C}{s} \times \dfrac{J}{C}$

*See Summary of Equations for more information.

If the watt power rating is not given it can be obtained by multiplying amps times volts. Also, since the time unit is in hours the time must be converted to the decimal equivalent of an hour if you want to know the cost of running an appliance for a number of minutes (x min/60 min).

Table 6.2 provides a summary of the electrical quantities and units.

EXAMPLE 6.7

What is the cost of operating a 100 W lightbulb for 1.00 h if the utility rate is $0.10 per kWh?

Solution

The power rating is given as 100 W, so the volt and amp units are not needed. Therefore,

$$IV = P = 100 \text{ W}$$
$$t = 1.00 \text{ h}$$
$$\text{rate} = \$0.10/\text{kWh}$$
$$\text{cost} = ?$$

$$\text{cost} = \frac{(\text{watts})(\text{time})(\text{rate})}{1,000 \dfrac{\text{watts}}{\text{kilowatt}}}$$

$$= \frac{(100 \text{ W})(1.00 \text{ h})(\$0.10/\text{kWh})}{1,000 \dfrac{\text{W}}{\text{kW}}}$$

$$= \frac{(100)(1.00)(0.10)}{1,000} \frac{\text{W}}{1} \times \frac{\text{h}}{1} \times \frac{\$}{\text{kWh}} \times \frac{\text{kW}}{\text{W}}$$

$$= \boxed{\$0.01}$$

The cost of operating a 100 W lightbulb at a rate of 10¢/kWh is 1¢/h.

EXAMPLE 6.8

An electric fan draws 0.5 A in a 120 V circuit. What is the cost of operating the fan if the rate is 10¢/kWh? (Answer: $0.006, which is 0.6 of a cent per hour)

Concepts Applied

Shocking Costs

You can predict the cost of running an electric appliance with just a few calculations. For example, suppose you want to know the cost of using a 1,300-watt hair dryer for 20 minutes if the utility charges 10 cents per kilowatt-hour. The equation would look like this:

$$\text{cost} = \frac{(1,300 \text{ W})(0.33 \text{ h})(\$0.10 / \text{kWh})}{1,000 \dfrac{\text{W}}{\text{kW}}}$$

Find answers to one or more of the following questions about the cost of running an electric appliance:

- What is your monthly electrical cost for watching television?
- What is the cost of drying your hair with a blow dryer?
- How much would you save by hanging out your clothes to dry rather than using an electric dryer?
- Compare the cost of using the following appliances: coffeemaker, toaster, can opener, vegetable steamer, microwave oven, and blender.
- How much does the electricity cost per month for the use of your desk lamp?
- Of all the electrical devices in a typical household, which three have the greatest monthly electrical cost?

Magnetism

The ability of a certain naturally occurring rock to attract iron has been known since at least 600 B.C. The early Greeks called this rock "Magnesian stone," since it was discovered near the ancient city of Magnesia in western Turkey. Knowledge about the iron-attracting properties of the Magnesian stone grew slowly. About A.D. 100, the Chinese learned to magnetize a piece of iron with a Magnesian stone, and sometime before A.D. 1000, they learned to use the magnetized iron or stone as a direction finder (compass). Today, the rock that attracts iron is known to be the black iron oxide mineral named *magnetite*.

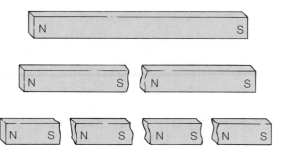

FIGURE 6.19

A bar magnet cut into halves always makes new, complete magnets with both a north and a south pole. The poles always come in pairs, and the separation of a pair into single poles, called monopoles, has never been accomplished.

FIGURE 6.18

Every magnet has ends, or poles, about which the magnetic properties seem to be concentrated. As this photo shows, more iron filings are attracted to the poles, revealing their location.

Magnetite is a natural magnet that strongly attracts iron and steel but also attracts cobalt and nickel. Such substances that are attracted to magnets are said to have *ferromagnetic properties*, or simply *magnetic* properties. Iron, cobalt, and nickel are considered to have magnetic properties, and most other common materials are considered not to have magnetic properties. Most of these nonmagnetic materials, however, are slightly attracted or slightly repelled by a strong magnet. In addition, certain rare earth elements, as well as certain metal oxides, exhibit strong magnetic properties.

Magnetic Poles

Every magnet has two **magnetic poles,** or ends, about which the force of attraction seems to be concentrated. Iron filings or other small pieces of iron are attracted to the poles of a magnet, for example, revealing their location (Figure 6.18). A magnet suspended by a string will turn, aligning itself in a north-south direction. The north-seeking pole is called the *north pole* of the magnet. The south-seeking pole is likewise named the *south pole* of the magnet. All magnets have both a north pole and a south pole, and neither pole can exist by itself. You cannot separate a north pole from a

south pole. If a magnet is broken into pieces, each new piece will have its own north and south poles (Figure 6.19).

You are probably familiar with the fact that two magnets exert forces on each other. For example, if you move the north pole of one magnet near the north pole of a second magnet, each will experience a repelling force. A repelling force also occurs if two south poles are moved close together. But if the north pole of one magnet is brought near the south pole of a second magnet, an attractive force occurs. The rule is, "*like magnetic poles repel and unlike magnetic poles attract.*"

A similar rule of "like charges repel and unlike charges attract" was used for electrostatic charges, so you might wonder if there is some similarity between charges and poles. The answer is they are not related. A magnet has no effect on a charged glass rod and the charged glass rod has no effect on either pole of a magnet.

Magnetic Fields

A magnet moved into the space near a second magnet experiences a magnetic force as it enters the **magnetic field** of the second magnet. A magnetic field can be represented by *magnetic field lines*. By convention, magnetic field lines are drawn to indicate how the *north pole* of a tiny imaginary magnet would point when in various places in the magnetic field. Arrowheads indicate the direction that the north pole would point, thus defining the direction of the magnetic field. The strength of the magnetic field is greater where the lines are closer together and weaker where they are farther apart. Figure 6.20 shows the magnetic field lines around the familiar bar magnet. Note that magnetic field lines emerge from the magnet at the north pole and enter the magnet at the south pole. Magnetic field lines always form closed loops.

The north end of a magnetic compass needle points north because the earth has a magnetic field. The earth's magnetic field is shaped and oriented as if there were a huge bar magnet inside the earth (Figure 6.21). The geographic North Pole is the axis of the earth's rotation, and this pole is used to determine the direction of true north on maps. A magnetic compass does not point to true north because the north magnetic

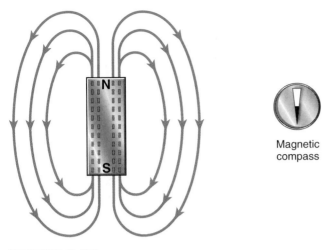

FIGURE 6.20

These lines are a map of the magnetic field around a bar magnet. The needle of a magnetic compass will follow the lines, with the north end showing the direction of the field.

Magnetic compass

pole and the geographic North Pole are in two different places. The difference is called the *magnetic declination*. The map in Figure 6.22 shows approximately how many degrees east or west of true north a compass needle will point in different locations. Magnetic declination must be considered when navigating with a compass. If you are navigating with a compass you might want to consider an up-to-date declination map. The magnetic north pole is continuously moving, so any magnetic declination map is probably a snapshot of how it used to be in the past.

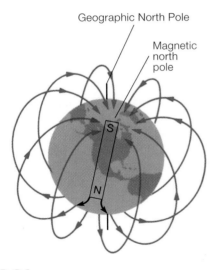

Geographic North Pole

Magnetic north pole

FIGURE 6.21

The earth's magnetic field. Note that the magnetic north pole and the geographic North Pole are not in the same place. Note also that the magnetic north pole acts as if the south pole of a huge bar magnet were inside the earth. You know that it must be a magnetic south pole, since the north end of a magnetic compass is attracted to it, and opposite poles attract.

Note in Figure 6.21 that the earth's magnetic field acts as if there is a huge bar magnet inside the earth with a *south* magnetic pole near the earth's geographic North Pole. This is not an error. The north pole of a magnet is attracted to the south pole of a second magnet, and the north pole of a compass needle points to the north. Therefore, the bar magnet must be arranged as shown. This apparent contradiction is a result of naming the magnetic poles after their "seeking" direction.

The typical compass needle pivots in a horizontal plane, moving to the left or right without up or down motion. Inspection of Figure 6.22, however, shows that the earth's magnetic field is horizontal to the surface only at the magnetic equator. A compass needle that is pivoted so that it moves only up and down will be horizontal only at the magnetic equator. Elsewhere, it shows the angle of the field from the horizontal, called the *magnetic dip*. The angle of dip is the vertical component of the earth's magnetic field. As you travel from the equator, the angle of magnetic dip increases from zero to a maximum of 90° at the magnetic poles.

The Source of Magnetic Fields

The observation that like magnetic poles repel and unlike magnetic poles attract might remind you of the forces involved with like and unlike charges. Recall that electric charges exist as single isolated units of positive protons and units of negative electrons. An object becomes electrostatically charged when charges are separated, and the object acquires an excess or deficiency of negative charges. You might wonder, by analogy, if the poles of a magnet are similarly made up of an excess or deficiency of magnetic poles. The answer is no; magnetic poles are different from electric charges. Positive and negative charges *can* be separated and isolated. But suppose that you try to separate and isolate the poles of a magnet by cutting a magnet into two halves. Cutting a magnet in half will produce two new magnets, each with north and south poles. You could continue cutting each half into new halves, but each time the new half will have its own north and south poles (Figure 6.19). It seems that no subdivision will ever separate and isolate a single magnetic pole, called a *monopole*. Magnetic poles always come in matched pairs of north and south, and a monopole has never been found. The two poles are always found to come together, and as it is understood today, magnetism is thought to be produced by *electric currents*, not an excess of monopoles. The modern concept of magnetism is electric in origin, and magnetism is understood to be a property of electricity.

The key discovery about the source of magnetic fields was reported in 1820 by a Danish physics professor named Hans Christian Oersted. Oersted found that a wire conducting an electric current caused a magnetic compass needle below the wire to move. When the wire was not connected to a battery the needle of the compass was lined up with the wire and pointed north as usual. But when the wire was connected to a battery, the compass needle moved perpendicular to the wire (Figure 6.23). Oersted had discovered that an electric current produces a magnetic field. An electric current is understood to be the movement of electric charges, so Oersted's discovery suggested that magnetism is a property of charges in motion.

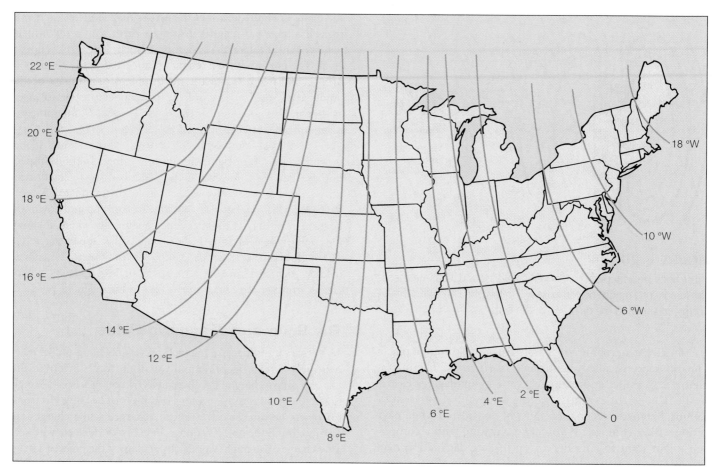

FIGURE 6.22

This magnetic declination map shows the approximate number of degrees east or west of the true geographic north that a magnetic-compass will point in various locations.

Permanent Magnets

The magnetic fields of bar magnets, horseshoe magnets, and other so-called permanent magnets are explained by the relationship between magnetism and moving charges. Electrons in atoms are moving around the nucleus, so they produce a magnetic field. Electrons also have a magnetic field associated with their spin. In most materials these magnetic fields cancel one another and neutralize the overall magnetic effect. In other materials, such as iron, cobalt, and nickel, the electrons are arranged and oriented in a complicated way that imparts a magnetic property to the atomic structure. These atoms are grouped in a tiny region called a **magnetic domain.** A magnetic domain is roughly 0.01 to 1 mm in length or width and does not have a fixed size (Figure 6.24). The atoms in each domain are magnetically aligned, contributing to the polarity of the domain. Each domain becomes essentially a tiny magnet with a north and south pole. In an unmagnetized piece of iron the domains are oriented in all possible directions and effectively cancel any overall magnetic effect. The net magnetism is therefore zero or near zero.

When an unmagnetized piece of iron is placed in a magnetic field, the orientation of the domain changes to align with the magnetic field, and the size of aligned domains may grow at the expense of unaligned domains. This explains why a "string" of iron paper clips is picked up by a magnet. Each paper clip has domains that become temporarily and slightly aligned by the magnetic field, and each paper clip thus acts as a temporary magnet while in the field of the magnet. In a strong magnetic field, the size of the aligned domains grows to such an extent that the paper clip becomes a "permanent magnet." The same result can be achieved by repeatedly stroking a paper clip with the pole of a magnet. The magnetic effect of a "permanent magnet" can be reduced or destroyed by striking, dropping, or

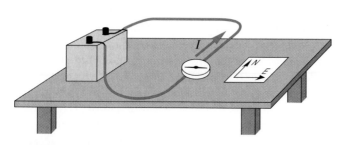

FIGURE 6.23

With the wire oriented along a north-south line, the compass needle deflects away from this line when there is a current in the wire.

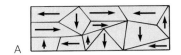

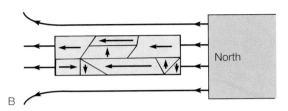

North

FIGURE 6.24

(A) In an unmagnetized piece of iron, the magnetic domains have a random arrangement that cancels any overall magnetic effect. (B) When an external magnetic field is applied to the iron, the magnetic domains are realigned, and those parallel to the field grow in size at the expense of the other domains, and the iron is magnetized.

heating the magnet to a sufficiently high temperature (770°C for iron). These actions randomize the direction of the magnetic domains, and the overall magnetic field disappears.

Earth's Magnetic Field

Earth's magnetic field is believed to originate deep within the earth. Like all other magnetic fields, Earth's magnetic field is believed to originate with moving charges. Earthquake waves and other evidence suggest that the earth has a solid inner core with a radius of about 1,200 km (about 750 mi), surrounded by a fluid outer core some 2,200 km (about 1,400 mi) thick. This core is probably composed of iron and nickel, which flows as the earth rotates, creating electric currents that result in Earth's magnetic field. How the electric currents are generated is not yet understood.

Other planets have magnetic fields, and there seems to be a relationship between the rate of rotation and the strength of the planet's magnetic field. Jupiter and Saturn rotate faster than Earth and have stronger magnetic fields than Earth. Venus and Mercury rotate more slowly than Earth and have weaker magnetic fields. This is indirect evidence that the rotation of a planet is associated with internal fluid movements, which somehow generate electric currents and produce a magnetic field.

In addition to questions about how the electric current is generated, there are puzzling questions from geologic evidence. Lava contains magnetic minerals that act like tiny compasses that are oriented to the earth's magnetic field when the lava is fluid but become frozen in place as the lava cools. Studies of these rocks by geologic dating and studies of the frozen magnetic mineral orientation show that Earth's magnetic field has undergone sudden reversals in polarity: the north magnetic pole becomes the south magnetic pole and vice versa. This has happened many times over the distant geologic past, and the most recent shift occurred about 780,000 years ago. The cause of such magnetic field reversals is unknown, but it must be related to changes in the flow patterns of the earth's fluid outer core of iron and nickel.

Electric Currents and Magnetism

As Oersted discovered, electric charges in motion produce a magnetic field. The direction of the magnetic field around a current-carrying wire can be determined by using a magnetic compass. The north-seeking pole of the compass needle will point in the direction of the magnetic field lines. If you move the compass around the wire the needle will always move to a position that is tangent to a circle around the wire. Evidently the magnetic field lines are closed concentric circles that are at right angles to the length of the wire (Figure 6.25).

Current Loops

The magnetic field around a current-carrying wire will interact with another magnetic field, one formed around a permanent magnet or one from a second current-carrying wire. The two fields interact, exerting forces just like the forces between the fields of two permanent magnets. The force could be increased by increasing the current, but there is a more efficient way to obtain a larger force. A current-carrying wire that is formed into a loop has perpendicular, circular field lines that pass through the inside of the loop in the same direction. This has the effect of concentrating the field lines, which increases the magnetic field intensity. Since the field lines all pass through the loop in the same direction, one side of the loop will have a north pole and the other side a south pole (Figure 6.26).

Many loops of wire formed into a cylindrical coil are called a *solenoid*. When a current passes through the loops of wire in a solenoid, each loop contributes field lines along the length of the cylinder (Figure 6.27). The overall effect is a magnetic field around the solenoid that acts just like the magnetic field of a bar magnet. This magnet, called an **electromagnet,** can be turned on or off by turning the current on or off. In addition, the strength of the electromagnet depends on the magnitude of the current and the number of loops (ampere-turns). The strength of the electromagnet can also be increased by placing a piece of soft iron in the coil. The domains of the iron become aligned by the influence of the magnetic field. This induced magnetism increases the overall

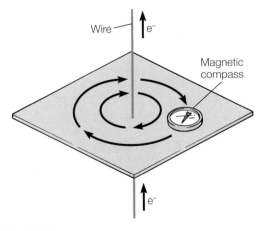

FIGURE 6.25

A magnetic compass shows the presence and direction of the magnetic field around a straight length of current-carrying wire.

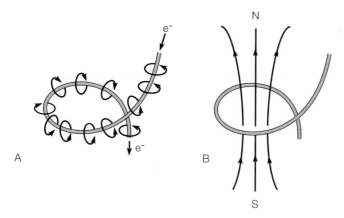

FIGURE 6.26

(A) Forming a wire into a loop causes the magnetic field to pass through the loop in the same direction. (B) This gives one side of the loop a north pole and the other side a south pole.

magnetic field strength of the solenoid as the magnetic field lines are gathered into a smaller volume within the core.

Applications of Electromagnets

The discovery of the relationship between an electric current, magnetism, and the resulting forces created much excitement in the 1820s and 1830s. This excitement was generated because it was now possible to explain some seemingly separate phenomena in terms of an interrelationship and because people began to see practical applications almost immediately. Within a year of Oersted's discovery, André Ampère had fully explored the

magnetic effects of currents, combining experiments and theory to find the laws describing these effects. Soon after Ampère's work, the possibility of doing mechanical work by sending currents through wires was explored. The electric motor, similar to motors in use today, was invented in 1834, only fourteen years after Oersted's momentous discovery.

The magnetic field produced by an electric current is used in many practical applications, including electrical meters, electromagnetic switches that make possible the remote or programmed control of moving mechanical parts, and electric motors. In each of these applications, an electric current is applied to an electromagnet.

Electric Meters

Since you cannot measure electricity directly, it must be measured indirectly through one of the effects that it produces. The strength of the magnetic field produced by an electromagnet is proportional to the electric current in the electromagnet. Thus, one way to measure a current is to measure the magnetic field that it produces. A device that measures currents from their magnetic fields is called a *galvanometer* (Figure 6.28). A galvanometer has a coil of wire that can rotate on pivots in the magnetic field of a permanent magnet. The coil has an attached pointer that moves across a scale and control springs that limit its motion and return the pointer to zero when there is no current. When there is a current in the coil the electromagnetic field is attracted and repelled by the field of the permanent magnet. The larger the current the greater the force and the more the coil will rotate until it reaches an equilibrium position with the control springs. The amount of movement of the coil (and thus the pointer) is proportional to the current in the coil. With

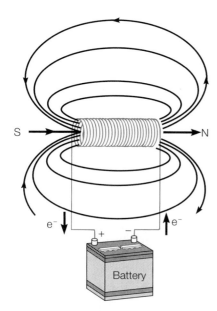

FIGURE 6.27

When a current is run through a cylindrical coil of wire, a solenoid, it produces a magnetic field like the magnetic field of a bar magnet.

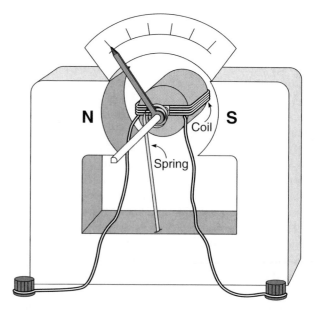

FIGURE 6.28

A galvanometer consists of a coil of wire, a permanent magnet, and a restoring spring to return the needle to zero when there is no current through the coil.

Lemon Battery

1. You can make a simple compass galvanometer that will detect a small electric current (Box Figure 6.2). All you need is a magnetic compass and some thin insulated wire (the thinner the better).

2. Wrap the thin insulated wire in parallel windings around the compass. Make as many parallel windings as you can, but leave enough room to see both ends of the compass needle. Leave the wire ends free for connections.

3. To use the galvanometer, first turn the compass so the needle is parallel to the wire windings. When a current passes through the coil of wire the magnetic field produced will cause the needle to move from its north-south position, showing the presence of a current. The needle will deflect one way or the other depending on the direction of the current.

4. Test your galvanometer with a "lemon battery." Roll a soft lemon on a table while pressing on it with the palm of your hand. Cut two slits in the lemon about 1 cm apart. Insert a 8-cm (approximate) copper wire in one slit and a same-sized length of a straightened paperclip in the other slit, making sure the metals do not touch inside the lemon. Connect the galvanometer to the two metals. Try the two metals in other fruits, vegetables, and liquids. Can you find a pattern?

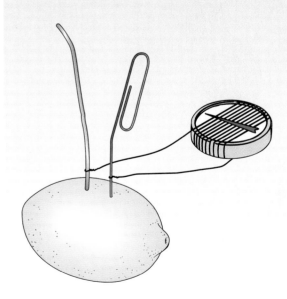

BOX FIGURE 6.2

You can use the materials shown here to create and detect an electric current.

certain modifications and applications, the galvanometer can be used to measure current (ammeter), potential difference (voltmeter), and resistance (ohmmeter).

Electromagnetic Switches

A *relay* is an electromagnetic switch device that makes possible the use of a low-voltage control current to switch a larger, high-voltage circuit on and off (Figure 6.29). A thermostat, for example, utilizes two thin, low-voltage wires in a glass tube of mercury. The glass tube of mercury is attached to a metal coil that expands and contracts with changes in temperature, tipping the attached glass tube. When the temperature changes enough to tip the glass tube, the mercury flows to the bottom end, which makes or breaks contact with the two wires, closing or opening the circuit. When contact is made, a weak current activates an electromagnetic switch, which closes the circuit on the large-current furnace or heat pump motor.

A solenoid is a coil of wire with a current. Some solenoids have a spring-loaded movable piece of iron inside. When a current flows in such a coil the iron is pulled into the coil by the magnetic field, and the spring returns the iron when the current is turned off. This device could be utilized to open a water valve, turning the hot or cold water on in a washing machine or dishwasher, for example. Solenoids are also used as mechanical switches on VCRs, automobile starters, and signaling devices such as bells and buzzers. The dot matrix computer printer works with a group of small solenoids that are activated by

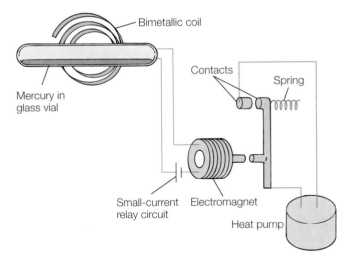

FIGURE 6.29

A schematic of a relay circuit. The mercury vial turns as changes in temperature expand or contract the coil, moving the mercury and making or breaking contact with the relay circuit. When the mercury moves to close the relay circuit, a small current activates the electromagnet, which closes the contacts on the large-current circuit.

electric currents from the computer. Seven or more of these small solenoids work together to strike a print ribbon, forming the letters or images as they rapidly move in and out.

Telephones and Loudspeakers

The mouthpiece of a typical telephone contains a cylinder of carbon granules with a thin metal diaphragm facing the front. When someone speaks into the telephone, the diaphragm moves in and out with the condensations and rarefactions of the sound wave (Figure 6.30). This movement alternately compacts and loosens the carbon granules, increasing and decreasing the electric current that increases and decreases with the condensations and rarefactions of the sound waves.

The moving electric current is fed to the earphone part of a telephone at another location. The current runs through a coil of wire that attracts and repels a permanent magnet attached to a speaker cone. When repelled forward, the speaker cone makes a condensation, and when attracted back the cone makes a rarefaction. The overall result is a series of condensations and rarefactions that, through the changing electric current, accurately match the sounds made by the other person.

The loudspeaker in a radio or stereo system works from changes in an electric current in a similar way, attracting and repelling a permanent magnet attached to the speaker cone. You can see the speaker cone in a large speaker moving back and forth as it creates condensations and rarefactions.

Electric Motors

An electric motor is an electromagnetic device that converts electrical energy to mechanical energy. Basically, a motor has two working parts, a stationary electromagnet called a *field magnet* and a cylindrical, movable electromagnet called an *armature.* The armature is on an axle and rotates in the magnetic field of the field magnet. The axle turns fan blades, compressors, drills, pulleys, or other devices that do mechanical work.

Different designs of electric motors are used for various applications, but the simple demonstration motor shown in Figure 6.31 can be used as an example of the basic operating principle. Both the field coil and the armature are connected to an electric current. The armature turns, and it receives the current through a *commutator* and *brushes.* The brushes are contacts that brush against the commutator as it rotates, maintaining contact. When the current is turned on, the field coil and the armature become electromagnets, and the unlike poles attract, rotating the armature. If the current is dc, the armature would turn no farther, stopping as it does in a galvanometer. But the commutator has insulated segments so when it turns halfway, the commutator segments switch brushes and there is a current through the armature in the *opposite* direction. This switches the armature poles, which are now repelled for another half-turn. The commutator again reverses the polarity, and the motion continues in one direction. An actual motor has many coils (called "windings") in the armature to obtain a useful force, and many commutator segments. This gives the motor a smoother operation with a greater turning force.

Electromagnetic Induction

So far, you have learned that (1) a moving charge and a current-carrying wire produce a magnetic field; (2) a second magnetic field exerts a force on a moving charge and exerts a force on a

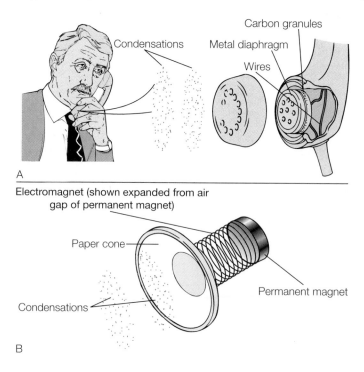

FIGURE 6.30

(A) Sound waves are converted into a changing electrical current in a telephone. (B) Changing electrical current can be changed to sound waves in a speaker by the action of an electromagnet pushing and pulling on a permanent magnet. The electromagnet is attached to a stiff paper cone or some other material that makes sound waves as it moves in and out.

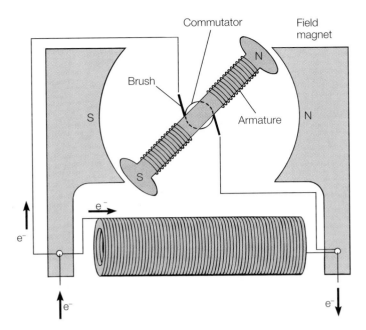

FIGURE 6.31
A schematic of a simple electric motor.

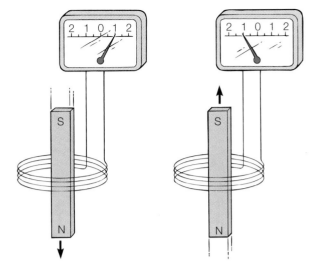

FIGURE 6.32
A current is induced in a coil of wire moved through a magnetic field. The direction of the current depends on the direction of motion.

current-carrying wire as their magnetic fields interact; and (3) the direction of the maximum force produced on a moving charge or moving charges is at right angles to their velocity and to the interacting magnetic field lines.

Soon after the discovery of these relationships by Oersted and Ampère, people began to wonder if the opposite effect was possible; that is, would a magnetic field produce an electric current? The discovery was made independently in 1831 by Joseph Henry in the United States and by Michael Faraday in England. They found that *if a loop of wire is moved in a magnetic field, or if the magnetic field is changed, a voltage is induced in the wire.* The voltage is called an *induced voltage,* and the resulting current in the wire is called an *induced current.* The overall interaction is called **electromagnetic induction.**

One way to produce electromagnetic induction is to move a bar magnet into or out of a coil of wire (Figure 6.32). A galvanometer shows that the induced current flows one way when the bar magnet is moved toward the coil and flows the other way when the bar magnet is moved away from the coil. The same effect occurs if you move the coil back and forth over a stationary magnet. Furthermore, no current is detected when the magnetic field and the coil of wire are not moving. Thus, electromagnetic induction depends on the relative motion of the magnetic field and the coil of wire. It does not matter which moves or changes, but one must move or change relative to the other for electromagnetic induction to occur.

Electromagnetic induction occurs when the loop of wire moves across magnetic field lines or when magnetic field lines move across the loop. The magnitude of the induced voltage is proportional to (1) the number of wire loops passing through the magnetic field lines, (2) the strength of the magnetic field, and (3) the rate at which magnetic field lines pass through the wire.

Simple Generator

1. Make a coil of wire from insulated bell wire (#18 copper wire) by wrapping fifty windings around a cardboard tube from a roll of paper. Tape the coil at several places so it does not come apart, and discard the cardboard tube.

2. Make a current-detecting instrument from a magnetic compass and some thin insulated wire (the thinner the better). Wrap the thin insulated wire in parallel windings around the compass. Make as many parallel windings as you can, but leave enough room to see both ends of the compass needle. Connect the wire ends to the coil you made in step 1.

3. Orient the compass so the needle is parallel to the wire around the compass. When a current passes through the coil of wire, the magnetic field produced will cause the needle to move, showing the presence of a current.

4. First, move a bar magnet into and out of the stationary coil of wire and observe the compass needle. Second, move the coil of wire back and forth over a stationary bar magnet and observe the compass needle.

5. Experiment with a larger coil of wire, bar magnets of greater or weaker strengths, and by moving the coil at varying speeds. See how many generalizations you can make concerning electromagnetic induction.

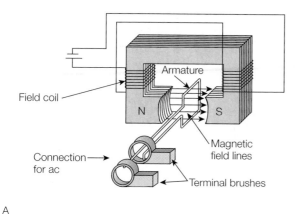

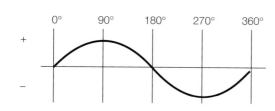

A

B

FIGURE 6.33

(A) Schematic of a simple alternator (ac generator) with one output loop. (B) Output of the single loop turning in a constant magnetic field, which alternates the induced current each half-cycle.

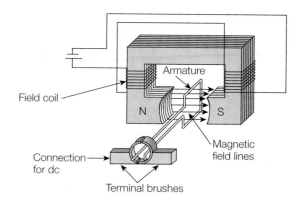

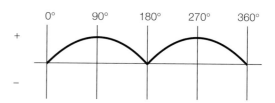

A

B

FIGURE 6.34

(A) Schematic of a simple dc generator with one output loop. (B) Output of the single loop turning in a constant magnetic field. The split ring (commutator) reverses the sign of the output when the voltage starts to reverse, so the induced current has half-cycle voltages of a constant sign, which is the definition of direct current.

Generators

Soon after the discovery of electromagnetic induction the **electric generator** was developed. The generator is essentially an axle with many wire loops that rotates in a magnetic field. The axle is turned by some form of mechanical energy, such as a water turbine or a steam turbine, which uses steam generated from fossil fuels or nuclear energy. The use of the electric generator began in the 1890s, when George Westinghouse designed and built a waterwheel-powered 75 kilowatt ac generator near Ouray, Colorado (Figure 6.33). Thomas Edison built a dc power plant in New York City about the same time (Figure 6.34). In 1896 a 3.7 megawatt power plant at Niagara Falls sent electrical power to factories in Buffalo, New York. The factories no longer had to be near a source of energy such as a waterfall, since electric energy could now be transmitted by wires.

Transformers

The production of electrical power began in the United States with generators built by George Westinghouse and Thomas Edison. A controversy arose, however, because Edison built dc generators, and Westinghouse built ac generators. Edison believed that alternating current was dangerous and argued for the use of direct current only. Alternating current eventually won because (1) the voltage of ac could be easily changed to meet

different applications, but the voltage of dc could *not* be easily changed, and (2) dc transmission suffered from excessive power losses in transmission while ac power was made possible by a **transformer,** a device that uses electromagnetic induction to increase or decrease ac voltage.

A transformer has two basic parts: (1) a *primary coil,* which is connected to a source of alternating current, and (2) a *secondary coil,* which is close by. Both coils are often wound on a single iron core but are always fully insulated from each other. When there is an alternating current through the primary coil, a magnetic field grows around the coil to a maximum size, collapses to zero, then grows to a maximum size with an opposite polarity. This happens 120 times a second as the alternating current oscillates at 60 hertz. The magnetic field is strengthened and directed by the iron core. The growing and collapsing magnetic field moves across the wires in the secondary coil, inducing a voltage in the secondary coil. The growing and collapsing magnetic field from the primary coil thus induces a voltage in the secondary coil, just as an induced voltage occurs in the wire loops of a generator.

The transformer increases or decreases the voltage in an alternating current because the magnetic field grows and collapses past the secondary coil, inducing a voltage. If a direct current is applied to the primary coil the magnetic field

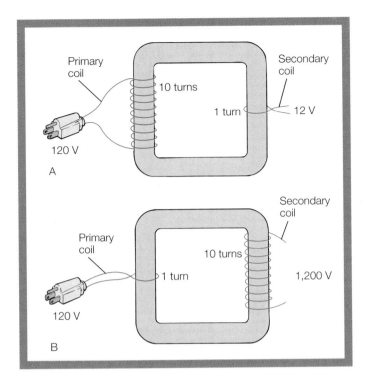

FIGURE 6.35

(A) This step-down transformer has ten turns on the primary for each turn on the secondary and reduces the voltage from 120 V to 12 V. (B) This step-up transformer increases the voltage from 120 V to 1,200 V, since there are ten turns on the secondary to each turn on the primary.

grows around the primary coil as the current is established but then becomes stationary. Recall that electromagnetic induction occurs when there is relative motion between the magnetic field lines and a wire loop. Thus, an induced voltage occurs from a direct current (1) only for an instant when the current is established and the growing field moves across the secondary coil and (2) only for an instant when the current is turned off and the field collapses back across the secondary coil. In order to use dc in a transformer, the current must be continually interrupted to produce a changing magnetic field.

When an alternating current or a continually interrupted direct current is applied to the primary coil, the magnitude of the induced voltage in the secondary coil is proportional to the ratio of wire loops in the two coils. If they have the same number of loops, the primary coil produces just as many magnetic field lines as are intercepted by the secondary coil. In this case, the induced voltage in the secondary coil will be the same as the voltage in the primary coil. Suppose, however, that the secondary coil has one-tenth as many loops as the primary coil. This means that the secondary loops will move across one-tenth as many field lines as the primary coil produces. As a result, the induced voltage in the secondary coil will be one-tenth the voltage in the primary coil. This is called

a *step-down transformer* because the voltage was stepped down in the secondary coil. On the other hand, more wire loops in the secondary coil will intercept more magnetic field lines. If the secondary coil has ten times *more* loops than the primary coil, then the voltage will be *increased* by a factor of 10. This is a *step-up transformer*. How much the voltage is stepped up or stepped down depends on the ratio of wire loops in the primary and secondary coils (Figure 6.35). Note that the *volts per wire loop* are the same in each coil. The relationship is

$$\frac{\text{volts}_{\text{primary}}}{(\text{number of loops})_{\text{primary}}} = \frac{\text{volts}_{\text{secondary}}}{(\text{number of loops})_{\text{secondary}}}$$

or

$$\frac{V_p}{N_p} = \frac{V_s}{N_s}$$

equation 6.8

EXAMPLE 6.9

A step-up transformer has five loops on its primary coil and twenty loops on its secondary coil. If the primary coil is supplied with an alternating current at 120 V, what is the voltage in the secondary coil?

Solution

$N_p = 5 \text{ loops}$ $\qquad \frac{V_p}{N_p} = \frac{V_s}{N_s} \quad \therefore \quad V_s = \frac{V_p N_s}{N_p}$

$N_s = 20 \text{ loops}$

$V_p = 120 \text{ V}$ $\qquad V_s = \frac{(120 \text{ V})(20 \text{ loops})}{5 \text{ loops}}$

$V_s = ?$

$\qquad\qquad = \frac{120 \times 20}{5} \frac{\text{V·loops}}{\text{loops}}$

$\qquad\qquad = \boxed{480 \text{ V}}$

A step-up or step-down transformer steps up or steps down the *voltage* of an alternating current according to the ratio of wire loops in the primary and secondary coils. Assuming no losses in the transformer, the *power input* on the primary coil equals the *power output* on the secondary coil. Since $P = IV$, you can see that when the voltage is stepped up the current is correspondingly decreased, as

$$\text{power input} = \text{power output}$$

$$\text{watts input} = \text{watts output}$$

$$(\text{amps} \times \text{volts})_{\text{in}} = (\text{amps} \times \text{volts})_{\text{out}}$$

or

$$V_p I_p = V_s I_s$$

equation 6.9

EXAMPLE 6.10

The step-up transformer in example 6.9 is supplied with an alternating current at 120 V and a current of 10.0 A in the primary coil. What current flows in the secondary circuit?

Solution

$V_p = 120$ V

$I_p = 10.0$ A

$V_s = 480$ V

$I_s = ?$

$V_p I_p = V_s I_s \quad \therefore \quad I_s \frac{V_p I_p}{V_s}$

$I_s = \frac{120 \text{ V} \times 10.0 \text{ A}}{480 \text{ V}}$

$= \frac{120 \times 10.0}{480} \frac{\text{V} \cdot \text{A}}{\text{V}}$

$= \boxed{2.5 \text{ A}}$

1. Make a coil of wire from insulated bell wire (#18 copper wire) by wrapping fifty windings around a narrow jar. Tape the coil at several places so it does not come apart.
2. Now make a second coil of wire from insulated bell wire and tape the coil as before.
3. Suspend both coils of wire on separate ring stands or some other support on a table top. The coils should hang so they will swing with the broad circle of the coil moving back and forth. Place a large magnet on supports so it is near the center of each coil (Box Figure 6.3).
4. Connect the two coils of wire.
5. Move one of the coils of wire and observe what happens to the second coil. The second coil should move, mirroring the movements of the first coil (if it does not move, find some stronger magnets).
6. Explain what happens in terms of magnetic fields and currents at the first coil and at the second coil.

Concepts Applied

Swinging Coils

The interactions between moving magnets and moving charges can be easily demonstrated with two large magnets and two coils of wire.

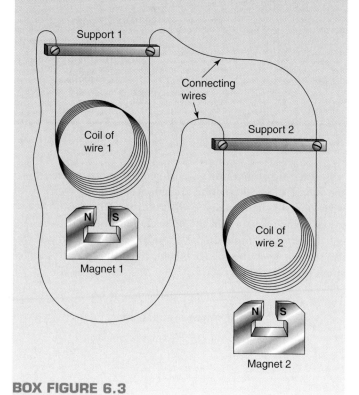

BOX FIGURE 6.3

Energy losses in transmission are reduced by stepping up the voltage. Recall that electrical resistance results in an energy loss and a corresponding absolute temperature increase in the conducting wire. If the current is large, there are many collisions between the moving electrons and positive ions of the wire, resulting in a large energy loss. Each collision takes energy from the electric field, diverting it into increased kinetic energy of the positive ions and thus increased temperature of the conductor. The energy lost to resistance is therefore reduced by *lowering* the current, which is what a transformer does by increasing the voltage. Hence electric power companies step up the voltage of generated power for economical transmission. A step-up transformer at a power plant, for example, might step up the voltage from 22,000 volts to 500,000 volts for transmission across the country to a city. This step up in voltage correspondingly reduces the current, lowering the resistance losses to a more acceptable 4 or 5 percent over long distances. A step-down transformer at a substation near the city reduces the voltage to several thousand volts for transmission around the city. Additional step-down transformers reduce this voltage to 120 volts for transmission to three or four houses (Figure 6.36).

A

B

FIGURE 6.36

Energy losses in transmission are reduced by increasing the voltage, so the voltage of generated power is stepped up at the power plant. (A) These transformers, for example, might step up the voltage from tens to hundreds of thousands of volts. After a step-down transformer reduces the voltage at a substation, still another transformer (B) reduces the voltage to 120 for transmission to three or four houses.

You may be familiar with many solid-state devices such as calculators, computers, word processors, digital watches, VCRs, digital stereos, and camcorders. All of these are called solid-state devices because they use a solid material, such as the semiconductor silicon, in an electric circuit in place of vacuum tubes. Solid-state technology developed from breakthroughs in the use of semiconductors during the 1950s, and the use of thin pieces of silicon crystal is common in many electric circuits today.

A related technology also uses thin pieces of a semiconductor such as silicon but not as a replacement for a vacuum tube. This technology is concerned with photovoltaic devices, also called *solar cells*, that generate electricity when exposed to light (Box Figure 6.4). A solar cell is unique in generating electricity since it produces electricity directly, without moving parts or chemical reactions, and potentially has a very long lifetime. This reading is concerned with how a solar cell generates electricity.

Light is an electromagnetic wave in a certain wavelength range. This wave is a pulse of electric and magnetic fields that are locked together, moving through space as they exchange energy back and forth, regenerating each other in an endless cycle until the wave is absorbed, giving up its energy. Light also has the ability to give its energy to an electron, knocking it out of a piece of metal. This phenomenon, called

A   B

BOX FIGURE 6.4

Solar cells are economical in remote uses such as (*A*) navigational aids, and (*B*) communications. The solar panels in both of these examples are oriented toward the south.

the *photoelectric effect*, is described in the next chapter.

Solid-state technology is mostly concerned with crystals and electrons in crystals. A *crystal* is a solid with an ordered, three-dimensional arrangement of atoms. A crystal of common table salt, sodium chloride, has an ordered arrangement that pro-duces a cubic structure. You can see this cubic structure of table salt if you look closely at a few grains. The regular geometric arrangement of atoms (or ions) in a crystal is called the *lattice,* or framework, of the crystal structure.

—*Continued top of next page*

Continued—

A highly simplified picture of a crystal of silicon has each silicon atom bonded with four other silicon atoms, with each pair of atoms sharing two electrons. Normally, this ties up all the electrons and none are free to move and produce an electric current. This happens in the dark in silicon crystals, and the silicon crystal is an insulator. Light, however, can break electrons free in a silicon crystal, so that they can move in a current. Silicon is therefore a semiconductor.

The conducting properties of silicon can be changed by *doping,* that is, artificially forcing atoms of other elements into the crystal lattice. Phosphorus, for example, has five electrons in its outermost shell compared to the four in a silicon atom. When phosphorus atoms replace silicon atoms in the crystal lattice, there are extra electrons not tied up in the two electron bonds. The extra electrons move easily through the crystal lattice, carrying a charge. Since the phosphorus-doped silicon carries a negative charge it is called an *n-type* semiconductor. The n means negative charge carrier.

A silicon crystal doped with boron will have atoms in the lattice with only three electrons in the outermost shell. This results in a deficiency, that is, electron "holes" that act as positive charges. A hole can move as an electron is attracted to it, but it leaves another hole elsewhere, where it moved from. Thus, a flow of electrons in one direction is equivalent to a flow of holes in the opposite direction. A hole, therefore, behaves as a positive charge. Since the boron-doped silicon carries a positive charge it is called a *p-type* semiconductor. The p means positive charge carrier.

The basic operating part of a silicon solar cell is typically an 8 cm wide and 3×10^{-1} mm (about one-hundredth of an inch) thick wafer cut from a silicon crystal. One side of the wafer is doped with boron to make p-silicon, and the other side is doped with phosphorus to make n-silicon. The place of contact between the two is called the p-n junction, which creates a *cell barrier.* The cell barrier forms as electrons are attracted from the n-silicon to the holes in the p-silicon. This creates a very thin zone of negatively charged p-silicon and positively charged n-silicon (Box Figure 6.5). Thus, an internal electric field is established at the p-n junction, and the field is the cell barrier. The barrier is about 3×10^{-5} mm (about one-millionth of an inch) thick.

A metal base plate is attached to the p-silicon side of the wafer, and a grid of metal contacts to the n-silicon side for electrical contacts. The grid is necessary to allow light into the cell. The entire cell is then coated with a transparent plastic covering.

The cell is thin, and light can penetrate through the p-n junction. Light strikes the p-silicon, freeing electrons. Low-energy free electrons might combine with a hole, but high-energy electrons cross the cell barrier into the n-silicon. The electron loses some of

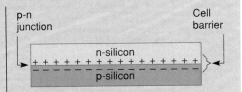

BOX FIGURE 6.5

The cell barrier forms at the p-n junction between the n-silicon and the p-silicon. The barrier creates a "one-way" door that accumulates negative charges in the n-silicon.

its energy, and the barrier prevents it from returning, creating an excess negative charge in the n-silicon and a positive charge in the p-silicon. This establishes a potential of about 0.5 volt, which will drive a 2 amp current (8 cm cell) through a circuit connected to the electrical contacts.

Solar cells are connected different ways for specific electrical energy requirements in about a 1 meter arrangement called a *module.* Several modules are used to make a solar *panel,* and panels are the units used to design a solar cell *array.*

Today, solar cells are essentially handmade and are economical only in remote power uses (navigational aids, communications, or irrigation pumps) and in consumer specialty items (solar-powered watches and calculators). Research continues on finding methods of producing highly efficient, highly reliable solar cells that are affordably priced.

Michael Faraday (1791–1867)

Michael Faraday was a British physicist and chemist who is often regarded as the greatest experimental scientist of the 1800s. He made pioneering contributions to electricity, inventing the electric motor, electric generator and the transformer, and discovering electromagnetic induction and the laws of electrolysis. He also discovered benzene and was the first to observe that the plane of polarization of light is rotated in a magnetic field.

Faraday was born in Newington, Surrey, on September 22, 1791. His father was a poor blacksmith, who went to London to seek work in the year that Faraday was born. Faraday received only a rudimentary education as a child and although he was literate, he gained little knowledge of mathematics. At the age of 14, he became an apprentice to a bookbinder in London and began to read voraciously. The article on electricity in the *Encyclopedia Britannica* fascinated him in particular, for it presented the view that electricity is a kind of vibration, an idea that was to remain with Faraday. He also read the works of Lavoisier and became interested in chemistry, and carried out what scientific experiments he could put together with his limited resources. He was aided by a manual dexterity gained from his trade, which also stood him in great stead in his later experimental work.

In 1810 Faraday was introduced to the City Philosophical Society and there received a basic grounding in science, attending lectures on most aspects of physics and chemistry and carrying out some experimental work. He also attended the Royal Institution, where he was enthralled by the lectures and demonstrations given by Humphry Davy (1778–1829). He made notes eagerly, assembling them at work into finely bound books. In 1812 Faraday came to the end of his apprenticeship and prepared to devote himself to his trade, not expecting to make a career in science. Almost immediately, however, there came an

extraordinary stroke of luck. Davy was temporarily blinded by an explosion in a chemistry experiment and asked Faraday to help him until he regained his sight. When he recovered, Faraday sent Davy the bound notes of his lectures. Impressed by the young man, Davy marked him out as his next permanent assistant at the Royal Institution and Faraday took up this post in 1813.

This was remarkably good fortune for Faraday, because Davy was a man of wide-ranging interests and great scientific insight as well as a brilliant exponent of ideas. Furthermore, Davy undertook a tour of France and Italy from 1813 to 1815 to visit the leading scientists of the day, including the pioneer of current electricity Alessandro Volta (1745–1827). Faraday accompanied Davy, gaining an immense amount of knowledge, and on his return to London threw himself wholeheartedly into scientific research.

Faraday remained at the Royal Institution and made most of his pioneering discoveries in chemistry and electricity there over the next 20 years. He became a great

popularizer of science with his lectures at the Royal Institution, which he began in 1825 and continued until 1862. His fame grew rapidly, soon eclipsing even that of Davy, who became embittered as a result. But the strain of his restless pursuit of knowledge told and in 1839 Faraday suffered a breakdown. He never totally recovered but at the instigation of Lord Kelvin (1824–1907), he returned to research in 1845 and made his important discoveries of the effect of magnetism on light and developed his field theory. In the 1850s, Faraday's mind began to lose its sharp grip, possibly as a result of low-grade poisoning caused by his chemical researches, and he abandoned research and then finally lecturing. He resigned from the Royal Institution in 1862 and retired to an apartment provided for him at Hampton Court, Middlesex, by Queen Victoria. He died there on August 25, 1867.

Faraday was mainly interested in chemistry during his early years at the Royal In-stitution. He investigated the effects of including precious metals in steel in 1818, producing high-quality alloys that later stimulated the production of special high-grade steels. Faraday's first serious chemical discoveries were made in 1820, when he prepared the chlorides of carbon—C_2Cl_6 from ethane and C_2Cl_4 from ethylene (ethene)—substitution reactions that anticipated the work a few years later by Jean Baptiste Dumas. In 1823 Faraday produced liquid chlorine by heating crystals of chlorine hydrate ($Cl_2 \cdot 8H_2O$) in an inverted U-tube, one limb of which was heated and the other placed in a freezing mixture (liquefaction resulted because of the high pressure of the gas cooled below its relatively high critical temperature). He then liquefied other gases, including sulfur dioxide, hydrogen sulfide, nitrous oxide (dinitrogen monoxide), chlorine dioxide, cyanogen, and hydrogen bromide. After

—*Continued top of next page*

the production of liquid carbon dioxide in 1835, Faraday used this coolant to liquefy such gases as ethylene (ethene), phosphine, silicon tetrafluoride, and boron trifluoride.

In the same year (1835) he made his greatest contribution to organic chemistry, the isolation of benzene from gas oils. He also worked out the empirical formula of naphthalene and prepared various sulfonic acids—later to have great importance in the industries devoted to dyestuffs and detergents. It was also at about this time that Faraday demonstrated the use of platinum as a catalyst and showed the importance in chemical reactions of surfaces and inhibitors—again foreshadowing a huge area of the modern chemical industry.

But Faraday's interest in science had been initiated by a fascination for electricity, and he eventually combined the knowledge that he gained of this subject with chemistry to produce the basic laws of electrolysis in 1834. His researches, summed up in Faraday's laws of electrolysis, established the link between electricity and chemical affinity, one of the most fundamental concepts in science. It was Faraday who coined the terms anode, cathode, cation, anion, electrode, and electrolyte. He postulated that during the electrolysis of an aqueous electrolyte, positively charged cations move toward the negatively charged cathode and negatively charged anions migrate to the positively charged anode. At each electrode the ions are discharged according to the following rules:

(a) the quantity of a substance produced is proportional to the amount of electricity passed;

(b) the relative quantities of different substances produced by the same amount of electricity are proportional to their equivalent weights (i.e., the atomic weight divided by the oxidation state or valence).

But his first major electrical discovery was made much earlier, in 1821, only a year after Hans Oersted (1777–1851) had discovered with a compass needle that a current of electricity flowing through a wire produces a magnetic field. Faraday was asked to investigate the phenomenon of electromagnetism by the editor of the *Philosophical Magazine,* who hoped that Faraday would elucidate the facts of the situation following the wild theories and opinions that Oersted's sensational discovery had aroused. Faraday conceived that circular lines of magnetic force are produced around the wire to explain the orientation of Oersted's compass needle, and therefore set about devising an apparatus that would demonstrate this by causing a magnet to revolve around an electric current. He succeeded in October 1821 with an elaborate device consisting of two vessels of mercury connected to a battery. Above the vessels and connected to each other were suspended a magnet and a wire; these were free to move and dipped just below the surface of the mercury. In the mercury were fixed a wire and a magnet, respectively. When the current was switched on, it flowed through both the fixed and free wires, generating a magnetic field in them. This caused the free magnet to revolve around the fixed wire, and the free wire to revolve around the fixed magnet.

This was a brilliant demonstration of the conversion of electrical energy into motive force, for it showed that either the conductor or the magnet could be made to move. In this experiment, Faraday demonstrated the basic principles governing the electric motor and although practical motors subsequently developed had a very different form to Faraday's apparatus, he is nevertheless usually credited with the invention of the electric motor.

Faraday's conviction that an electric current gave rise to lines of magnetic force arose from his idea that electricity was a form of vibration and not a moving fluid. He believed that electricity was a state of varying strain in the molecules of the conductor, and this gave rise to a similar strain in the medium surrounding the conductor. It was reasonable to consider therefore that the transmitted strain might set up a similar strain in the molecules of another nearby conductor—that a magnetic field might bring about an electric current in the reverse of the electromagnetic effect discovered by Oersted.

Faraday hunted for this effect from 1824 onward, expecting to find that a magnetic field would induce a steady electric current in a conductor. In 1824, François Arago (1786–1853) found that a rotating non-magnetic disk, specifically of copper, caused the deflection of a magnetic needle placed above it. This was in fact a demonstration of electromagnetic induction, but nobody at that time could explain Arago's wheel (as it was called). Faraday eventually succeeded in producing induction in 1831. In August of that year, he wound two coils around an iron bar and connected one to a battery and the other to a galvanometer. Nothing happened when the current flowed through the first coil, but Faraday noticed that the galvanometer gave a kick whenever the current was switched on or off. Faraday found an immediate explanation with his lines of force. If the lines of force were cut—that is, if the magnetic field changed—then an electric current would be induced in a conductor placed within the magnetic field. The iron core in fact helped to concentrate the magnetic field, as Faraday later came to understand, and a current was induced in the second coil by the magnetic field momentarily set up as current entered or left the first coil. With this device, Faraday had discovered the transformer, a modern transformer being no different in essence even though the alternating current required had not then been discovered.

Faraday is thus also credited with the simultaneous discovery of electromagnetic induction, though the same discovery had been made in the same way by Joseph Henry (1797–1878) in 1830. However, busy teaching, Henry had not been able to publish his findings before Faraday did, although both men are now credited with the independent discovery of induction.

Faraday's insight enabled him to make another great discovery soon afterwards. He realized that the motion of the copper wheel relative to the magnet in Arago's experiment caused an electric current to flow in the disk,

—Continued top of next page

which in turn set up a magnetic field and de-flected the magnet. He set about constructing a similar device in which the current produced could be led off, and in October 1831 built the first electric generator. This consisted of a copper disk that was rotated between the poles of a magnet; Faraday touched wires to the edge and center of the disk and connected them to a galvanometer, which registered a steady current. This was the first electric generator, and generators employing coils and magnets in the same way as modern generators were developed by others over the next two years.

Faraday's next discoveries in electricity, apart from his major contribution to electrochemistry, were to show in 1832 that an electrostatic charge gives rise to the same effects as current electricity, thus proving that there is no basic difference between them. Then in 1837 he investigated electrostatic force and demonstrated that it consists of a field of curved lines of force, and that different substances take up different amounts of electric charge when subjected to an electric field. This led Faraday to conceive of specific inductive capacity. In 1838, he proposed a theory of electricity based on his discoveries that elaborated his idea of varying strain in molecules. In a good conductor, a rapid buildup and breakdown of strain took place, transferring energy quickly from one molecule to the next. This also accounted for the decomposition of compounds in electrolysis. At the same time, Faraday rejected the notion that electricity involved the movement of any kind of electrical fluid. In this, he was

wrong (because the motion of electrons is involved) but in that this motion causes a rapid transfer of electrical energy through a conductor, Faraday's ideas were valid.

Faraday's theory was not taken seriously by many scientists, but his concept of the line of force was developed mathematically by Kelvin. In 1845, he suggested that Faraday investigate the action of electricity on polarized light. Faraday had in fact already carried out such experiments with no success, but this could have been because electrical forces were not strong. Faraday now used an electromagnet to give a strong magnetic field instead and found that it causes the plane of polarization to rotate, the angle of rotation being proportional to the strength of the magnetic field.

Several further discoveries resulted from this experiment. Faraday realized that the glass block used to transmit the beam of light must also transmit the magnetic field, and he noticed that the glass tended to set itself at right angles to the poles of the magnet rather than lining up with it as an iron bar would. Faraday showed that the differing responses of substances to a magnetic field depended on the distribution of the lines of force through them, and not on the induction of different poles. He called materials that are attracted to a magnetic field paramagnetic, and those that are repulsed diamagnetic. Faraday then went on to point out that the energy of a magnet is in the field around it and not in the magnet itself, and he extended this basic conception of field theory to electrical and gravitational systems.

Finally Faraday considered the nature of light and in 1846 arrived at a form of the electromagnetic theory of light that was later developed by James Clerk Maxwell (1831–1879). In a brilliant demonstration of both his intuition and foresight, Faraday said "The view which I am so bold to put forth considers radiation as a high species of vibration in the lines of force which are known to connect particles, and also masses of matter, together. It endeavors to dismiss the ether but not the vibrations." It was a bold view, for no scientist until Albert Einstein (1879–1955) was to take such a daring step.

Michael Faraday was a scientific genius of a most extraordinary kind. Without any mathematical ability at all, he succeeded in making the basic discoveries on which virtually all our uses of electricity depend and also in conceiving the fundamental nature of magnetism and, to a degree, of electricity and light. He owed this extraordinary degree of insight to an amazing talent for producing valid pictorial interpretations of the workings of nature. Faraday himself was a modest man, content to serve science as best he could without undue reward, and he declined both a knighthood and the Presidency of the Royal Society. Characteristically, he also refused to take part in the preparation of poison gas for use in the Crimean War. His many achievements are honored in the use of his name in science, the farad being the SI unit of capacitance and the Faraday constant being the quantity of electricity required to liberate a standard amount of substance in electrolysis.

From the Hutchinson *Dictionary of Scientific Biography*. © Research Machines plc [2003] All Rights Reserved. Helicon Publishing is a division of Research Machines.

Summary

The first electrical phenomenon recognized was the charge produced by friction, which today is called *static electricity*. By the early 1900s, the *electron theory of charge* was developed from studies of the *atomic nature of matter*. These studies led to the understanding that matter is made of *atoms*, which are composed of *negatively charged electrons* moving about a central *nucleus*, which contains *positively charged protons*. The two kinds of charges interact as *like charges produce a repellant force*, and *unlike charges produce an attractive force*. An object acquires an *electric charge* when it has an excess or deficiency of electrons, which is called an *electrostatic charge*.

A *quantity of charge* (q) is measured in units of *coulombs* (C), the charge equivalent to the transfer of 6.24×10^{18} charged particles such as the electron. The *fundamental charge* of an electron or proton is 1.60×10^{-19} coulomb. The *electrical forces* between two charged objects can be calculated from the relationship between the quantity of charge and the distance between two charged objects. The relationship is known as *Coulomb's law*.

A charged object in an electric field has *electric potential energy* that is related to the charge on the object and the work done to move it into a field of like charge. The resulting *electric potential difference* (V) is a ratio of the work done (W) to move a quantity of charge (q). In units, a joule of work done to move a coulomb of charge is called a *volt*.

A flow of electric charge is called an *electric current* (I). A current requires some device, such as a generator or battery, to maintain a potential difference. The device is called a *voltage source.* An *electric circuit* contains (1) a voltage source, (2) a *continuous path* along which the current flows, and (3) a device such as a lamp or motor where work is done, called a *voltage drop.* Current (I) is measured as the *rate of flow* of charge, the quantity of charge (q) through a conductor in a period of time (t). The unit of current in coulomb/second is called an *ampere* or *amp* for short (A).

Current occurs in a conductor when a *potential difference* is applied and an *electric field* travels through the conductor at near the speed of light. The electrons drift *very slowly,* accelerated by the electric field. The field moves the electrons in one direction in a *direct current* (dc) and moves them back and forth in an *alternating current* (ac).

Materials have a property of opposing or reducing an electric current called *electrical resistance* (R). Resistance is a ratio between the potential difference (V) between two points and the resulting current (I), or $R = V/I$. The unit is called the *ohm* (Ω), and $1.00 \; \Omega = 1.00$ volt/1.00 amp. The relationship between voltage, current, and resistance is called *Ohm's law.*

Disregarding the energy lost to resistance, the *work* done by a voltage source is equal to the work accomplished in electrical devices in a circuit. The *rate of doing work is power, or work per unit time,* $P = W/t$. *Electrical power* can be calculated from the relationship of $P = IV$, which gives the power unit of *watts.*

Magnets have two poles about which their attraction is concentrated. When free to turn, one pole moves to the north and the other to the south. The north-seeking pole is called the *north pole* and the south-seeking pole is called the *south pole. Like poles repel* one another and *unlike poles attract.*

The property of magnetism is *electric in origin,* produced by charges in motion. *Permanent magnets* have tiny regions called *magnetic domains,* each with its own north and south poles. An *unmagnetized* piece of iron has randomly arranged domains. When *magnetized,* the domains become aligned and contribute to the overall magnetic effect.

A *current-carrying wire* has magnetic field lines of closed, *concentric circles* that are at right angles to the length of wire. The *direction of* the magnetic field depends on the direction of the current. A coil of many loops is called a *solenoid* or *electromagnet.* The electromagnet is the working part in electrical meters, electromagnetic switches, and the electric motor.

When a loop of wire is moved in a magnetic field, or if a magnetic field is moved past a wire loop, a voltage is induced in the wire loop. The interaction is called *electromagnetic induction.* An electric generator is a rotating coil of wire in a magnetic field. The coil is rotated by mechanical energy, and electromagnetic induction induces a voltage, thus converting mechanical energy to electrical energy. A *transformer* steps up or steps down the voltage of an alternating current. The ratio of input and output voltage is determined by the number of loops in the primary and secondary coils. Increasing the voltage decreases the current, which makes long-distance transmission of electrical energy economically feasible.

Summary of Equations

6.1

$$\text{Quantity of charge} = (\text{number of electrons})(\text{electron charge})$$

$$q = ne$$

6.2

$$\text{Electrical force} = (\text{constant}) \times \frac{\text{charge on one object} \times \text{charge on second object}}{\text{distance between objects squared}}$$

$$F = k\frac{q_1 q_2}{d^2}$$

where $k = 9.00 \times 10^9$ newton·meters^2/coulomb^2

6.3

$$\text{Electric potential} = \frac{\text{work to create potential}}{\text{charge moved}}$$

$$V = \frac{W}{q}$$

6.4

$$\text{Electric current} = \frac{\text{quantity of charge}}{\text{time}}$$

$$I = \frac{q}{t}$$

6.5

$$\text{Volts} = \text{current} \times \text{resistance}$$

$$V = IR$$

6.6

$$\text{Electrical power} = (\text{amps})(\text{volts})$$

$$P = IV$$

6.7

$$\text{Cost} = \frac{(\text{watts})(\text{time})(\text{rate})}{1,000 \; \text{W/kW}}$$

6.8

$$\frac{\text{volts}_{\text{primary}}}{(\text{number of loops})_{\text{primary}}} = \frac{\text{volts}_{\text{secondary}}}{(\text{number of loops})_{\text{secondary}}}$$

$$\frac{V_p}{N_p} = \frac{V_s}{N_s}$$

6.9

$$(\text{volts}_{\text{primary}})(\text{current}_{\text{primary}}) = (\text{volts}_{\text{secondary}})(\text{current}_{\text{secondary}})$$

$$V_p I_p = V_s I_s$$

KEY TERMS

alternating current (p. **152**)

amp (p. **150**)

ampere (p. **150**)

coulomb (p. **145**)

Coulomb's law (p. **146**)

direct current (p. **152**)

electric circuit (p. **149**)

electric current (p. **149**)

electric field (p. **147**)

electric generator (p. **166**)

electrical conductors (p. **145**)

electrical resistance (p. **152**)

electromagnet (p. **161**)

electromagnetic induction
   (p. **165**)

electrostatic charge (p. **143**)

force field (p. **147**)

fundamental charge (p. **145**)

magnetic domain (p. **160**)

magnetic field (p. **158**)

magnetic poles (p. **158**)

ohm (p. **152**)

Ohm's law (p. **153**)

transformer (p. **166**)

volt (p. **148**)

watt (p. **155**)

APPLYING THE CONCEPTS

1. How does an electron acquire a negative charge?
   a. from an imbalance of subatomic particles
   b. by induction or contact with charged objects
   c. from the friction of certain objects rubbing together
   d. Charge is a fundamental property of an electron.

2. An object that acquires an excess of electrons becomes a (an)
   a. ion.
   b. negatively charged object.
   c. positively charged object.
   d. electrical conductor.

3. Which of the following is most likely to acquire an electrostatic charge?
   a. electrical conductor
   b. electrical nonconductor
   c. Both are equally likely.
   d. None of the above is correct.

4. A quantity of electric charge is measured in a unit called a (an)
   a. coulomb.
   b. volt.
   c. amp.
   d. watt.

5. The unit that describes the potential difference that occurs when a certain amount of work is used to move a certain quantity of charge is called the
   a. ohm.
   b. volt.
   c. amp.
   d. watt.

6. Which of the following units are measures of *rates?*
   a. amp and volt
   b. coulomb and joule
   c. volt and watt
   d. amp and watt

7. An electric current is measured in units of
   a. coulomb.
   b. volt.
   c. amp.
   d. watt.

8. If you consider a current to be positive charges that flow from the positive to the negative terminal of a battery, you are using which description of a current?
   a. electron current
   b. conventional current
   c. proton current
   d. alternating current

9. In an electric current the electrons are moving
   a. at a very slow rate.
   b. at the speed of light.
   c. faster than the speed of light.
   d. at a speed described as supersonic.

10. In which of the following currents is there no electron movement from one end of a conducting wire to the other end?
    a. electron current
    b. direct current
    c. alternating current
    d. None of the above is correct.

11. If you multiply amps × volts, the answer will be in units of
    a. resistance.
    b. work.
    c. current.
    d. power.

12. The unit of resistance is the
    a. watt.
    b. ohm.
    c. amp.
    d. volt.

13. Which of the following is a measure of electrical work?
    a. kilowatt
    b. C
    c. kWh
    d. C/s

14. Considering a direct current in a thick wire, a thin wire of the same length, material, and temperature has
    a. less electrical resistance.
    b. more electrical resistance.
    c. the same electrical resistance.
    d. none of the above.

15. If an electric charge is somehow caused to disappear, the electric field that surrounds it will
    a. immediately cease to exist.
    b. collapse inward at some speed.
    c. continue to exist until neutralized.
    d. move off into space until it finds another charge.

16. The earth's north magnetic pole
    a. is located at the geographic North Pole.
    b. is a magnetic south pole.
    c. has always had the same orientation.
    d. None of the above is correct.

17. A permanent magnet has magnetic properties because
    a. the magnetic fields of its electrons are balanced.
    b. of an accumulation of monopoles in the ends.
    c. the magnetic domains are aligned.
    d. all of the above.

18. A current-carrying wire has a magnetic field around it because
    a. a moving charge produces a magnetic field of its own.
    b. the current aligns the magnetic domains in the metal of the wire.
    c. the metal was magnetic before the current was established and the current enhanced the magnetic effect.
    d. none of the above.

19. If you reverse the direction that a current is running in a wire, the magnetic field around the wire
    a. is oriented as it was before.
    b. is oriented in an opposite direction.
    c. flips to become aligned parallel to the length of the wire.
    d. ceases to exist.

20. A step-up transformer steps up (the)
    a. power.
    b. current.
    c. voltage.
    d. all of the above.

Answers

1. d 2. b 3. b 4. a 5. b 6. d 7. c 8. b 9. a 10. c 11. d 12. b 13. c 14. b 15. b
16. b 17. c 18. a 19. b 20. c

1. Explain why a balloon that has been rubbed sticks to a wall for a while.
2. Explain what is happening when you walk across a carpet and receive a shock when you touch a metal object.
3. Why does a positively or negatively charged object have multiples of the fundamental charge?
4. Explain how you know that it is an electric field, not electrons, that moves rapidly through a circuit.
5. Is a kWh a unit of power or a unit of work? Explain.
6. What is the difference between ac and dc?
7. What is a magnetic pole? How are magnetic poles named?
8. How is an unmagnetized piece of iron different from the same piece of iron when it is magnetized?
9. Explain why the electric utility company increases the voltage of electricity for long-distance transmission.
10. Describe how an electric generator is able to generate an electric current.
11. Why does the north pole of a magnet point to the geographic North Pole if like poles repel?
12. Explain what causes an electron to move toward one end of a wire when the wire is moved across a magnetic field.

PARALLEL EXERCISES

The exercises in groups A and B cover the same concepts. Solutions to group A exercises are located in appendix D.

Group A

1. A rubber balloon has become negatively charged from being rubbed with a wool cloth, and the charge is measured as 1.00×10^{-14} C. According to this charge, the balloon contains an excess of how many electrons?
2. One rubber balloon with a negative charge of 3.00×10^{-14} C is suspended by a string and hangs 2.00 cm from a second rubber balloon with a negative charge of 2.00×10^{-12} C. (a) What is the direction of the force between the balloons? (b) What is the magnitude of the force?
3. A dry cell does 7.50 J of work through chemical energy to transfer 5.00 C between the terminals of the cell. What is the electric potential between the two terminals?
4. An electric current through a wire is 6.00 C every 2.00 s. What is the magnitude of this current?
5. A 1.00 A electric current corresponds to the charge of how many electrons flowing through a wire per second?
6. There is a current of 4.00 A through a toaster connected to a 120.0 V circuit. What is the resistance of the toaster?
7. What is the current in a 60.0 Ω resistor when the potential difference across it is 120.0 V?

Group B

1. An inflated rubber balloon is rubbed with a wool cloth until an excess of a billion electrons is on the balloon. What is the magnitude of the charge on the balloon?
2. What is the force between two balloons with a negative charge of 1.6×10^{-10} C if the balloons are 5.0 cm apart?
3. How much energy is available from a 12 V storage battery that can transfer a total charge equivalent to 100,000 C?
4. A wire carries a current of 2.0 A. At what rate is the charge flowing?
5. What is the magnitude of the least possible current that could theoretically exist?
6. There is a current of 0.83 A through a lightbulb in a 120 V circuit. What is the resistance of this lightbulb?
7. What is the voltage across a 60.0 Ω resistor with a current of 3 1/3 amp?

8. A lightbulb with a resistance of 10.0 Ω allows a 1.20 A current to flow when connected to a battery. (a) What is the voltage of the battery? (b) What is the power of the lightbulb?

9. A small radio operates on 3.00 V and has a resistance of 15.0 Ω. At what rate does the radio use electric energy?

10. A 1,200 W hair dryer is operated on a 120 V circuit for 15 min. If electricity costs $0.10/kWh, what was the cost of using the blow dryer?

11. An automobile starter rated at 2.00 hp draws how many amps from a 12.0 V battery?

12. An average-sized home refrigeration unit has a 1/3 hp fan motor for blowing air over the inside cooling coils, a 1/3 hp fan motor for blowing air over the outside condenser coils, and a 3.70 hp compressor motor. (a) All three motors use electric energy at what rate? (b) If electricity costs $0.10/kWh, what is the cost of running the unit per hour? (c) What is the cost for running the unit 12 hours a day for a 30-day month?

13. A 15 ohm toaster is turned on in a circuit that already has a 0.20 hp motor, three 100 W lightbulbs, and a 600 W electric iron that are on. Will this trip a 15 A circuit breaker? Explain.

14. A power plant generator produces a 1,200 V, 40 A alternating current that is fed to a step-up transformer before transmission over the high lines. The transformer has a ratio of 200 to 1 wire loops. (a) What is the voltage of the transmitted power? (b) What is the current?

15. A step-down transformer has an output of 12 V and 0.5 A when connected to a 120 V line. Assuming no losses: (a) What is the ratio of primary to secondary loops? (b) What current does the transformer draw from the line? (c) What is the power output of the transformer?

16. A step-up transformer on a 120 V line has 50 loops on the primary and 150 loops on the secondary, and draws a 5.0 A current. Assuming no losses: (a) What is the voltage from the secondary? (b) What is the current from the secondary? (c) What is the power output?

8. A 10.0 Ω lightbulb is connected to a 12.0 V battery. (a) What current flows through the bulb? (b) What is the power of the bulb?

9. A lightbulb designed to operate in a 120.0 V circuit has a resistance of 192 Ω. At what rate does the bulb use electric energy?

10. What is the monthly energy cost of leaving a 60 W bulb on continuously if electricity costs 10¢ per kWh?

11. An electric motor draws a current of 11.5 A in a 240 V circuit. What is the power of this motor in W?

12. A swimming pool requiring a 2.0 hp motor to filter and circulate the water runs for 18 hours a day. What is the monthly electrical cost for running this pool pump if electricity costs 10¢ per kWh?

13. Is it possible for two people to simultaneously operate 1,300 W hair dryers on the same 120 V circuit without tripping a 15 A circuit breaker? Explain.

14. A step-up transformer has a primary coil with 100 loops and a secondary coil with 1,500 loops. If the primary coil is supplied with a household current of 120 V and 15 A, (a) what voltage is produced in the secondary circuit? (b) What current flows in the secondary circuit?

15. The step-down transformer in a local neighborhood reduces the voltage from a 7,200 V line to 120 V. (a) If there are 125 loops on the secondary, how many are on the primary coil? (b) What current does the transformer draw from the line if the current in the secondary is 36 A? (c) What are the power input and output?

16. A step-down transformer connected to a 120 V electric generator has 30 loops on the primary for each loop in the secondary. (a) What is the voltage of the secondary? (b) If the transformer has a 90 A current in the primary, what is the current in the secondary? (c) What are the power input and output?

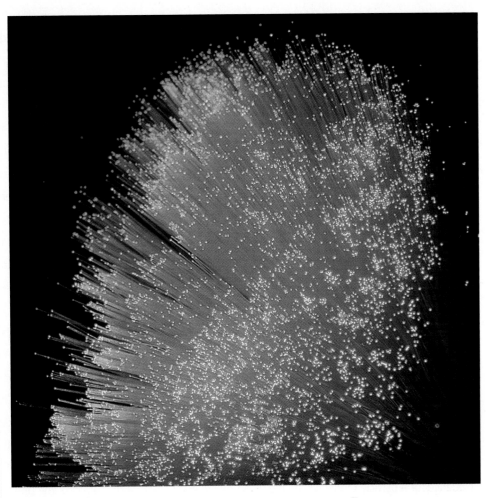

This fiber optics bundle carries pulses of light from an infrared laser to carry much more information than could be carried by electrons moving through wires. This is part of a dramatic change underway, a change that will first find a hybrid "optoelectronics" replacing the more familiar "electronics" of electrons and wires.

CHAPTER

7

Light

You use light and your eyes more than any other sense to learn about your surroundings. All of your other senses—touch, taste, sound, and smell—involve matter, but the most information is provided by light. Yet, light seems more mysterious than matter. You can study matter directly, measuring its dimensions, taking it apart, and putting it together to learn about it. Light, on the other hand, can only be studied indirectly in terms of how it behaves (Figure 7.1). Once you understand its behavior, you know everything there is to know about light. Anything else is thinking about what the behavior means.

The behavior of light has stimulated thinking, scientific investigations, and debate for hundreds of years. The investigations and debate have occurred because light cannot be directly observed, which makes the exact nature of light very difficult to pin down. For example, you know that light moves energy from one place to another place. You can feel energy from the sun as sunlight warms you, and you know that light has carried this energy across millions of miles of empty space. The ability of light to move energy like this could be explained (1) as energy transported by waves, just as sound waves carry energy from a source, or (2) as the kinetic energy of a stream of moving particles, which give up their energy when they strike a surface. The movement of energy from place to place could be explained equally well by a wave model of light or by a particle model of light. When two possibilities exist like this in science, experiments are designed and measurements are made to support one model and reject the other. Light, however, presents a baffling dilemma. Some experiments provide evidence that light consists of waves and not a stream of moving particles. Yet other experiments provide evidence of just the opposite, that light is a stream of particles and not a wave. Evidence for accepting a wave or particle model seems to depend on which experiments are considered.

The purpose of using a model is to make new things understandable in terms of what is already known. When these new things concern light, three models are useful in visualizing separate behaviors. Thus, the electromagnetic wave model will be used to describe how light is created at a source. Another model, a model of light as a ray, a small beam of light, will be used to discuss some common properties of light such as reflection and the refraction, or bending, of light. Finally, properties of light that provide evidence for a particle model will be discussed before ending with a discussion of the present understanding of light.

Sources of Light

The sun and other stars, lightbulbs, and burning materials all give off light. When something produces light it is said to be **luminous.** The sun is a luminous object that provides almost all of the *natural* light on the earth. A small amount of light does reach the earth from the stars but not really enough to see by on a moonless night. The moon and planets shine by reflected light and do not produce their own light, so they are not luminous.

Burning has been used as a source of *artificial* light for thousands of years. A wood fire and a candle flame are luminous because of their high temperatures. When visible light is given off as a result of high temperatures, the light source is said to be **incandescent.** A flame from any burning source, an ordinary lightbulb, and the sun are all incandescent sources because of high temperatures.

How do incandescent objects produce light? One explanation is given by the electromagnetic wave model. This model describes a relationship between electricity, magnetism, and light. The model pictures an electromagnetic wave as forming whenever an electric charge is *accelerated* by some external force. The acceleration produces a wave consisting of electrical

and magnetic fields that become isolated from the accelerated charge, moving off into space (Figure 7.2). As the wave moves through space, the two fields exchange energy back and forth, continuing on until they are absorbed by matter and give up their energy.

The frequency of an electromagnetic wave depends on the acceleration of the charge; the greater the acceleration, the higher the frequency of the wave that is produced. The complete range of frequencies is called the *electromagnetic spectrum* (Figure 7.3). The spectrum ranges from radio waves at the low-frequency end of the spectrum to gamma rays at the high-frequency end. Visible light occupies only a small part of the middle portion of the complete spectrum.

Visible light is emitted from incandescent sources at high temperatures, but actually electromagnetic radiation is given off from matter at *any* temperature. This radiation is called **blackbody radiation,** which refers to an idealized material (the *blackbody*) that perfectly absorbs and perfectly emits electromagnetic radiation. From the electromagnetic wave model, the radiation originates from the acceleration of charged particles near the surface of an object. The frequency of the blackbody

FIGURE 7.1

Light, sounds, and odors can identify the pleasing environment of this garden, but light provides the most information. Sounds and odors can be identified and studied directly, but light can only be studied indirectly, that is, in terms of how it behaves. As a result, the behavior of light has stimulated scientific investigations and debate for hundreds of years. Perhaps you have wondered about light and its behaviors. What is light?

radiation is determined by the energy available for accelerating charged particles, that is, the temperature of the object. Near absolute zero, there is little energy available and no radiation is given off. As the temperature of an object is increased, more energy is available, and this energy is distributed over a range of values, so more than one frequency of radiation is emitted. A

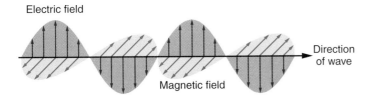

FIGURE 7.2

The electric and magnetic fields in an electromagnetic wave vary together. Here the fields are represented by arrows that indicate the strength and direction of the fields. Note the fields are perpendicular to each other and to the direction of the wave.

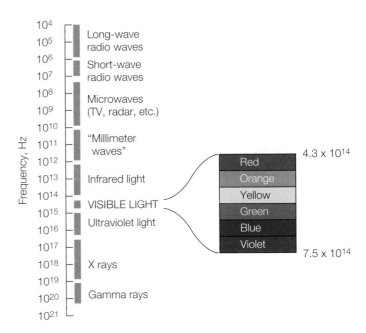

FIGURE 7.3

The electromagnetic spectrum. All electromagnetic waves have the same fundamental character and the same speed in a vacuum, but many aspects of their behavior depend on their frequency.

graph of the frequencies emitted from the range of available energy is thus somewhat bell-shaped. The steepness of the curve and the position of the peak depend on the temperature (Figure 7.4). As the temperature of an object increases, there is an increase in the *amount* of radiation given off, and the peak radiation emitted progressively *shifts* toward higher and higher frequencies.

At room temperature the radiation given off from an object is in the infrared region, invisible to the human eye. When the temperature of the object reaches about 700°C (about 1,300°F), the peak radiation is still in the infrared region, but the peak has shifted enough toward the higher frequencies that a little visible light is emitted as a dull red glow. As the temperature of the object continues to increase, the amount of radiation increases, and the peak continues to shift toward shorter wavelengths. Thus, the object begins to glow brighter, and the color changes from red, to orange, to yellow, and eventually to white. The association of this color change with temperature is noted in the referent description of an object being "red hot," "white hot," and so forth.

The incandescent flame of a candle or fire results from the blackbody radiation of carbon particles in the flame. At a blackbody temperature of 1,500°C (about 2,700°F), the carbon particles emit visible light in the red to yellow frequency range. The tungsten filament of an incandescent lightbulb is heated to about 2,200°C (about 4,000°F) by an electric current. At this temperature, the visible light emitted is in the reddish, yellow-white range.

The radiation from the sun, or sunlight, comes from the sun's surface, which has a temperature of about 5,700°C

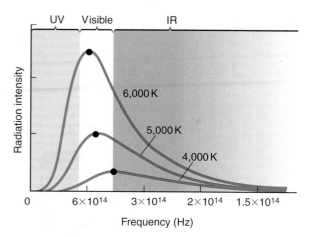

FIGURE 7.4

Three different objects emitting blackbody radiation at three different temperatures. The frequency of the peak of the curve (shown by dot) shifts to higher frequency at higher temperatures.

(about 10,000°F). As shown in Figure 7.5, the sun's radiation has a broad spectrum centered near the yellow-green frequency. Your eye is most sensitive to this frequency of sunlight. The spectrum of sunlight before it travels through the earth's atmosphere is infrared (about 51 percent), visible light (about 40 percent), and ultraviolet (about 9 percent). Sunlight originated as energy released from nuclear reactions in the sun's core. This energy requires about a million years to work its way up to the surface. At the surface, the energy from the core accelerates charged particles, which then emit light like tiny antennas. The sunlight requires about eight minutes to travel the distance from the sun's surface to the earth.

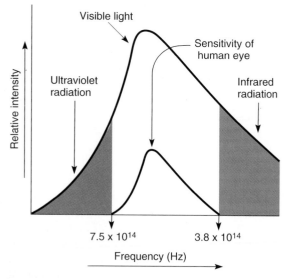

FIGURE 7.5

Sunlight is about 9 percent ultraviolet radiation, 40 percent visible light, and 51 percent infrared radiation before it travels through the earth's atmosphere.

Properties of Light

You can see luminous objects from the light they emit, and you can see nonluminous objects from the light they reflect, but you cannot see the path of the light itself. For example, you cannot see a flashlight beam unless you fill the air with chalk dust or smoke. The dust or smoke particles reflect light, revealing the path of the beam. This simple observation must be unknown to the makers of science fiction movies, since they always show visible laser beams zapping through the vacuum of space.

Some way to represent the invisible travels of light is needed in order to discuss some of its properties. Throughout history a **light ray model** has been used to describe the travels of light. The meaning of this model has changed over time, but it has always been used to suggest that "something" travels in *straight-line paths*. The light ray is a line that is drawn to represent the straight-line travel of light. A line is drawn to represent this imaginary beam to illustrate the law of reflection (as from a mirror) and the law of refraction (as through a lens). There are limits to using a light ray for explaining some properties of light, but it works very well in explaining mirrors, prisms, and lenses.

Light Interacts with Matter

A ray of light travels in a straight line from a source until it encounters some object or particles of matter (Figure 7.6). What happens next depends on several factors, including (1) the smoothness of the surface, (2) the nature of the material, and (3) the angle at which the light ray strikes the surface.

The *smoothness* of the surface of an object can range from perfectly smooth to extremely rough. If the surface is perfectly smooth, rays of light undergo *reflection*, leaving the surface parallel to each other. A mirror is a good example of a very smooth surface that reflects light this way (Figure 7.7A). If a surface is not smooth, the light rays are reflected in many random directions as *diffuse reflection* takes place (Figure 7.7B).

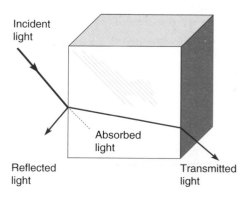

FIGURE 7.6

Light that interacts with matter can be reflected, absorbed, or transmitted through transparent materials. Any combination of these interactions can take place, but a particular substance is usually characterized by what it mostly does to light.

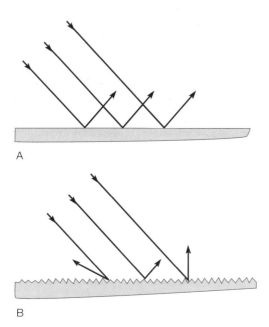

FIGURE 7.7

(*A*) Rays reflected from a perfectly smooth surface are parallel to each other. (*B*) Diffuse reflection from a rough surface causes rays to travel in many random directions.

FIGURE 7.8

Light travels in a straight line, and the color of an object depends on which wavelengths of light the object reflects. Each of these flowers absorbs the colors of white light and reflects the color that you see.

Rough and irregular surfaces and dust in the air make diffuse reflections. It is diffuse reflection that provides light in places not in direct lighting, such as under a table or under a tree. Such shaded areas would be very dark without the diffuse reflection of light.

Some *materials* allow much of the light that falls on them to move through the material without being reflected. Materials that allow transmission of light through them are called *transparent*. Glass and clear water are examples of transparent materials. Many materials do not allow transmission of any light and are called *opaque*. Opaque materials reflect light, absorb light, or some combination of partly absorbing and partly reflecting light (Figure 7.8). The light that is reflected varies with wavelength and gives rise to the perception of color, which will be discussed shortly. Absorbed light gives up its energy to the material and may be reemitted at a different wavelength, or it may simply show up as a temperature increase.

The *angle* of the light ray to the surface and the nature of the material determine if the light is absorbed, transmitted through a transparent material, or reflected. Vertical rays of light, for example, are mostly transmitted through a transparent material with some reflection and some absorption. If the rays strike the surface at some angle, however, much more of the light is reflected, bouncing off the surface. Thus, the glare of reflected sunlight is much greater around a body of water in the late afternoon than when the sun is directly overhead.

Light that interacts with matter is reflected, transmitted, or absorbed, and all combinations of these interactions are possible. Materials are usually characterized by which of these interactions they *mostly* do, but this does not mean that other interactions are not occurring too. For example, a window glass is usually characterized as a transmitter of light. Yet, the glass always *reflects* about 4 percent of the light that strikes it. The reflected light usually goes unnoticed during the day because of the bright light that is transmitted from the outside. When it is dark outside you notice the reflected light as the window glass now appears to act much like a mirror. A one-way mirror is another example of both reflection and transmission occurring (Figure 7.9). A mirror is usually characterized as a reflector of light. A one-way mirror, however, has a very thin silvering that reflects most of the light but still transmits a little. In a lighted room, a one-way mirror appears to reflect light just as any other mirror does. But a person behind the mirror in a dark room can see into the lighted room by means of the transmitted light. Thus, you know that this mirror transmits as well as reflects light. One-way mirrors are used to unobtrusively watch for shoplifters in many businesses.

Reflection

Most of the objects that you see are visible from diffuse reflection. For example, consider some object such as a tree that you see during a bright day. Each *point* on the tree must reflect light in all directions, since you can see any part of the tree from any angle (Figure 7.10). As a model, think of bundles of light rays entering your eye, which enable you to see the tree. This means that you can see any part of the tree from any angle because different bundles of reflected rays will enter your eye from different parts of the tree.

Light rays that are diffusely reflected move in all possible directions, but rays that are reflected from a smooth surface, such as a mirror, leave the mirror in a definite direction. Suppose you

(A)

(B)

FIGURE 7.9

(A) A one-way mirror reflects most of the light and transmits some light. You can see such a mirror around the top of the walls in this store. (B) Here is the view from behind the mirror.

FIGURE 7.10

Bundles of light rays are reflected diffusely in all directions from every point on an object. Only a few light rays are shown from only one point on the tree in this illustration. The light rays that move to your eyes enable you to see the particular point from which they were reflected.

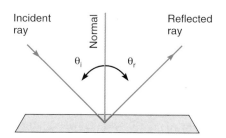

FIGURE 7.11

The law of reflection states that the angle of incidence (θ_i) is equal to the angle of reflection (θ_r). Both angles are measured from the *normal*, a reference line drawn perpendicular to the surface at the point of reflection.

ray, called the *reflected ray*, and (3) a reference line that is perpendicular to the reflecting surface and is located at the point where the incident ray struck the surface. This line is called the *normal*. The angle between the incident ray and the normal is called the *angle of incidence*, θ_i, and the angle between the reflected ray and the normal is called the *angle of reflection*, θ_r (Figure 7.11). The *law of reflection*, which was known to the ancient Greeks, is that the *angle of incidence equals the angle of reflection*, or

$$\theta_i = \theta_r$$

equation 7.1

look at a tree in a mirror. There is only one place on the mirror where you look to see any one part of the tree. Light is reflecting off the mirror from all parts of the tree, but the only rays that reach your eyes are the rays that are reflected at a certain angle from the place where you look. The relationship between the light rays moving from the tree and the direction in which they are reflected from the mirror to reach your eyes can be understood by drawing three lines: (1) a line representing an original ray from the tree, called the *incident ray*, (2) a line representing a reflected

Figure 7.12 shows how the law of reflection works when you look at a flat mirror. Light is reflected from all points on the block, and of course only the rays that reach your eyes are detected. These rays are reflected according to the law of reflection, with the angle of reflection equaling the angle of incidence. If you move your head slightly, then a different bundle of rays reaches your eyes. Of all the bundles of rays that reach your eyes, only two rays from a point are shown in the illustration. After these two rays are reflected, they continue to spread apart at the

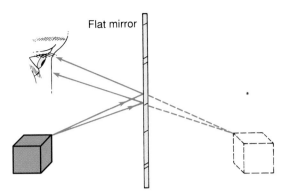

FIGURE 7.12

Light rays leaving a point on the block are reflected according to the law of reflection, and those reaching your eye are seen. After reflecting, the rays continue to spread apart at the same rate. You interpret this to be a block the same distance behind the mirror. You see a virtual image of the block, because light rays do not actually move from the image.

same rate that they were spreading before reflection. Your eyes and brain do not know that the rays have been reflected, and the diverging rays appear to come from behind the mirror, as the dashed lines show. The image, therefore, appears to be the same distance *behind* the mirror as the block is from the front of the mirror. Thus, a mirror image is formed where the rays of light *appear* to originate. This is called a **virtual image.** A virtual image is the result of your eyes' and brain's interpretations of light rays, not actual light rays originating from an image. Light rays that do originate from the other kind of image are called a **real image.** A real image is like the one displayed on a movie screen, with light originating from the image. A virtual image cannot be displayed on a screen, since it results from an interpretation.

Curved mirrors are either *concave,* with the center part curved inward, or *convex,* with the center part bulging outward. A concave mirror can be used to form an enlarged virtual image, such as a shaving or makeup mirror, or it can be used to form a real image, as in a reflecting telescope. Convex mirrors, for example, the mirrors on the sides of trucks and vans are often used to increase the field of vision. Convex mirrors are also used above an aisle in a store to show a wide area.

Stand facing the mirror and tell your helper exactly where to place a piece of masking tape on the mirror to mark the top of your head. Place a second piece of tape on the mirror to mark the top of your belt, or some other part of your clothing.

Now we need some measurements on you and on the mirror. First, measure and record the actual distance from the top of your head to the top of your belt or the other clothing marked. Second, measure and record the distance between the two pieces of masking tape on the mirror.

Now step back, perhaps doubling your distance from the mirror, and repeat the procedure. What can you conclude about your distance from the mirror and the image you see? Is there a difference in how much you can see? How tall a mirror do you need to see yourself from head to toe? Finally, is there some way you can cause your image to wink the same eye as you?

Refraction

You may have observed that an object that is partly in the air and partly in water appears to be broken, or bent, where the air and water meet. When a light ray moves from one transparent material to another, such as from water through air, the ray undergoes a change in the direction of travel at the boundary between the two materials. This change of direction of a light ray at the boundary is called **refraction.** The amount of change can be measured as an angle from the normal, just as it was for the angle of reflection. The incoming ray is called the *incident ray* as before, and the new direction of travel is called the *refracted ray.* The angles of both rays are measured from the normal (Figure 7.13).

Refraction results from a *change in speed* when light passes from one transparent material into another. The speed of light in a vacuum is 3.00×10^8 m/s, but it is slower when moving

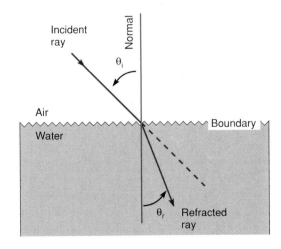

FIGURE 7.13

A ray diagram shows refraction at the boundary as a ray moves from air through water. Note that θ_i does not equal θ_r in refraction.

Chapter Seven: Light **185**

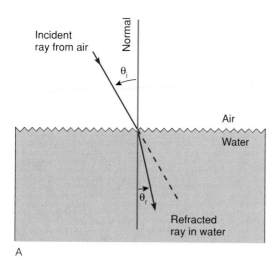

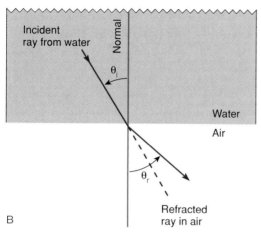

FIGURE 7.14

(A) A light ray moving to a new material with a slower speed of light is refracted toward the normal ($\theta_i > \theta_r$). (B) A light ray moving to a new material with a faster speed is refracted away from the normal ($\theta_i < \theta_r$).

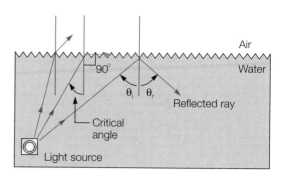

FIGURE 7.15

When the angle of incidence results in an angle of refraction of 90°, the refracted light ray is refracted along the water surface. The angle of incidence for a material that results in an angle of refraction of 90° is called the *critical angle*. When the incident ray is at this critical angle or greater, the ray is reflected internally. The critical angle for water is about 49°, and for a diamond it is about 25°.

through a transparent material. In water, for example, the speed of light is reduced to about 2.30×10^8 m/s. The speed of light has a magnitude that is specific for various transparent materials.

When light moves from one transparent material to another transparent material with a *slower* speed of light, the ray is refracted *toward* the normal (Figure 7.14A). For example, light travels through air faster than through water. Light traveling from air into water is therefore refracted toward the normal as it enters the water. On the other hand, if light has a *faster* speed in the new material, it is refracted *away* from the normal. Thus, light traveling from water into the air is refracted away from the normal as it enters the air (Figure 7.14B).

The magnitude of refraction depends on (1) the angle at which light strikes the surface and (2) the ratio of the speed of light in the two transparent materials. An incident ray that is perpendicular (90°) to the surface is not refracted at all. As the angle of incidence is increased, the angle of refraction is also increased. There is a limit, however, that occurs when the angle of refraction reaches 90°, or along the water surface. Figure 7.15 shows rays of light traveling from water to air at various angles. When the incident ray is

about 49°, the angle of refraction that results is 90°, along the water surface. This limit to the angle of incidence that results in an angle of refraction of 90° is called the *critical angle* for a water-to-air surface (Figure 7.15). At any incident angle greater than the critical angle, the light ray does not move from the water to the air but is *reflected* back from the surface as if it were a mirror. This is called **total internal reflection** and implies that the light is trapped inside if it arrived at the critical angle or beyond. Faceted transparent gemstones such as the diamond are brilliant because they have a small critical angle and thus reflect much light internally. Total internal reflection is also important in fiber optics, which is discussed in the reading at the end of this chapter.

Concepts Applied

Internal Reflection

Seal a flashlight in a clear plastic bag to waterproof it, then investigate the critical angle and total internal reflection in a swimming pool, play pool, or large tub of water. In a darkened room or at night, shine the flashlight straight up from beneath the water, then at different angles until it shines almost horizontally beneath the surface. Report your observation of the critical angle for the water used.

As was stated earlier, refraction results from a change in speed when light passes from one transparent material into another. The ratio of the speeds of light in the two materials determines the magnitude of refraction at any given angle of incidence. The greatest speed of light possible, according to current theory, occurs when light is moving through a vacuum. The speed of light in a vacuum is accurately known to nine decimals but is usually rounded to 3.00×10^8 m/s for general discussion. The speed of light in a vacuum is a very important constant in physical science, so it is given a symbol of its own, c. The ratio of c to the speed of light in some transparent material, v, is called the **index of refraction**, n, of that material or

$$n = \frac{c}{v}$$

equation 7.2

The indexes of refraction for some substances are listed in Table 7.1. The values listed are constant physical properties and can be used to identify a specific substance. Note that a larger value means a greater refraction at a given angle. Of the materials listed, diamond refracts light the most and air the least. The index for air is nearly 1, which means that light is slowed only slightly in air.

EXAMPLE 7.1

What is the speed of light in a diamond?

Solution

The relationship between the speed of light in a material (v), the speed of light in a vacuum ($c = 3.00 \times 10^8$ m/s), and the index of refraction is given in equation 7.2. The index of refraction of a diamond is found in Table 7.1 ($n = 2.42$).

$$n_{diamond} = 2.42$$
$$c = 3.00 \times 10^8 \text{ m/s}$$
$$v = ?$$

$$n = \frac{c}{v} \quad \therefore \quad v = \frac{c}{n}$$

$$v = \frac{3.00 \times 10^8 \text{ m/s}}{2.42}$$

$$= \boxed{1.24 \times 10^8 \text{ m/s}}$$

Note that Table 7.1 shows that colder air at 0°C (32°F) has a higher index of refraction than warmer air at 30°C (86°F), which means that light travels faster in warmer air. This difference explains the "wet" highway that you sometimes see at a distance in the summer. The air near the road is hotter on a clear, calm day.

TABLE 7.1

Index of refraction

Substance	$n = c/v$
Glass	1.50
Diamond	2.42
Ice	1.31
Water	1.33
Benzene	1.50
Carbon tetrachloride	1.46
Ethyl alcohol	1.36
Air (0°C)	1.00029
Air (30°C)	1.00026

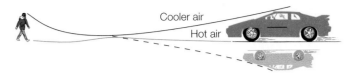

FIGURE 7.16

Mirages are caused by hot air near the ground refracting, or bending light rays upward into the eyes of a distant observer. The observer believes he is seeing an upside down image reflected from water on the highway.

Light rays traveling toward you in this hotter air are refracted upward as they enter the cooler air. Your brain interprets this refracted light as *reflected* light. Light traveling downward from other cars is also refracted upward toward you, and you think you are seeing cars "reflected" from the wet highway (Figure 7.16). When you reach the place where the "water" seemed to be, it disappears, only to appear again farther down the road.

Sometimes convection currents produce a mixing of warmer air near the road with the cooler air just above. This mixing refracts light one way, then the other, as the warmer and cooler air mix. This produces a shimmering or quivering that some people call "seeing heat." They are actually seeing changing refraction, which is a *result* of heating and convection. In addition to causing distant objects to quiver, the same effect causes the point source of light from stars to appear to twinkle. The light from closer planets does not twinkle because the many light rays from the disklike sources are not refracted together as easily as the fewer rays from the point sources of stars. The light from planets will appear to quiver, however, if the atmospheric turbulence is great.

Concepts Applied

Seeing Around Corners

Place a coin in an empty cup. Position the cup so the coin appears to be below the rim, just out of your line of sight. Do not move from this position as your helper slowly pours water into the cup. Explain why the coin becomes visible, then appears to rise in the cup. Use a sketch such as one of those in Figure 7.13 to help with your explanation.

Dispersion and Color

Electromagnetic waves travel with the speed of light with a whole spectrum of waves of various frequencies and wavelengths. The speed of electromagnetic waves (c) is related to the wavelength (λ) and the frequency (f) by a form of the wave equation, or

$$c = \lambda f$$

equation 7.3

Visible light is the part of the electromagnetic spectrum that your eyes can detect, a narrow range of wavelength from

FIGURE 7.17

The flowers appear to be red because they reflect light in the 7.9×10^{-7} m to 6.2×10^{-7} m range of wavelengths.

TABLE 7.2

Range of wavelengths and frequencies of the colors of visible light

Color	Wavelength (in meters)	Frequency (in hertz)
Red	7.9×10^{-7} to 6.2×10^{-7}	3.8×10^{14} to 4.8×10^{14}
Orange	6.2×10^{-7} to 6.0×10^{-7}	4.8×10^{14} to 5.0×10^{14}
Yellow	6.0×10^{-7} to 5.8×10^{-7}	5.0×10^{14} to 5.2×10^{14}
Green	5.8×10^{-7} to 4.9×10^{-7}	5.2×10^{14} to 6.1×10^{14}
Blue	4.9×10^{-7} to 4.6×10^{-7}	6.1×10^{14} to 6.6×10^{14}
Violet	4.6×10^{-7} to 3.9×10^{-7}	6.6×10^{14} to 7.7×10^{14}

about 7.90×10^{-7} m to 3.90×10^{-7} m. In general, this range of visible light can be subdivided into ranges of wavelengths that you perceive as colors (Figure 7.17). These are the colors of the rainbow, and there are six distinct colors that blend one into another. These colors are *red, orange, yellow, green, blue,* and *violet.* The corresponding ranges of wavelengths and frequencies of these colors are given in Table 7.2.

In general, light is interpreted to be white if it has the same mixture of colors as the solar spectrum. That sunlight is made up of component colors was first investigated in detail by Isaac Newton. While a college student, Newton became interested in grinding lenses, light, and color. At the age of twenty-three, Newton visited a local fair and bought several triangular glass prisms and proceeded to conduct a series of experiments with a beam of sunlight in his room. In 1672, he reported the results of his experiments with prisms and color, concluding that white light is a mixture of all the independent colors. Newton found that a beam of sunlight falling on a glass prism in a darkened room produced a band of colors he called a *spectrum.* Further, he found that a second glass prism would not subdivide each separate color but would combine all the colors back into white sunlight. Newton concluded that sunlight consists of a mixture of the six colors.

EXAMPLE 7.2

The colors of the spectrum can be measured in units of wavelength, frequency, or energy, which are alternative ways of describing colors of light waves. The human eye is most sensitive to light with a wavelength of 5.60×10^{-7} m, which is a yellow-green color. What is the frequency of this wavelength?

Solution

The relationship between the wavelength (λ), frequency (f), and speed of light in a vacuum (c), is found in equation 7.3, $c = \lambda f$.

$$c = 3.00 \times 10^8 \text{ m/s}$$
$$\lambda = 5.60 \times 10^{-7} \text{ m}$$
$$f = ?$$

$$c = \lambda f \quad \therefore \quad f = \frac{c}{\lambda}$$

$$f = \frac{3.00 \times 10^8 \frac{\text{m}}{\text{s}}}{5.60 \times 10^{-7} \text{ m}}$$

$$= \frac{3.00 \times 10^8}{5.60 \times 10^{-7}} \frac{\cancel{\text{m}}}{\text{s}} \times \frac{1}{\cancel{\text{m}}}$$

$$= 5.40 \times 10^{14} \frac{1}{\text{s}}$$

$$= \boxed{5.40 \times 10^{14} \text{ Hz}}$$

Historians tell us there are many early stories and legends about the development of ancient optical devices. The first glass vessels were made about 1500 B.C., so it is possible that samples of clear, transparent glass were available soon after. One legend claimed that the ancient Chinese invented eyeglasses as early as 500 B.C. A burning glass (lens) was mentioned in an ancient Greek play written about 424 B.C. Several writers described how Archimedes saved his hometown of Syracuse with a burning glass in about 214 B.C. Syracuse was besieged by Roman ships when Archimedes supposedly used the burning glass to focus sunlight on the ships, setting them on fire. It is not known if this story is true or not, but it is known that the Romans indeed did have burning glasses. Glass spheres, which were probably used to start fires, have been found in Roman ruins, including a convex lens recovered from the ashes of Pompeii.

Today, lenses are no longer needed to start fires, but they are common in cameras, scanners, optical microscopes, eyeglasses, lasers, binoculars, and many other optical devices. Lenses are no longer just made from glass, and today many are made from a transparent, hard plastic that is shaped into a lens.

The primary function of a lens is to form an image of a real object by refracting incoming parallel light rays. Lenses have two basic shapes, with the center of a surface either bulging in or bulging out. The outward bulging shape is thicker at the center than around the outside edge and this is called a *convex lens* (Box Figure 7.1A). The other basic lens shape is just the opposite, thicker around the outside edge than at the center, and is called a *concave lens* (Box Figure 7.1B).

Convex lenses are used to form images in magnifiers, cameras, eyeglasses, projectors, telescopes, and microscopes (Box Figure 7.2). Concave lenses are used in some eyeglasses and in combinations with the convex lens to correct for defects. The convex lens is the most commonly used lens shape.

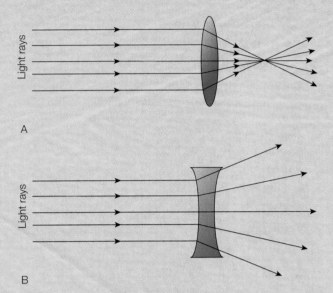

BOX FIGURE 7.1

(*A*) Convex lenses are called converging lenses because they bring together, or converge, parallel rays of light. (*B*) Concave lenses are called diverging lenses because they spread apart, or diverge, parallel rays of light.

Your eyes are optical devices with convex lenses. Box Figure 7.3 shows the basic structure. First, a transparent hole called the *pupil* allows light to enter the eye. The size of the pupil is controlled by the *iris*, the colored part that is a muscular diaphragm. The *lens* focuses a sharp image on the back surface of the eye, the *retina*. The retina is made up of millions of light-sensitive structures, and nerves carry electrical signals from the retina to the optic nerve, then to the brain.

The lens is a convex, pliable material held in place and changed in shape by the attached *ciliary muscle*. When the eye is focused on a distant object the ciliary muscle is completely relaxed. Looking at a closer object requires the contraction of the ciliary muscles to change the curvature of the lens. This adjustment of focus by the action of the

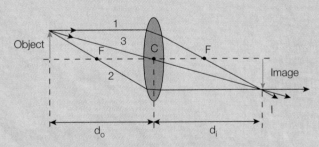

BOX FIGURE 7.2

A convex lens forms an inverted image from refracted light rays of an object outside the focal point. Convex lenses are mostly used to form images in cameras, file or overhead projectors, magnifying glasses, and eyeglasses.

—*Continued top of next page*

Continued—

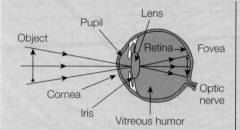

BOX FIGURE 7.3

Light rays from a distant object are focused by the lens onto the retina, a small area on the back of the eye.

ciliary muscle is called *accommodation*. The closest distance an object can be seen without a blurred image is called the *near point*, and this is the limit to accommodation.

The near point moves outward with age as the lens becomes less pliable. By middle age the near point may be twice this distance or greater, creating the condition known as farsightedness. The condition of farsightedness, or *hyperopia*, is a problem associated with aging (called presbyopia). Hyperopia can be caused at an early age by an eye that is too short or by problems with the cornea or lens that focus the image behind the retina. Farsightedness can be corrected with a convex lens as shown in Box Figure 7.4F.

Nearsightedness, or *myopia*, is a problem caused by an eye that is too long or problems with the cornea or lens that focus the image in front of the retina. Nearsightedness can be corrected with a concave lens as shown in Box Figure 7.4D.

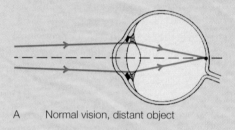

A   Normal vision, distant object

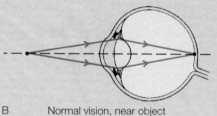

B   Normal vision, near object

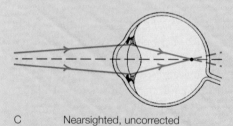

C   Nearsighted, uncorrected

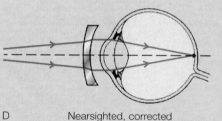

D   Nearsighted, corrected

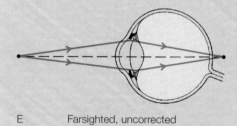

E   Farsighted, uncorrected

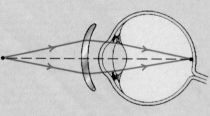

F   Farsighted, corrected

BOX FIGURE 7.4

(*A*) The relaxed, normal eye forms an image of distant objects on the retina. (*B*) For close objects, the lens of the normal eye changes shape to focus the image on the retina. (*C*) In a nearsighted eye, the image of a distant object forms in front of the retina. (*D*) A diverging lens corrects for nearsightedness. (*E*) In a farsighted eye, the image of a nearby object forms beyond the retina. (*F*) A converging lens corrects for farsightedness.

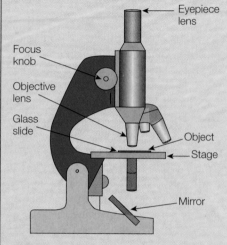

BOX FIGURE 7.5

A simple microscope uses a system of two lenses, which are an objective lens that makes an enlarged image of the specimen, and an eyepiece lens that makes an enlarged image of that image.

The microscope is an optical device used to make things look larger. It is essentially a system of two lenses, one to produce an image of the object being studied, and the other to act as a magnifying glass and enlarge that image. The power of the microscope is basically determined by the *objective lens*, which is placed close to the specimen on the stage of the microscope. Light is projected up through the specimen, and the objective lens makes an enlarged image of the specimen inside the tube between the two lenses. The *eyepiece lens* is adjusted up and down to make a sharp enlarged image of the image produced by the objective lens (Box Figure 7.5).

Telescopes are optical instruments used to provide enlarged images of near and distant objects. There are two major types of telescopes, *refracting* telescopes that use two lenses, and *reflecting* telescopes that use combinations of mirrors, or a mirror and a lens. The refracting telescope has two lenses, with the objective lens forming a reduced image, which is viewed with an eyepiece lens to enlarge that image. In reflecting telescopes,

—Continued top of next page

mirrors are used instead of lenses to collect the light (Box Figure 7.6).

Finally, the *digital camera* is one of the more recently developed light-gathering and photograph-taking optical instruments. This camera has a group of small photocells, with perhaps thousands lined up on the focal plane behind a converging lens. An image falls on the array, and each photocell stores a charge that is proportional to the amount of light falling on the

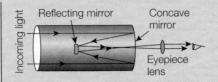

BOX FIGURE 7.6
This illustrates how the path of light moves through a simple reflecting astronomical telescope. Several different designs and mirror placements are possible.

cell. A microprocessor measures the amount of charge registered by each photocell and considers it as a pixel, a small bit of the overall image. A shade of gray or a color is assigned to each pixel and the image is ready to be enhanced, transmitted to a screen, printed, or magnetically stored for later use.

A glass prism separates sunlight into a spectrum of colors because the index of refraction is different for different wavelengths of light. The same processes that slow the speed of light in a transparent substance have a greater effect on short wavelengths than they do on longer wavelengths. As a result, violet light is refracted most, red light is refracted least, and the other colors are refracted between these extremes. This results in a beam of white light being separated, or dispersed, into a spectrum when it is refracted. Any transparent material in which the index of refraction varies with wavelength has the property of *dispersion*. The dispersion of light by ice crystals sometimes produces a colored halo around the sun and the moon.

Concepts Applied

Colors and Refraction

A convex lens is able to magnify by forming an image with refracted light. This application is concerned with magnifying, but it is really more concerned with experimenting to find an explanation.

Here are three pairs of words:

SCIENCE BOOK

RAW HIDE

CARBON DIOXIDE

Hold a cylindrical solid glass rod over the three pairs of words, using it as a magnifying glass. A clear, solid, and transparent plastic rod or handle could also be used as a magnifying glass.

Notice that some words appear inverted but others do not. Does this occur because red letters are refracted differently than blue letters?

Make some words with red and blue letters to test your explanation. What is your explanation for what you observed?

Evidence for Waves

The nature of light became a topic of debate toward the end of the 1600s as Isaac Newton published his *particle theory* of light. He believed that the straight-line travel of light could be better explained as small particles of matter that traveled at great speed from a source of light. Particles, reasoned Newton, should follow a straight line according to the laws of motion. Waves, on the other hand, should bend as they move, much as water waves on a pond bend into circular shapes as they move away from a disturbance. About the same time that Newton developed his particle theory of light, Christian Huygens (pronounced "hi-ganz") was concluding that light is not a stream of particles but rather a longitudinal wave.

Both theories had advocates during the 1700s, but the majority favored Newton's particle theory. By the beginning of the 1800s, new evidence was found that favored the wave theory, evidence that could not be explained in terms of anything but waves.

Interference

In 1801, Thomas Young published evidence of a behavior of light that could only be explained in terms of a wave model of light. Young's experiment is illustrated in Figure 7.18A. Light from a single source is used to produce two beams of light that are in phase, that is, having their crests and troughs together as they move away from the source. This light falls on a card with two slits, each less than a millimeter in width. The light is diffracted from each slit, moving out from each as an expanding arc. Beyond the card the light from one slit crosses over the light from the other slit to produce a series of bright lines on a screen. Young had produced a phenomenon of light called **interference**, and interference can only be explained by waves.

The pattern of bright lines and dark zones is called an *interference pattern* (Figure 7.18B). The light moved from each slit in phase, crest to crest and trough to trough. Light from both slits traveled the same distance directly across to the screen, so

A rainbow is a spectacular, natural display of color that is supposed to have a pot of gold under one end. Understanding the why and how of a rainbow requires information about water droplets and knowledge of how light is reflected and refracted. This information will also explain why the rainbow seems to move when you move—making it impossible to reach the end to obtain that mythical pot of gold.

First, note the pattern of conditions that occur when you see a rainbow. It usually appears when the sun is shining low in one part of the sky and rain is falling in the opposite part. With your back to the sun, you are looking at a zone of raindrops that are all showing red light, another zone that are all showing violet light, with zones of the other colors between (ROYGBV). For a rainbow to form like this requires a surface that refracts and reflects the sunlight, a condition met by spherical raindrops.

Water molecules are put together in such a way that they have a positive side and a negative side, and this results in strong molecular attractions. It is the strong attraction of water molecules for one another that results in the phenomenon of surface tension. Surface tension is the name given to the surface of water acting as if it is covered by an ultra-thin elastic membrane that is contracting. It is surface tension that pulls raindrops into a spherical shape as they fall through the air.

Box Figure 7.7 shows one thing that can happen when a ray of sunlight strikes a single spherical raindrop near the top of the drop. At this point some of the sunlight is reflected, and some is refracted into the raindrop. The refraction disperses the light into its spectrum colors, with the violet light being refracted most and red the least. The refracted light travels through the drop to the opposite side, where some of it might be reflected back into the drop. The reflected part travels back through the drop again, leaving the front surface of the raindrop. As it leaves, the light is refracted for a second time. The combined refraction, reflection, and second refraction is the source of the zones of colors you see in a rainbow. This also explains why you see a rainbow in the part of the sky opposite from the sun.

The light from any one raindrop is one color, and that color comes from all drops on the arc of a circle that is a certain angle between the incoming sunlight and the refracted light. Thus the raindrops in the red region refract red light toward your eyes at an angle of 42°, and all other colors are refracted over your head by these drops. Raindrops in the violet region refract violet light toward your eyes at an angle of 40° and the red and other colors toward your feet. Thus the light from any one drop is seen as one color, and all drops showing this color are on the arc of a circle. An arc is formed because

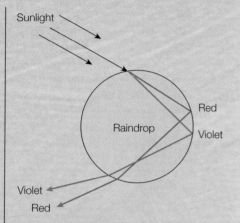

BOX FIGURE 7.7

Light is refracted when it enters a raindrop and when it leaves. The part that leaves the front surface of the raindrop is the source of the light in thousands upon thousands of raindrops from which you see zones of color—a rainbow.

the angle between the sunlight and the refracted light of a color is the same for each of the spherical drops.

There is sometimes a fainter secondary rainbow, with colors reversed, that forms from sunlight entering the bottom of the drop, reflecting twice, and then refracting out the top. The double reflection reverses the colors and the angles are 50° for the red and 54° for the violet.

---

they arrived in phase. The crests from the two slits are superimposed here, and constructive interference produces a bright line in the center of the pattern. But for positions above and below the center, the light from the two slits must travel different distances to the screen. At a certain distance above and below the bright center line, light from one slit had to travel a greater distance and arrives one-half wavelength after light from the other slit. Destructive interference produces a zone of darkness at these positions. Continuing up and down the screen, a bright line of constructive interference will occur at each position where the distance traveled by light from the two slits differs by any whole number of wavelengths. A dark zone of destructive interference will occur at each position where the distance traveled by light from the two slits differs by any half-wavelength. Thus, bright lines occur above and below the center bright line

at positions representing differences in paths of 1, 2, 3, 4, and so on wavelengths. Similarly, zones of darkness occur above and below the center bright line at positions representing differences in paths of ½, 1½, 2½, 3½, and so on wavelengths. Young found all of the experimental data such as this in full agreement with predictions from a wave theory of light. About fifteen years later, A. J. Fresnel (pronounced "fray-nel") demonstrated mathematically that diffraction as well as other behaviors of light could be fully explained with the wave theory. In 1821 Fresnel determined that the wavelength of red light was about 8×10^{-7} m and of violet light about 4×10^{-7} m, with other colors in between these two extremes. The work of Young and Fresnel seemed to resolve the issue of considering light to be a stream of particles or a wave, and it was generally agreed that light must be waves.

A laser is a device that produces a coherent beam of single-frequency, in-phase light. The beam comes from atoms that have been stimulated by electricity. Most ordinary light sources produce incoherent light; light that is emitted randomly and at different frequencies. The coherent light from a laser has the same frequency, phase, and direction, so it does not tend to spread out and it can be very intense. This has made possible a number of specialized applications, and the list of uses continues to grow.

There are different kinds of lasers in use and new ones are under development. One common type of laser is a gas-filled tube with mirrors at both ends. The mirror at one end is only partly silvered, which allows light to escape as the laser beam. The distance between the mirrors matches the resonant frequency of the light produced, so the trapped light will set up an optical standing wave. An electric discharge produces fast electrons that raise the energy level of the electrons of the specific gas atoms in the tube. The electrons of the energized gas atoms emit a particular frequency of light as they drop back to a lower level, and this emitted light sets up the standing wave. The standing wave stimulates other atoms of the gas, resulting in the emission of more light at the same frequency and phase.

Lasers are everywhere today and have connections with a wide variety of technologies. At the supermarket a laser and detector unit reads the bar code on each grocery item. The laser sends the pattern to a computer, which sends a price to the register as well as tracking the store inventory. A low-powered laser and detector also reads your CD music disc or MP3 disc and can be used to make a three-dimensional image. Most laser printers use a laser, and laser makes the operational part of a fiber optics communication system. Stronger lasers are used for cutting, drilling, and welding. Lasers are used extensively in many different medical procedures, from welding a detached retina to bloodless surgery.

Polarization

Huygens' wave theory and Newton's particle theory could explain some behaviors of light satisfactorily, but there were some behaviors that neither (original) theory could explain. Both theories failed to explain some behaviors of light, such as light moving through certain transparent crystals. For example, a slice of the mineral tourmaline transmits what appears to be a low-intensity greenish light. But if a second slice of tourmaline is placed on the first and rotated, the transmitted light passing through both slices begins to dim. The transmitted light is practically zero when the second slice is rotated 90°. Newton suggested that this behavior had something to do with "sides" or

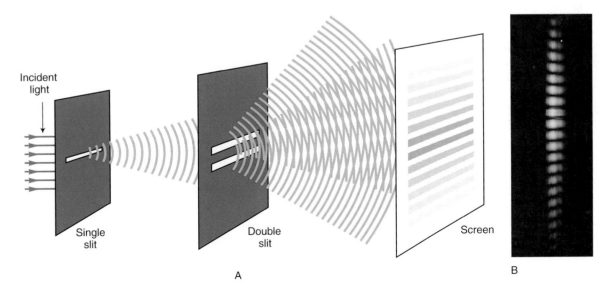

FIGURE 7.18

(A) The arrangement for Young's double-slit experiment. Sunlight passing through the first slit is coherent and falls on two slits close to each other. Light passing beyond the two slits produces an interference pattern on a screen. (B) The double-slit pattern of a small-diameter beam of light from a helium-neon laser.

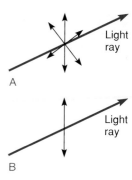

FIGURE 7.19

(A) Unpolarized light has transverse waves vibrating in all possible directions perpendicular to the direction of travel. (B) Polarized light vibrates only in one plane. In this illustration, the wave is vibrating in a vertical direction only.

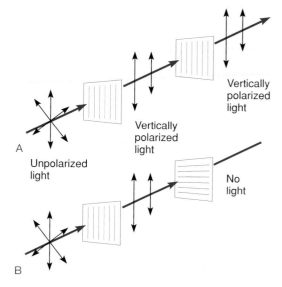

FIGURE 7.20

(A) Two crystals that are aligned both transmit vertically polarized light that looks like any other light. (B) When the crystals are crossed, no light is transmitted.

"poles" and introduced the concept of what is now called the *polarization* of light.

The waves of Huygens' wave theory were longitudinal, moving like sound waves, with wave fronts moving in the direction of travel. A longitudinal wave could not explain the polarization behavior of light. In 1817, Young modified Huygens' theory by describing the waves as *transverse*, vibrating at right angles to the direction of travel. This modification helped explain the polarization behavior of light transmitted through the two crystals and provided firm evidence that light is a transverse wave. As shown in Figure 7.19A, **unpolarized light** is assumed to consist of transverse waves vibrating in all conceivable random directions. Polarizing materials, such as the tourmaline crystal, transmit light that is vibrating in one direction only, such as the vertical direction in Figure 7.19B. Such a wave is said to be **polarized**, or *plane-polarized*, since it vibrates only in one plane. The single crystal polarized light by transmitting only waves that vibrate parallel to a certain direction while selectively absorbing waves that vibrate in all other directions. Your eyes cannot tell the difference between unpolarized and polarized light, so the light transmitted through a single crystal looks just like any other light. When a second crystal is placed on the first, the amount of light transmitted depends on the alignment of the two crystals (Figure 7.20). When the two crystals are *aligned*, the polarized light from the first crystal passes through the second with little absorption. When the crystals are *crossed* at 90°, the light transmitted by the first is vibrating in a plane that is absorbed by the second crystal, and practically all the light is absorbed. At some other angle, only a fraction of the polarized light from the first crystal is transmitted by the second.

You can verify whether or not a pair of sunglasses is made of polarizing material by rotating a lens of one pair over a lens of a second pair. Light is transmitted when the lenses are aligned but mostly absorbed at 90° when the lenses are crossed.

Light is completely polarized when all the waves are removed except those vibrating in a single direction. Light is partially polarized when some of the waves are in a particular orientation, and any amount of polarization is possible. There

are several means of producing partially or completely polarized light, including (1) selective absorption, (2) reflection, and (3) scattering.

Selective absorption is the process that takes place in certain crystals, such as tourmaline, where light in one plane is transmitted and all the other planes are absorbed. A method of manufacturing a polarizing film was developed in the 1930s by Edwin H. Land. The film is called *Polaroid*. Today Polaroid is made of long chains of hydrocarbon molecules that are aligned in a film. The long-chain molecules ideally absorb all light waves that are parallel to their lengths and transmit light that is perpendicular to their lengths. The direction that is *perpendicular* to the oriented molecular chains is thus called the polarization direction or the *transmission axis*.

Reflected light with an angle of incidence between 1° and 89° is partially polarized as the waves parallel to the reflecting surface are reflected more than other waves. Complete polarization, with all waves parallel to the surface, occurs at a particular angle of incidence. This angle depends on a number of variables, including the nature of the reflecting material. Figure 7.21 illustrates polarization by reflection. Polarizing sunglasses reduce the glare of reflected light because they have vertically oriented transmission axes. This absorbs the horizontally oriented reflected light. If you turn your head from side to side so as to rotate your sunglasses while looking at a reflected glare, you will see the intensity of the reflected light change. This means that the reflected light is partially polarized.

The phenomenon called *scattering* occurs when light is absorbed and reradiated by particles about the size of gas molecules that make up the air. Sunlight is initially unpolarized. When it strikes a molecule, electrons are accelerated and vibrate horizontally and vertically. The vibrating charges reradiate

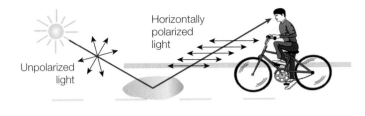

FIGURE 7.21

Light that is reflected becomes partially or fully polarized in a horizontal direction, depending on the incident angle and other variables.

polarized light. Thus, if you look at the blue sky with a pair of polarizing sunglasses and rotate them, you will observe that light from the sky is polarized. Bees are believed to be able to detect polarized skylight and use it to orient the direction of their flights. Violet and blue light have the shortest wavelengths of visible light, and red and orange light have the largest. The violet and blue rays of sunlight are scattered the most. At sunset the path of sunlight through the atmosphere is much longer than when the sun is more directly overhead. Much of the blue and violet have been scattered away as a result of the longer path through the atmosphere at sunset. The remaining light that comes through is mostly red and orange, so these are the colors you see at sunset.

Concepts Applied

Polarization

1. Describe how you could test two pairs of polarizing sunglasses to make sure they are polarized.
2. Look through the glass of a car windshield while rotating a lens of a pair of polarizing sunglasses. What evidence do you find that the windshield is or is not polarized? If it is polarized, can you determine the direction of polarization and a reason for this?
3. Look at the sky through one lens of a pair of polarizing sunglasses as you rotate the lens. What evidence do you find that light from the sky is or is not polarized? Is there any relationship between the direction of polarization and the position of the sun?
4. Position yourself with a wet street or puddle of water between you and the sun when it is low in the sky. Rotate the lens of a pair of polarizing sunglasses as you look at the glare reflected off the water. Explain how these sunglasses are able to eliminate reflected glare.

Evidence for Particles

The evidence from diffraction, interference, and polarization of light was very important in the acceptance of the wave theory because there was simply no way to explain these behaviors with a particle theory. Then, in 1850, J. L. Foucault was able to prove that light travels much more slowly in transparent materials than it does in air. This was in complete agreement with the wave theory and completely opposed to the particle theory. By the end of the 1800s, Maxwell's theoretical concept of electric and magnetic fields changed the concept of light from mechanical waves to waves of changing electric and magnetic fields. Further evidence removed the necessity for ether, the material supposedly needed for waves to move through. Light was now seen as electromagnetic waves that could move through empty space. By this time it was possible to explain all behaviors of light moving through empty space or through matter with a wave theory. Yet, there were nagging problems that the wave theory could not explain. In general, these problems concerned light that is absorbed by or emitted from matter.

Photoelectric Effect

Light is a form of energy, and it gives its energy to matter when it is absorbed. Usually the energy of absorbed light results in a temperature increase, such as the warmth you feel from absorbed sunlight. Sometimes, however, the energy from absorbed light results in other effects. In some materials, the energy is acquired by electrons, and some of the electrons acquire sufficient energy to jump out of the material. The movement of electrons as a result of energy acquired from light is known as the **photoelectric effect.** The photoelectric effect is put to a practical use in a solar cell, which transforms the energy of light into an electric current (Figure 7.22).

The energy of light can be measured with great accuracy. The kinetic energy of electrons after they absorb light can also be measured with great accuracy. When these measurements were made of the light and electrons involved in the photoelectric effect, some unexpected results were observed. Monochromatic light, that is, light of a single, fixed frequency, was used to produce the photoelectric effect. First, a low-intensity, or dim, light was used, and the numbers and energy of the ejected electrons were measured. Then a high-intensity light was used, and the numbers and energy of the ejected electrons were again measured. Measurement showed that (1) low-intensity light caused fewer electrons to be ejected, and high-intensity light caused many to be ejected, and (2) all electrons ejected from low- or high-intensity light ideally had the *same* kinetic energy. Surprisingly, the kinetic energy of the ejected electrons was found to be *independent* of the light intensity. This was contrary to what the wave theory of light would predict, since a stronger light should mean that waves with more energy have more energy to give to the electrons. Here is a behavior involving light that the wave theory could not explain.

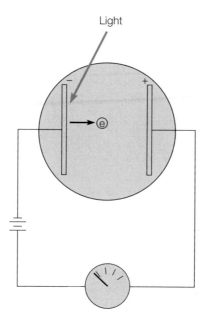

FIGURE 7.22

A setup for observing the photoelectric effect. Light strikes the negatively charged plate, and electrons are ejected. The ejected electrons move to the positively charged plate and can be measured as a current in the circuit.

This relationship says that higher-frequency light (e.g., blue light at 6.50×10^{14} Hz) has more energy than lower-frequency light (e.g., red light at 4.00×10^{14} Hz). The energy of such high- and low-frequency light can be verified by experiment.

The photon theory also explained the photoelectric effect. According to this theory, light is a stream of moving photons. It is the number of photons in this stream that determines if the light is dim or intense. A high-intensity light has many, many photons, and a low-intensity light has only a few photons. At any particular fixed frequency, all the photons would have the same energy, the product of the frequency and Planck's constant (hf). When a photon interacts with matter, it is absorbed and gives up all of its energy. In the photoelectric effect, this interaction takes place between photons and electrons. When an intense light is used, there are more photons to interact with the electrons, so more electrons are ejected. The energy given up by each photon is a function of the frequency of the light, so at a fixed frequency, the energy of each photon, hf, is the same, and the acquired kinetic energy of each ejected electron is the same. Thus, the photon theory explains the measured experimental results of the photoelectric effect.

Quantization of Energy

In addition to the problem of the photoelectric effect, there were problems with blackbody radiation, light emitted from hot objects. The experimental measurements of light emitted through blackbody radiation did not match predictions made from theory. In 1900, Max Planck (pronounced "plonk"), a German physicist, found that he could fit the experimental measurements and theory together by assuming that the vibrating molecules that emitted the light could only have a *discrete amount* of energy. Instead of energy existing through a continuous range of amounts, Planck found that the vibrating molecules could only have energy in multiples of energy in certain amounts, or **quanta** (meaning "discrete amounts"; *quantum* is singular, and *quanta* plural).

Planck's discovery of quantized energy states was a radical, revolutionary development, and most scientists, including Planck, did not believe it at the time. Planck, in fact, spent considerable time and effort trying to disprove his own discovery. It was, however, the beginning of the quantum theory, which was eventually to revolutionize physics.

Five years later, in 1905, Albert Einstein applied Planck's quantum concept to the problem of the photoelectric effect. Einstein described the energy in a light wave as quanta of energy called **photons.** Each photon has an energy E that is related to the frequency f of the light through Planck's constant h, or

$$E = hf \qquad \textbf{equation 7.4}$$

The value of Planck's constant is 6.63×10^{-34} J·s.

EXAMPLE 7.3

What is the energy of a photon of red light with a frequency of 4.00×10^{14} Hz?

Solution

The relationship between the energy of a photon (E) and its frequency (f) is found in equation 7.4. Planck's constant is given as 6.63×10^{-34} J·s.

$f = 4.00 \times 10^{14}$ Hz
$h = 6.63 \times 10^{-34}$ J·s
$E = ?$

$$E = hf$$
$$= (6.63 \times 10^{-34} \text{ J·s})\left(4.00 \times 10^{14} \frac{1}{s}\right)$$
$$= (6.63 \times 10^{-34})(4.00 \times 10^{14}) \text{ J·s} \times \frac{1}{s}$$
$$= 2.65 \times 10^{-19} \frac{\text{J·s}}{s}$$
$$= \boxed{2.65 \times 10^{-19} \text{ J}}$$

EXAMPLE **7.4**

What is the energy of a photon of violet light with a frequency of 7.00×10^{14} Hz? (Answer: 4.64×10^{-19} J)

FIGURE 7.23

It would seem very strange if there were not a sharp distinction between objects and waves in our everyday world. Yet this appears to be the nature of light.

The photoelectric effect is explained by considering light to be photons with quanta of energy, not a wave of continuous energy. This is not the only evidence about the quantum nature of light, and more will be presented in the next chapter. But, as you can see, there is a dilemma. The electromagnetic wave theory and the photon theory seem incompatible. Some experiments cannot be explained by the wave theory and seem to support the photon theory. Other experiments are contradictions, providing seemingly equal evidence to reject the photon theory in support of the wave theory.

The Present Theory

Today, light is considered to have a dual nature, sometimes acting like a wave and sometimes acting like a particle. A wave model is useful in explaining how light travels through space and how it exhibits such behaviors as refraction, interference, and diffraction. A particle model is useful in explaining how light is emitted from and absorbed by matter, exhibiting such behaviors as blackbody radiation and the photoelectric effect. Together, both of these models are part of a single theory of light, a theory that pictures light as having both particle and wave properties. Some properties are more useful when explaining some observed behaviors, and other properties are more useful when explaining other behaviors.

Frequency is a property of a wave, and the energy of a photon is a property of a particle. Both frequency and the energy of a photon are related in equation 7.4, $E = hf$. It is thus possible to describe light in terms of a frequency (or wavelength) or in terms of a quantity of energy. Any part of the electromagnetic spectrum can thus be described by units of frequency, wavelength, or energy, which are alternative means of describing light. The radio radiation parts of the spectrum are low-frequency, low-energy, and long-wavelength radiations. Radio radiations have more wave properties and practically no particle properties, since the energy levels are low. Gamma radiation, on the other hand, is high-frequency, high-energy, and short-wavelength radiation. Gamma radiation has more particle properties, since the extremely short wavelengths have very high energy levels. The more familiar part of the spectrum, visible light, is between these two extremes and exhibits both wave and particle properties, but it never exhibits both properties at the same time in the same experiment.

Part of the problem in forming a concept or mental image of the exact nature of light is understanding this nature in terms of what is already known. The things you already know about are observed to be particles, or objects, or they are observed to be waves. You can see objects that move through the air, such as baseballs or footballs, and you can see waves on water or in a field of grass. There is nothing that acts like a moving object in some situations but acts like a wave in other situations. Objects are objects, and waves are waves, but objects do not become waves, and waves do not become objects. If this dual nature did exist it would seem very strange. Imagine, for example, holding a book at a certain height above a lake (Figure 7.23). You can make measurements and calculate the kinetic energy the book will have when dropped into the lake. When it hits the water, the book disappears, and water waves move away from the point of impact in a circular pattern that moves across the water. When the waves reach another person across the lake, a book identical to the one you dropped pops up out of the water as the waves disappear. As it leaves the water across the lake, the book has the same kinetic energy that your book had when it hit the water in front of you. You and the other person could measure things about either book, and you could measure things about the waves, but you could not measure both at the same time. You might say that this behavior is not only strange but impossible. Yet, it is an analogy to the observed behavior of light.

As stated, light has a dual nature, sometimes exhibiting the properties of a wave and sometimes exhibiting the properties of moving particles but never exhibiting both properties at the same time. Both the wave and the particle nature are accepted as being part of one model today, with the understanding that the exact nature of light is not describable in terms of anything that is known to exist in the everyday-sized world. Light is an extremely small-scale phenomenon that must be different, without a sharp distinction between a particle and a wave. Evidence about this strange nature of an extremely small-scale phenomenon will be considered again in the next chapter as a basis for introducing the quantum theory of matter.

A compact disc (CD) is a laser-read (also called *optically read*) data storage device on which music, video, or any type of computer data can be stored. Three types in popular use today are the CD audio, DVD (or MP3), and the CD-ROM used to store any type of computer data. All three kinds of players rotate between 200 and 500 revolutions per minute, but the drive changes speed to move the head at a constant linear velocity over the recording track, faster near the inner hub and slower near the outer edge of the disc. Furthermore, the CD drive reads from the inside out, so the disc will slow as it is played.

The CD itself is a 12 cm diameter, 1.3 mm thick sandwich of a hard plastic core, a mirrorlike layer of metallic aluminum, and a tough, clear plastic overcoating that protects the thin layer of aluminum. The CD records digitized data; music, video, or computer data that have been converted into a string of binary numbers. First, a master disc is made. The binary numbers are translated into a series of pulses that are fed to a laser. The laser is focused onto a photosensitive material on a spinning master disc. Whenever there is a pulse in the signal, the laser burns a small oval pit into the surface, making a pattern of pits and bumps on the track of the master disc. The laser beam is incredibly small, making marks about a micron or so in diameter. A micron is one-millionth of a meter, so you can fit a tremendous number of data tracks onto the disc, which has each track spaced 1.6 microns apart. Next, the CD audio or CD-ROM discs are made by using the master disc as a mold. Soft plastic is pressed against the master disc in a vacuum-forming machine so the small physical marks—the pits and bumps made by the laser—are pressed into the plastic. This makes a record of the strings of binary numbers that were etched into the master disc by the strong but tiny laser beam. During playback, a low-powered laser beam is reflected off the track to read the binary marks on it. The optical sensor head contains a tiny diode laser, a lens, mirrors, and tracking devices that can move the head in three directions. The head moves side to side to keep the head over a single track (within 1.6 micron), it moves up and down to keep the laser beam in focus, and it moves forward and backward as a fine adjustment to maintain a constant linear velocity.

The advantages of the CD audio and video discs over conventional records or tapes include more uniform and accurate frequency response, a complete absence of background noise, and absence of wear—since nothing mechanical touches the surface of the disc when it is played. The advantages of the CD-ROM over the traditional magnetic floppy disks or magnetic hard disks include storage capacity, long-term reliability, and the impossibility of a head crash with a resulting loss of data.

The disadvantage of the CD audio and CD-ROM discs is the lack of ability to do writing or rewriting. Rewritable optical media are available, and these are called CD-R and CD-RW.

A CD-R records data to a disc by using a laser to burn spots into an organic dye. Such a "burned" spot reflects less light than an area that was not heated by the laser. This is designed to mimic the way light reflects from pits and bumps of a normal CD, except this time the string of binary numbers are burned (nonreflective) and not burned areas (reflective). Since this is similar to how data on a normal CD is represented, a CD-R disc can generally be used in a CD audio or CD-ROM player as if it were a pressed CD. The dyes in a CD-R disc are photosensitive organic compounds that are similar to those used in making photographs. The color of a CD-R disc is a combined result of the kind of dye that was used in the recording layer combined with the type of reflective coating used. Some of these dye and reflective coating combinations appear green, some appear blue, and others appear to be gold. Once a CD-R disc is burned it cannot be rewritten or changed.

The CD-RW is designed to have the ability to do writing or rewriting. It uses a different technology, but again mimics the way light reflects from the pits and bumps of a pressed audio CD. Instead of a dye-based recording layer the CD-RW uses a compound made from silver, indium, antimony, and tellurium. This layer has a property that permits rewriting the information on a disc. The nature of this property is that when it is heated to a certain temperature and cooled it becomes crystalline. However, when it is heated to a higher temperature and cooled it becomes noncrystalline. A crystalline surface reflects a laser beam while a noncrystalline surface absorbs the laser beam. The CD-RW is again designed to mimic the way light reflects from the pits and bumps of a normal CD, except this time the string of binary numbers are noncrystalline (nonreflective) and crystalline areas (reflective). In order to write, erase, and read the CD-RW recorder must have three different laser powers. It must have (1) a high power to heat spots to about 600°C, which cool rapidly and make noncrystalline spots that are less reflective. It must have (2) a medium power to erase data by heating the media to about 200°C, which allows the media to crystallize and have a uniform reflectivity. Finally, it must have (3) a low setting that is used for finding and reading nonreflective and the more reflective areas of a disc. The writing and rewriting of a CD-RW can be repeated hundreds of times. A problem is that not all CD players can read the less reflective data bumps of a CD-RW unless they have dual-wavelength heads that meet multi-read specifications.

The communication capacities of fiber optics are enormous. When fully developed and exploited, a single fiber the size of a human hair could in principle carry all the telephone conversations and all the radio and television broadcasts in the United States at one time without interfering with one another. This enormous capacity is possible because of a tremendous range of frequencies in a beam of light. This range determines the information-carrying capacity, which is called *bandwidth*. The traditional transmission of telephone communication depends on the movement of electrons in a metallic wire, and a single radio station usually broadcasts on a single frequency. Light has a great number of different frequencies, as you can imagine when looking at Table 7.2. An optic fiber has a bandwidth about a million times greater than the frequency used by a radio station.

Physically, an optical fiber has two main parts. The center part, called the *core*, is practically pure silica glass with an index of refraction of about 1.6. Larger fibers, about the size of a human hair or larger, transmit light by total internal reflection. Light rays internally reflect back and forth about a thousand times or so in each 15 cm length of such an optical fiber core. Smaller fibers, about the size of a light wavelength, act as waveguides, with rays moving straight through the core. A layer of transparent material called the *cladding* is used to cover the core. The cladding has a lower index of refraction, perhaps 1.50, and helps keep the light inside the core.

Presently fiber optics systems are used to link telephone offices across the United States and other nations, and in a submarine cable that links many nations. The electrical signals of a telephone conversation are converted to light signals, which travel over optical fibers. The light signals are regenerated by devices called *repeaters* until the signal reaches a second telephone office. Here, the light signals are transformed to electrical signals, which are then transmitted by more traditional means to homes and businesses. Today the laying of fiber optics lines to homes and businesses is a goal of nearly every major telephone company in the United States.

One way of transmitting information over a fiber optics system is to turn the transmitting light on and off rapidly. Turning the light on and off is not a way of sending ordinary codes or signals, but is a rapid and effective way of sending *digital information*. Digital information is information that has been converted into digits, that is, into a string of numbers. Fax machines, CD-ROM players, computers, and many other modern technologies use digitized information. Closely examine a black-and-white newspaper photograph, and you will see that it is made up of many tiny dots that vary in density from place to place. The image can be digitized by identifying the location of every tiny piece of the image, which is called a *pixel*, with a string of numbers. Once digitized, an image can be manipulated by a computer, reproduced by a printer, sent over a phone line by a fax machine, or stored for future use. Most fax machines transmit images with a resolution of about 240 pixels per inch, which is also called 240 dots per inch (dpi).

How is a telephone conversation digitized? The mouthpiece part of a telephone converts the pressure fluctuations of a voice sound wave into electric current fluctuations. This fluctuating electric current is digitized by a logic circuit that samples the amplitude of the wave at regular intervals, perhaps thousands of times per second. The measured amplitude of a sample (i.e., how far the sample is above or below the baseline) is now described by numbers. The succession of sample heights becomes a succession of numbers, and the frequency and amplitude pattern of a voice is now digitized.

Digitized information is transmitted and received by using the same numbering system that is used by a computer. A computer processes information by recognizing if a particular circuit carries an electric current or not, that is, if a circuit is on or off. The computer thus processes information by counting in twos, using what is called the *binary system*. Unlike our ordinary decimal system, in which we count by tens, the binary system has only two symbols, the digits 0 and 1. For example, the decimal number 4 has a binary notation of 0100; the decimal number 5 is 0101, 6 is 0110, 7 is 0111, 8 is 1000, and so on. In good sound-recording systems, a string of 16 binary digits is used to record the magnitude of each sample.

Digitized information is transmitted by translating the binary numbers, the 0s and 1s, into a series of pulses that are fed to a laser (with 0 meaning off, and 1 meaning on). Millions of bits per second can be transmitted over the fiber optics system to a receiver at the other end. The receiver recognizes the very rapid on-and-off patterns as binary numbers, which are converted into the original waveform. The waveform is now transformed into electric current fluctuations, just like the electric current fluctuations that occurred at the other end of the circuit. These fluctuations are amplified and converted back into the pressure fluctuations of the original sound waveform by the earpiece of the receiving telephone.

Currently, new optical fiber systems have increased almost tenfold the number of conversations that can be carried through the much smaller, lighter-weight fiber optics. Extension of the optical fiber to the next generation of computers will be the next logical step, with hundreds of available television channels, two-way videophone conversations, and unlimited data availability. The use of light for communications will have come a long way since the first use of signaling mirrors and flashing lamps from mountain tops and church steeples.

Thomas Young (1773–1829)

Thomas Young was a British physicist and physician who discovered the principle of interference of light, showing it to be caused by light waves. He also made important discoveries in the physiology of vision and is also remembered for Young's modulus, the ratio of stress to strain in elasticity. In addition, Young was also an Egyptologist and was instrumental in the deciphering of hieroglyphics.

Young was born in Milverton, Somerset, on June 13, 1773. He was an infant prodigy, learning to read by the age of two, whereupon he read the whole Bible twice through. Young was largely self-taught, although he did attend school. He developed great ability at languages, taking particular interest in the ancient languages of the Middle East, and mastered mathematics, physics, and chemistry while still a youth.

Young's early work was concerned with the physiology of vision. In 1793, he recognized that the mechanism of accommodation in focusing the eye on near or distant objects is due to a change of shape in the lens of the eye. Young confirmed this view in 1801 after obtaining experimental proof of it. He also showed that astigmatism, from which he himself suffered, is due to irregular curvature of the cornea. In 1801, Young was also the first to recognize that color sensation is due to the presence in the retina of structures that respond to the three colors red, green, and violet, showing that color blindness is due to the inability of one or more of these structures to respond to light. Young's work in this field was elaborated into a proper theory of vision by Hermann Helmholtz (1821–1894) and James Clerk Maxwell (1831–1879).

Young also made a thorough study of optics at the same time. There was great controversy as to the nature of light—whether it consisted of streams of particles or of waves. In Britain, the particulate theory was strongly favored because it had been advanced by Isaac Newton (1642–1727) a century before. In 1800, Young reopened the debate by suggesting that the wave theory of

Christiaan Huygens (1629–1695) gave a more convincing explanation for the phenomena of reflection, refraction, and diffraction. He assumed that light waves are propagated in a similar way to sound waves and are longitudinal vibrations but with a different medium and frequency. He also proposed that different colors consist of different frequencies.

In the following year, Young announced his discovery of the principle of interference. He explained that the bright bands of fringes in effects such as Newton's rings result from light waves interfering so that they reinforce each other. This was a hypothetical deduction and over the next two years, Young obtained experimental proof for the principle of interference by passing light through extremely narrow openings and observing the interference patterns produced.

Young's discovery of interference was convincing proof that light consists of waves, but it did not confirm that the waves are longitudinal. The discovery of the polarization of light by Etienne Malus (1775–1812) in 1808 led Young to suggest in 1817 that light waves may contain a transverse component. In 1821, Augustin Fresnel (1788–1827) proved that light waves are entirely transverse, and the wave nature of light was finally

established. In mechanics, Young established many important concepts in a course of lectures published in 1807. He was the first to use the terms "energy" for the product of the mass of a body with the square of its velocity and the expression "labor expended" (i.e., work done) for the product of the force exerted on a body "with the distance through which it moved." He also stated that these two products are proportional to each other. He introduced absolute measurements in elasticity by defining the modulus as the weight that would double the length of a rod of unit cross-section. Today it is usually referred to as Young's modulus, and it equals stress divided by strain in the elastic region of the loading of a material. Young also published in 1805 a theory of capillary action and accounted for the angle of contact in surface-tension effects for liquids.

From 1815 onward, Young published papers on Egyptology, mostly concerning the reading of tablets involving hieroglyphics. He was one of the first to interpret the writings on the Rosetta Stone, which was found at the mouth of the Nile in 1799. The hieroglyphic vocabulary that he established of approximately 200 signs has stood the test of time remarkably well.

Thomas Young was a scientist who possessed an extraordinary range of talents and a rare degree of insight, and he was able to initiate important paths of investigation that others were to take up and complete. His discoveries in vision are fundamental to an understanding of visual problems and to color reproduction, and his discovery of interference was a vital step to a full explanation of the nature of light.

Young demonstrated the interference of light by passing monochromatic light through a narrow slit S (he originally used a small hole) and then letting it pass through a pair of closely spaced slits. Reinforcement and cancellation of the wave trains as they reached the screen produced characteristic interference fringes of alternate light and dark bands.

Source: From the Hutchinson *Dictionary of Scientific Biography*. © Research Machines plc [2003] All Rights Reserved. Helicon Publishing is a division of Research Machines.

Summary

Electromagnetic radiation is emitted from all matter with a temperature above absolute zero, and as the temperature increases, more radiation and shorter wavelengths are emitted. Visible light is emitted from matter hotter than about 700°C, and this matter is said to be *incandescent*. The sun, a fire, and the ordinary lightbulb are incandescent sources of light.

The behavior of light is shown by a light ray model that uses straight lines to show the straight-line path of light. Light that interacts with matter is *reflected* with parallel rays, moves in random directions by *diffuse reflection* from points, or is *absorbed*, resulting in a temperature increase. Matter is *opaque*, reflecting light, or *transparent*, transmitting light.

In reflection, the incoming light, or *incident ray*, has the same angle as the *reflected ray* when measured from a perpendicular from the point of reflection, called the *normal*. That the two angles are equal is called the *law of reflection*. The law of reflection explains how a flat mirror forms a *virtual image*, one from which light rays do not originate. Light rays do originate from the other kind of image, a *real image*.

Light rays are bent, or *refracted*, at the boundary when passing from one transparent medium to another. The amount of refraction depends on the *incident angle* and the *index of refraction*, a ratio of the speed of light in a vacuum to the speed of light in the medium. When the refracted angle is 90°, *total internal reflection* takes place. This limit to the angle of incidence is called the *critical angle*, and all light rays with an incident angle at or beyond this angle are reflected internally.

Each color of light has a range of wavelengths that forms the *spectrum* from red to violet. A glass prism has the property of *dispersion*, separating a beam of white light into a spectrum. Dispersion occurs because the index of refraction is different for each range of colors, with short wavelengths refracted more than larger ones.

A wave model of light can be used to explain interference and polarization. *Interference* occurs when light passes through two small slits or holes and produces an *interference pattern* of bright lines and dark zones. *Polarized light* vibrates in one direction only, in a plane. Light can be polarized by certain materials, by reflection, or by scattering. Polarization can only be explained by a transverse wave model.

A wave model fails to explain observations of light behaviors in the *photoelectric effect* and *blackbody radiation*. Max Planck found that he could modify the wave theory to explain blackbody radiation by assuming that vibrating molecules could only have discrete amounts, or *quanta*, of energy and found that the quantized energy is related to the frequency and a constant known today as *Planck's constant*. Albert Einstein applied Planck's quantum concept to the photoelectric effect and described a light wave in terms of quanta of energy called *photons*. Each photon has an energy that is related to the frequency and Planck's constant.

Today, the properties of light are explained by a model that incorporates both the wave and the particle nature of light. Light is considered to have both wave and particle properties and is not describable in terms of anything known in the everyday-sized world.

Summary of Equations

7.1

$$\text{angle of incidence} = \text{angle of reflection}$$

$$\theta_i = \theta_r$$

7.2

$$\text{index of refraction} = \frac{\text{speed of light in vacuum}}{\text{speed of light in material}}$$

$$n = \frac{c}{v}$$

7.3

$$\text{speed of light in vacuum} = (\text{wavelength})(\text{frequency})$$

$$c = \lambda f$$

7.4

$$\text{energy of photon} = \left(\begin{array}{c}\text{Planck's}\\\text{constant}\end{array}\right)(\text{frequency})$$

$$E = hf$$

KEY TERMS

blackbody radiation (p. **180**)

incandescent (p. **180**)

index of refraction (p. **187**)

interference (p. **191**)

light ray model (p. **182**)

luminous (p. **180**)

photoelectric effect (p. **195**)

photons (p. **196**)

polarized (p. **194**)

quanta (p. **196**)

real image (p. **185**)

refraction (p. **185**)

total internal reflection (p. **186**)

unpolarized light (p. **194**)

virtual image (p. **185**)

APPLYING THE CONCEPTS

1. A luminous object is an object that
   a. reflects a dim blue-green light in the dark.
   b. produces light of its own by any method.
   c. shines by reflected light only, such as the moon.
   d. an object that glows only in the absence of light.

2. An object is hot enough to emit a dull red glow. When this object is heated even more, it will
   a. emit shorter-wavelength, higher-frequency radiation.
   b. emit longer-wavelength, lower-frequency radiation.
   c. emit the same wavelengths as before, but with more energy.
   d. emit more of the same wavelengths with more energy.

3. The difference in the light emitted from a candle, an incandescent lightbulb, and the sun is basically from differences in
   a. energy sources.
   b. materials.
   c. temperatures.
   d. phases of matter.

4. Before it travels through the earth's atmosphere, sunlight is
   a. infrared radiation.
   b. visible light.
   c. ultraviolet radiation.
   d. all of the above.

5. You are able to see in shaded areas, such as under a tree, because light has undergone
   a. refraction.
   b. incident bending.
   c. a change in speed.
   d. diffuse reflection.

6. An image that is not produced by light rays coming from the image, but is the result of your brain's interpretations of light rays is called a(an)
   a. real image.
   b. imagined image.
   c. virtual image.
   d. phony image.

7. Light traveling at some angle as it moves from water into the air is refracted away from the normal as it enters the air, so the fish you see under water is actually (draw a sketch if needed)
   a. above the refracted image.
   b. below the refracted image.
   c. beside the refracted image.
   d. in the same place as the refracted image.

8. The ratio of the speed of light in a vacuum to the speed of light in some transparent materials is called
   a. the critical angle.
   b. total internal reflection.
   c. the law of reflection.
   d. the index of refraction.

9. Any part of the electromagnetic spectrum, including the colors of visible light, can be measured in units of
   a. wavelength.
   b. frequency.
   c. energy.
   d. any of the above.

10. A prism separates the colors of sunlight into a spectrum because
    a. each wavelength of light has its own index of refraction.
    b. longer wavelengths are refracted more than shorter wavelengths.
    c. red light is refracted the most, violet the least.
    d. all of the above.

11. Light moving through a small pinhole does not make a shadow with a distinct, sharp edge because of
    a. refraction.
    b. diffraction.
    c. polarization.
    d. interference.

12. Which of the following can only be explained by a wave model of light?
    a. reflection
    b. refraction
    c. interference
    d. photoelectric effect

13. The polarization behavior of light is best explained by considering light to be
    a. longitudinal waves.
    b. transverse waves.
    c. particles.
    d. particles with ends, or poles.

14. The sky appears to be blue when the sun is high in the sky because
    a. blue is the color of air, water, and other fluids in large amounts.
    b. red light is scattered more than blue.
    c. blue light is scattered more than the other colors.
    d. none of the above.

15. The photoelectric effect proved to be a problem for a wave model of light because
    a. the number of electrons ejected varied directly with the intensity of the light.
    b. the light intensity had no effect on the energy of the ejected electrons.
    c. the energy of the ejected electrons varied inversely with the intensity of the light.
    d. the energy of the ejected electrons varied directly with the intensity of the light.

16. Max Planck made the revolutionary discovery that the energy of vibrating molecules involved in blackbody radiation existed only in
    a. multiples of certain fixed amounts.
    b. amounts that smoothly graded one into the next.
    c. the same, constant amount of energy in all situations.
    d. amounts that were never consistent from one experiment to the next.

17. Einstein applied Planck's quantum discovery to light and found
    a. a direct relationship between the energy and frequency of light.
    b. that the energy of a photon divided by the frequency of the photon always equaled a constant known as Planck's constant.
    c. that the energy of a photon divided by Planck's constant always equaled the frequency.
    d. all of the above.

18. Today, light is considered to be
    a. tiny particles of matter that move through space, having no wave properties.
    b. electromagnetic waves only, with no properties of particles.
    c. a small-scale phenomenon without a sharp distinction between particle and wave properties.
    d. something that is completely unknown.

Answers

1. b 2. a 3. c 4. d 5. d 6. c 7. b 8. d 9. d 10. a 11. b 12. c 13. b 14. c 15. b 16. a 17. d 18. c

1. What determines if an electromagnetic wave emitted from an object is a visible light wave or a wave of infrared radiation?
2. What model of light does the polarization of light support? Explain.
3. Which carries more energy, red light or blue light? Should this mean anything about the preferred color of warning and stop lights? Explain.
4. What model of light is supported by the photoelectric effect? Explain.
5. What happens to light that is absorbed by matter?
6. One star is reddish, and another is bluish. Do you know anything about the relative temperatures of the two stars? Explain.
7. When does total internal reflection occur? Why does this occur in the diamond more than other gemstones?
8. Why does a highway sometimes appear wet on a hot summer day when it is not wet?
9. How can you tell if a pair of sunglasses is polarizing or not?
10. What conditions are necessary for two light waves to form an interference pattern of bright lines and dark areas?
11. Explain why the intensity of reflected light appears to change if you tilt your head from side to side while wearing polarizing sunglasses.
12. Why do astronauts in orbit around Earth see a black sky with stars that do not twinkle but see a blue Earth?
13. What was so unusual about Planck's findings about blackbody radiation? Why was this considered revolutionary?
14. Why are both the photon model and the electromagnetic wave model accepted today as a single theory? Why was this so difficult for people to accept at first?

The exercises in groups A and B cover the same concepts. Solutions to group A exercises are located in appendix D.

Group A

1. What is the speed of light while traveling through (a) water and (b) ice?
2. How many minutes are required for sunlight to reach the earth if the sun is 1.50×10^8 km from the earth?
3. How many hours are required before a radio signal from a space probe near the planet Pluto reaches Earth, 6.00×10^9 km away?
4. A light ray is reflected from a mirror with an angle 10° to the normal. What was the angle of incidence?
5. Light travels through a transparent substance at 2.20×10^8 m/s. What is the substance?
6. The wavelength of a monochromatic light source is measured to be 6.00×10^{-7} m in a diffraction experiment. (a) What is the frequency? (b) What is the energy of a photon of this light?
7. At a particular location and time, sunlight is measured on a 1 square meter solar collector with an intensity of 1,000.0 W. If the peak intensity of this sunlight has a wavelength of 5.60×10^{-7} m, how many photons are arriving each second?
8. A light wave has a frequency of 4.90×10^{14} cycles per second. (a) What is the wavelength? (b) What color would you observe (see Table 7.2)?
9. What is the energy of a gamma photon of frequency 5.00×10^{20} Hz?
10. What is the energy of a microwave photon of wavelength 1.00 mm?
11. What is the speed of light traveling through glass?
12. What is the frequency of light with a wavelength of 5.00×10^{-7} m?

Group B

1. (a) What is the speed of light while traveling through a vacuum? (b) while traveling through air at 30°C? (c) while traveling through air at 0°C?
2. How much time is required for reflected sunlight to travel from the Moon to Earth if the distance between Earth and the Moon is 3.85×10^5 km?
3. How many minutes are required for a radio signal to travel from Earth to a space station on Mars if the planet Mars is 7.83×10^7 km from Earth?
4. An incident light ray strikes a mirror with an angle of 30° to the surface of the mirror. What is the angle of the reflected ray?
5. The speed of light through a transparent substance is 2.00×10^8 m/s. What is the substance?
6. A monochromatic light source used in a diffraction experiment has a wavelength of 4.60×10^{-7} m. What is the energy of a photon of this light?
7. In black-and-white photography, a photon energy of about 4.00×10^{-19} J is needed to bring about the changes in the silver compounds used in the film. Explain why a red light used in a darkroom does not affect the film during developing.
8. The wavelength of light from a monochromatic source is measured to be 6.80×10^{-7} m. (a) What is the frequency of this light? (b) What color would you observe?
9. How much greater is the energy of a photon of ultraviolet radiation ($\lambda = 3.00 \times 10^{-7}$ m) than the energy of an average photon of sunlight ($\lambda = 5.60 \times 10^{-7}$ m)?
10. At what rate must electrons in a wire vibrate to emit microwaves with a wavelength of 1.00 mm?
11. What is the speed of light in ice?
12. What is the frequency of a monochromatic light used in a diffraction experiment that has a wavelength of 4.60×10^{-7} m?

13. What is the energy of a photon of orange light with a frequency of 5.00×10^{14} Hz?

14. What is the energy of a photon of blue light with a frequency of 6.50×10^{14} Hz?

15. At a particular location and time, sunlight is measured on a one-square-meter solar collector with an intensity of 1,000.0 watts. If the peak intensity of this sunlight has a wavelength of 5.60×10^{-7} m, how many photons are arriving each second?

13. What is the energy of a photon of red light with a frequency of 4.3×10^{14} Hz?

14. What is the energy of a photon of ultraviolet radiation with a wavelength of 3.00×10^{-7} m?

15. At a particular location and time, sunlight is measured on a one-square-meter solar collector with an intensity of 500.0 watts. If the peak intensity of this sunlight has a wavelength of 5.60×10^{-7} m, how many photons are arriving each second?

Conversion Factors

Length
1 in = 2.54 cm
1 cm = 0.394 in
1 ft = 30.5 cm
1 m = 39.4 in = 3.281 ft
1 km = 0.621 mi
1 mi = 5280 ft = 1.609 km
1 light-year = 9.461×10^{15} m

Mass
1 lb = 453.6 g (where g = 9.8 m/s^2)
1 kg = 2.205 lb (where g = 9.8 m/s^2)
1 atomic mass unit u = 1.66061×10^{-27} kg

Volume
1 liter = 1.057 quarts
1 in^3 = 16.39 cm^3
1 gallon = 3.786 liter
1 ft^3 = 0.02832 m^3

Energy
1 cal = 4.184 J
1 J = 0.738 ft·lb = 0.239 cal
1 ft·lb = 1.356 J
1 Btu = 252 cal = 778 ft·lb
1 kWh = 3.60×10^6 J = 860 kcal
1 hp = 550 ft·lb/s = 746 W
1 W = 0.738 ft·lb/s
1 Btu/h = 0.293 W
Absolute zero (0K) = $-273.15°C$
1 J = 6.24×10^{18} eV
1 eV = 1.6022×10^{-19} J

Speed
1 km/h = 0.2778 m/s = 0.6214 mi/h
1 m/s = 3.60 km/h = 2.237 mi/h = 3.281 ft/s
1 mi/h = 1.61 km/h = 0.447 m/s = 1.47 ft/s
1 ft/s = 0.3048 m/s = 0.6818 mi/h

Force
1 N = 0.2248 lb
1 lb = 4.448 N

Pressure
1 atm = 1.013 bar = 1.013×10^5 N/m^2 = 14.7 lb/in^2
1 lb/in^2 = 6.90×10^3 N/m^2

Metric Prefixes

Prefix	Symbol	Meaning	Unit Multiplier
exa-	E	quintillion	10^{18}
peta-	P	quadrillion	10^{15}
tera-	T	trillion	10^{12}
giga-	G	billion	10^9
mega-	M	million	10^6
kilo-	k	thousand	10^3
hecto-	h	hundred	10^2
deka-	da	ten	10^1
unit			
deci-	d	one-tenth	10^{-1}
centi-	c	one-hundredth	10^{-2}
milli-	m	one-thousandth	10^{-3}
micro-	μ	one-millionth	10^{-6}
nano-	n	one-billionth	10^{-9}
pico-	p	one-trillionth	10^{-12}
femto-	f	one-quadrillionth	10^{-15}
atto-	a	one-quintillionth	10^{-18}

Physical Constants

Quantity	Approximate Value
Gravity (earth)	$g = 9.8$ m/s^2
Gravitational law constant	$G = 6.67 \times 10^{-11}$ N·m^2/kg^2
Earth radius (mean)	6.38×10^6 m
Earth mass	5.97×10^{24} kg
Earth-sun distance (mean)	1.50×10^{11} m
Earth-moon distance (mean)	3.84×10^8 m
Fundamental charge	1.60×10^{-19} C
Coulomb law constant	$k = 9.00 \times 10^9$ N·m^2/C^2
Electron rest mass	9.11×10^{-31} kg
Proton rest mass	1.6726×10^{-27} kg
Neutron rest mass	1.6750×10^{-27} kg
Bohr radius	5.29×10^{-11} m
Avogadro's number	6.022045×10^{23}/mol
Planck's constant	6.62×10^{-34} J·s
Speed of light (vacuum)	3.00×10^8 m/s
Pi	$\pi = 3.1415926536$

Periodic Table of the Elements

Period																		

Alkali Metals / **Alkaline Earth Metals** / **Transition Elements** / **Halogens** / **Noble gases**

Group labels: IA (1), IIA (2), IIIB (3), IVB (4), VB (5), VIB (6), VIIB (7), VIIIB (8), VIIIB (9), VIIIB (10), IB (11), IIB (12), IIIA (13), IVA (14), VA (15), VIA (16), VIIA (17), VIIIA (18)

Period 1
- Hydrogen 1 H 1.008
- Helium 2 He 4.003

Period 2
- Lithium 3 Li 6.941
- Beryllium 4 Be 9.012
- Boron 5 B 10.81
- Carbon 6 C 12.01
- Nitrogen 7 N 14.01
- Oxygen 8 O 16.00
- Fluorine 9 F 19.00
- Neon 10 Ne 20.18

Period 3
- Sodium 11 Na 22.99
- Magnesium 12 Mg 24.31
- Aluminum 13 Al 26.98
- Silicon 14 Si 28.09
- Phosphorus 15 P 30.97
- Sulfur 16 S 32.07
- Chlorine 17 Cl 35.45
- Argon 18 Ar 39.95

Period 4
- Potassium 19 K 39.10
- Calcium 20 Ca 40.08
- Scandium 21 Sc 44.96
- Titanium 22 Ti 47.88
- Vanadium 23 V 50.94
- Chromium 24 Cr 52.00
- Manganese 25 Mn 54.94
- Iron 26 Fe 55.85
- Cobalt 27 Co 58.93
- Nickel 28 Ni 58.69
- Copper 29 Cu 63.55
- Zinc 30 Zn 65.39
- Gallium 31 Ga 69.72
- Germanium 32 Ge 72.61
- Arsenic 33 As 74.92
- Selenium 34 Se 78.96
- Bromine 35 Br 79.90
- Krypton 36 Kr 83.80

Period 5
- Rubidium 37 Rb 85.47
- Strontium 38 Sr 87.62
- Yttrium 39 Y 88.91
- Zirconium 40 Zr 91.22
- Niobium 41 Nb 92.91
- Molybdenum 42 Mo 95.94
- Technetium 43 Tc (98)
- Ruthenium 44 Ru 101.1
- Rhodium 45 Rh 102.9
- Palladium 46 Pd 106.4
- Silver 47 Ag 107.9
- Cadmium 48 Cd 112.4
- Indium 49 In 114.8
- Tin 50 Sn 118.7
- Antimony 51 Sb 121.8
- Tellurium 52 Te 127.6
- Iodine 53 I 126.9
- Xenon 54 Xe 131.3

Period 6
- Cesium 55 Cs 132.9
- Barium 56 Ba 137.3
- Lanthanum 57 La 138.9
- Hafnium 72 Hf 178.5
- Tantalum 73 Ta 180.9
- Tungsten 74 W 183.8
- Rhenium 75 Re 186.2
- Osmium 76 Os 190.2
- Iridium 77 Ir 192.2
- Platinum 78 Pt 195.1
- Gold 79 Au 197.0
- Mercury 80 Hg 200.6
- Thallium 81 Tl 204.4
- Lead 82 Pb 207.2
- Bismuth 83 Bi 209.0
- Polonium 84 Po (209)
- Astatine 85 At (210)
- Radon 86 Rn (222)

Period 7
- Francium 87 Fr (223)
- Radium 88 Ra (226)
- Actinium 89 Ac (227)
- Rutherfordium 104 Rf (261)
- Dubnium 105 Db (262)
- Seaborgium 106 Sg (266)
- Bohrium 107 Bh (264)
- Hassium 108 Hs (277)
- Meitnerium 109 Mt (268)
- Darmstadtium 110 Ds (281)
- Unununium 111 Uuu (272)
- Ununbium 112 Uub (285)
- Ununquadium 114 Uuq (289)

Inner Transition Elements

Lanthanides (6)
- Cerium 58 Ce 140.1
- Praseodymium 59 Pr 140.9
- Neodymium 60 Nd 144.2
- Promethium 61 Pm (145)
- Samarium 62 Sm 150.4
- Europium 63 Eu 152.0
- Gadolinium 64 Gd 157.3
- Terbium 65 Tb 158.9
- Dysprosium 66 Dy 162.5
- Holmium 67 Ho 164.9
- Erbium 68 Er 167.3
- Thulium 69 Tm 168.9
- Ytterbium 70 Yb 173.0
- Lutetium 71 Lu 175.0

Actinides (7)
- Thorium 90 Th 232.0
- Protactinium 91 Pa 231.0
- Uranium 92 U 238.0
- Neptunium 93 Np (237)
- Plutonium 94 Pu (244)
- Americium 95 Am (243)
- Curium 96 Cm (247)
- Berkelium 97 Bk (247)
- Californium 98 Cf (251)
- Einsteinium 99 Es (252)
- Fermium 100 Fm (257)
- Mendelevium 101 Md (258)
- Nobelium 102 No (259)
- Lawrencium 103 Lr (262)

Legend:
- Metals
- Semiconductors
- Nonmetals

Values in parentheses are the mass numbers of the most stable or best-known isotopes.

Names and symbols for elements 111–114 are under review.

Key
- element name — Hydrogen
- atomic number — 1
- symbol of element — H
- atomic weight — 1.008

591

Appendix A
Mathematical Review

Working with Equations

Many of the problems of science involve an equation, a shorthand way of describing patterns and relationships that are observed in nature. Equations are also used to identify properties and to define certain concepts, but all uses have well-established meanings, symbols that are used by convention, and allowed mathematical operations. This appendix will assist you in better understanding equations and the reasoning that goes with the manipulation of equations in problem-solving activities.

Background

In addition to a knowledge of rules for carrying out mathematical operations, an understanding of certain quantitative ideas and concepts can be very helpful when working with equations. Among these helpful concepts are (1) the meaning of inverse and reciprocal, (2) the concept of a ratio, and (3) fractions.

The term *inverse* means the opposite, or reverse, of something. For example, addition is the opposite, or inverse, of subtraction, and division is the inverse of multiplication. A *reciprocal* is defined as an inverse multiplication relationship between two numbers. For example, if the symbol n represents any number (except zero), then the reciprocal of n is $1/n$. The reciprocal of a number $(1/n)$ multiplied by that number (n) always gives a product of 1. Thus, the number multiplied by 5 to give 1 is $1/5$ ($5 \times 1/5 = 5/5 = 1$). So $1/5$ is the reciprocal of 5, and 5 is the reciprocal of $1/5$. Each number is the *inverse* of the other.

The fraction $1/5$ means 1 divided by 5, and if you carry out the division it gives the decimal 0.2. Calculators that have a $1/x$ key will do the operation automatically. If you enter 5, then press the $1/x$ key, the answer of 0.2 is given. If you press the $1/x$ key again, the answer of 5 is given. Each of these numbers is a reciprocal of the other.

A *ratio* is a comparison between two numbers. If the symbols m and n are used to represent any two numbers, then the ratio of the number m to the number n is the fraction m/n. This expression means to divide m by n. For example, if m is 10 and n is 5, the ratio of 10 to 5 is $10/5$, or $2:1$.

Working with *fractions* is sometimes necessary in problem-solving exercises, and an understanding of these operations is needed to carry out unit calculations. It is helpful in many of these operations to remember that a number (or a unit) divided by itself is equal to 1, for example,

$$\frac{5}{5} = 1 \qquad \frac{\text{inch}}{\text{inch}} = 1 \qquad \frac{5 \text{ inches}}{5 \text{ inches}} = 1$$

When one fraction is divided by another fraction, the operation commonly applied is to "invert the denominator and multiply." For example, 2/5 divided by 1/2 is

$$\frac{\dfrac{2}{5}}{\dfrac{1}{2}} = \frac{2}{5} \times \frac{2}{1} = \frac{4}{5}$$

What you are really doing when you invert the denominator of the larger fraction and multiply is making the denominator $(1/2)$ equal to 1. Both the numerator $(2/5)$ and the denominator $(1/2)$ are multiplied by $2/1$, which does not change the value of the overall expression. The complete operation is

$$\frac{\dfrac{2}{5}}{\dfrac{1}{2}} \times \frac{\dfrac{2}{1}}{\dfrac{2}{1}} = \frac{\dfrac{2}{5} \times \dfrac{2}{1}}{\dfrac{1}{2} \times \dfrac{2}{1}} = \frac{\dfrac{4}{5}}{\dfrac{2}{2}} = \frac{\dfrac{4}{5}}{1} = \frac{4}{5}$$

Symbols and Operations

The use of symbols seems to cause confusion for some students because it seems different from their ordinary experiences with arithmetic. The rules are the same for symbols as they are for numbers, but you cannot do the operations with the symbols until you know what values they represent. The operation signs, such as $+$, $\div$, $\times$, and $-$ are used with symbols to indicate the operation that you *would* do if you knew the values. Some of the mathematical operations are indicated several ways. For example, $a \times b$, $a \cdot b$, and ab all indicate the same thing, that a is to be multiplied by b. Likewise, $a \div b$, a/b, and $a \times 1/b$ all indicate that a is to be divided by b. Since it is not possible to carry out the operations on symbols alone, they are called *indicated operations*.

Operations in Equations

An equation is a shorthand way of expressing a simple sentence with symbols. The equation has three parts: (1) a left side, (2) an equal sign $(=)$, which indicates the equivalence of the two sides, and (3) a right side. The left side has the same value and units as the right side, but the two sides may have a very different appearance. The two sides may also have the symbols that indicate mathematical operations ($+$, $-$, $\times$, and so forth) and may be in certain forms that indicate operations (a/b, ab, and so forth). In any case, the equation is a complete expression that states the left side has the same value and units as the right side.

Equations may contain different symbols, each representing some unknown quantity. In science, the term "solve the equation" means to perform certain operations with one symbol (which represents some variable) by itself on one side of the equation. This single symbol is usually, but not necessarily, on the left side and is not present on the other side. For example, the equation $F = ma$ has the symbol F on the left side. In science, you would say that this equation is solved for F. It could also be solved for m or for a, which will be considered shortly. The equation $F = ma$ is solved for F, and the *indicated operation* is to multiply m by a because they are in the form ma, which means the same thing as $m \times a$. This is the only indicated operation in this equation.

A solved equation is a set of instructions that has an order of indicated operations. For example, the equation for the relationship between a Fahrenheit and Celsius temperature, solved for °C, is $C = 5/9(F - 32)$. A list of indicated operations in this equation is as follows:

1. Subtract 32° from the given Fahrenheit temperature.
2. Multiply the result of (1) by 5.
3. Divide the result of (2) by 9.

Why are the operations indicated in this order? Because the bracket means 5/9 of the *quantity* $(F - 32°)$. In its expanded form, you can see that $5/9(F - 32°)$ actually means $5/9(F) - 5/9(32°)$. Thus, you cannot multiply by 5 or divide by 9 until you have found the quantity $(F - 32°)$. Once you have figured out the order of operations, finding the answer to a problem becomes almost routine as you complete the needed operations on both the numbers and the units.

Solving Equations

Sometimes it is necessary to rearrange an equation to move a different symbol to one side by itself. This is known as solving an equation for an unknown quantity. But you cannot simply move a symbol to one side of an equation. Since an equation is a statement of equivalence, the right side has the same value as the left side. If you move a symbol, you must perform the operation in a way that the two sides remain equivalent. This is accomplished by "canceling out" symbols until you have the unknown on one side by itself. One key to understanding the canceling operation is to remember that a fraction with the same number (or unit) over itself is equal to 1. For example, consider the equation $F = ma$, which is solved for F. Suppose you are considering a problem in which F and m are given, and the unknown is a. You need to solve the equation

for a so it is on one side by itself. To eliminate the m, you do the *inverse* of the indicated operation on m, dividing both sides by m. Thus,

$$F = ma$$

$$\frac{F}{m} = \frac{ma}{m}$$

$$\frac{F}{m} = a$$

Since m/m is equal to 1, the a remains by itself on the right side. For convenience, the whole equation may be flipped to move the unknown to the left side,

$$a = \frac{F}{m}$$

Thus, a quantity that indicated a multiplication (ma) was removed from one side by an inverse operation of dividing by m.

Consider the following inverse operations to "cancel" a quantity from one side of an equation, moving it to the other side:

If the Indicated Operation of the Symbol You Wish to Remove Is:	Perform This Inverse Operation on Both Sides of the Equation
multiplication	division
division	multiplication
addition	subtraction
subtraction	addition
squared	square root
square root	square

EXAMPLE A.1

The equation for finding the kinetic energy of a moving body is $KE = 1/2mv^2$. You need to solve this equation for the velocity, v.

Solution

The order of indicated operations in the equation is as follows:

1. Square v.
2. Multiply v^2 by m.
3. Divide the result of (2) by 2.

To solve for v, this order is *reversed* as the "canceling operations" are used:

Step 1: Multiply both sides by 2

$$KE = \frac{1}{2}mv^2$$

$$2KE = \frac{2}{2}mv^2$$

$$2KE = mv^2$$

Step 2: Divide both sides by m

$$\frac{2KE}{m} = \frac{mv^2}{m}$$

$$\frac{2KE}{m} = v^2$$

Step 3: Take the square root of both sides

$$\sqrt{\frac{2KE}{m}} = \sqrt{v^2}$$

$$\sqrt{\frac{2KE}{m}} = v$$

or

$$v = \sqrt{\frac{2KE}{m}}$$

The equation has been solved for v, and you are now ready to substitute quantities and perform the needed operations (see example 1.3 in chapter 1 for information on this topic).

Significant Figures

The numerical value of any measurement will always contain some uncertainty. Suppose, for example, that you are measuring one side of a square piece of paper as shown in Figure A.1. You

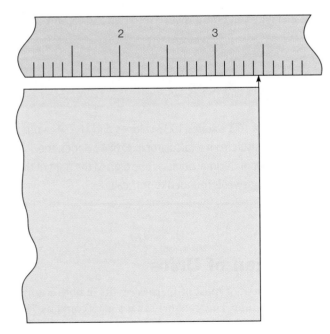

FIGURE A.1

How wide is this sheet of paper? Write your answer before reading the text _____.

could say that the paper is *about* 3.5 cm wide and you would be correct. This measurement, however, would be unsatisfactory for many purposes. It does not approach the true value of the length and contains too much uncertainty. It seems clear that the paper width is larger than 3.4 cm but shorter than 3.5 cm. But how much larger than 3.4 cm? You cannot be certain if the paper is 3.44, 3.45, or 3.46 cm wide. As your best estimate, you might say that the paper is 3.45 cm wide. Everyone would agree that you can be certain about the first two numbers (3.4) and they should be recorded. The last number (0.05) has been estimated and is not certain. The two certain numbers, together with one uncertain number, represent the greatest accuracy possible with the ruler being used. The paper is said to be 3.45 cm wide.

A *significant figure* is a number that is believed to be correct with some uncertainty only in the last digit. The value of the width of the paper, 3.45 cm, represents three significant figures. As you can see, the number of significant figures can be determined by the degree of accuracy of the measuring instrument being used. But suppose you need to calculate the area of the paper. You would multiply 3.45 cm $\times$ 3.45 cm and the product for the area would be 11.9025 cm^2. This is a greater precision than you were able to obtain with your measuring instrument. The result of a calculation can be no more accurate than the values being treated. Because the measurement had only three significant figures (two certain, one uncertain), then the answer can have only three significant figures. The area is correctly expressed as 11.9 cm^2.

There are a few simple rules that will help you determine how many significant figures are contained in a reported measurement:

1. All digits reported as a direct result of a measurement are significant.
2. Zero is significant when it occurs between nonzero digits. For example, 607 has three significant figures, and the zero is one of the significant figures.
3. In figures reported as *larger than the digit 1*, the digit zero is not significant when it follows a nonzero digit to indicate place. For example, in a report that "23,000 people attended the rock concert," the digits 2 and 3 are significant but the zeros are not significant. In this situation the 23 is the measured part of the figure, and the three zeros tell you an estimate of how many attended the concert, that is, 23 thousand. If the figure is a measurement rather than an estimate, then it is written *with a decimal point after the last zero* to indicate that the zeros *are* significant. Thus 23,000 has *two* significant figures (2 and 3), but 23,000. has *five* significant figures. The figure 23,000 means "about 23 thousand," but 23,000. means 23,000. and not 22,999 or 23,001.
4. In figures reported as *smaller than the digit 1*, zeros after a decimal point that come before nonzero digits *are not* significant and serve only as place holders. For example, 0.0023 has two significant figures: 2 and 3. Zeros alone after a decimal point or zeros after a nonzero digit indicate a measurement, however, so these zeros *are* significant. The figure 0.00230, for example, has three significant figures since the 230 means 230 and not 229 or 231. Likewise, the figure 3.000 cm has four significant figures because the presence of the three zeros means that the measurement was actually 3.000 and not 2.999 or 3.001.

Multiplication and Division

When you multiply or divide measurement figures, the answer may have no more significant figures than the *least* number of significant figures in the figures being multiplied or divided. This simply means that an answer can be no more accurate than the least accurate measurement entering into the calculation, and that you cannot improve the accuracy of a measurement by doing a calculation. For example, in multiplying 54.2 mi/h × 4.0 h to find out the total distance traveled, the first figure (54.2) has three significant figures but the second (4.0) has only two significant figures. The answer can contain only two significant figures since this is the weakest number of those involved in the calculation. The correct answer is therefore 220 mi, not 216.8 mi. This may seem strange since multiplying the two numbers together gives the answer of 216.8 mi. This answer, however, means a greater accuracy than is possible, and the accuracy cannot be improved over the weakest number involved in the calculation. Since the weakest number (4.0) has only two significant figures the answer must also have only two significant figures, which is 220 mi.

The result of a calculation is *rounded* to have the same least number of significant figures as the least number of a measurement involved in the calculation. When rounding numbers the last significant figure is increased by 1 if the number after it is 5 or larger. If the number after the last significant figure is 4 or less, the nonsignificant figures are simply dropped. Thus, if two significant figures are called for in the answer of the previous example, 216.8 is rounded up to 220 because the last number after the two significant figures is 6 (a number larger than 5). If the calculation result had been 214.8, the rounded number would be 210 miles.

Note that *measurement figures* are the only figures involved in the number of significant figures in the answer. Numbers that are counted or **defined** are not included in the determination of significant figures in an answer. For example, when dividing by 2 to find an average, the 2 is ignored when considering the number of significant figures. Defined numbers are defined exactly and are not used in significant figures. Since 1 kilogram is *defined* to be exactly 1,000 grams, such a conversion is not a measurement.

Addition and Subtraction

Addition and subtraction operations involving measurements, as with multiplication and division, cannot result in an answer that implies greater accuracy than the measurements had before the calculation. Recall that the last digit to the right in a measurement is uncertain, that is, it is the result of an estimate. The answer to an addition or subtraction calculation can have this uncertain number *no farther from the decimal place than it was in the weakest number involved in the calculation.* Thus, when 8.4 is added to 4.926, the weakest number is 8.4, and the uncertain number is .4, one place to the right of the decimal. The sum of 13.326 is therefore rounded to 13.3, reflecting the placement of this weakest doubtful figure.

The rules for counting zeros tell us that the numbers 203 and 0.200 both have three significant figures. Likewise, the numbers 230 and 0.23 only have two significant figures. Once you remember the rules, the counting of significant figures is straightforward. On the other hand, sometimes you find a number that seems to make it impossible to follow the rules. For example, how would you write 3,000 with two significant figures? There are several special systems in use for taking care of problems such as this, including the placing of a little bar over the last significant digit. One of the convenient ways of showing significant figures for difficult numbers is to use scientific notation, which is also discussed elsewhere in this appendix. The convention for writing significant figures is to display one digit to the left of the decimal. The exponents are not considered when showing the number of significant figures in scientific notation. Thus if you want to write three thousand showing one significant figure, you would write 3×10^3. To show two significant figures it is 3.0×10^3 and for three significant figures it becomes 3.00×10^3. As you can see, the correct use of scientific notation leaves little room for doubt about how many significant figures are intended.

EXAMPLE A.2

In a problem it is necessary to multiply 0.0039 km by 15.0 km. The result from a calculator is 0.0585 km². The least number of significant figures involved in this calculation is *two* (0.0039 is two significant figures; 15.0 is three—read the rules again to see why). The calculator result is therefore rounded off to have only two significant figures, and the answer is recorded as 0.059 km².

EXAMPLE A.3

The quantities of 10.3 calories, 10.15 calories, and 16.234 calories are added. The result from a calculator is 36.684 calories. The smallest number of decimal points is *one* digit to the right of the decimal, so the answer is rounded to 16.7 calories.

Conversion of Units

The measurement of most properties results in both a numerical value and a unit. The statement that a glass contains 50 cm³ of a liquid conveys two important concepts—the numerical value of 50 and the referent unit of cubic centimeters. Both the numerical value and the unit are necessary to communicate correctly the volume of the liquid.

When working with calculations involving measurement units, *both* the numerical value and the units are treated mathematically. As in other mathematical operations, there are general rules to follow.

1. Only properties with like units may be added or subtracted. It should be obvious that adding quantities such as 5 dollars and 10 dimes is meaningless. You must first convert to like units before adding or subtracting.
2. Like or unlike units may be multiplied or divided and treated in the same manner as numbers. You have used this rule when dealing with area (length $\times$ length = length^2, for example, cm $\times$ cm = cm^2) and when dealing with volume (length $\times$ length $\times$ length = length^3, for example, cm $\times$ cm $\times$ cm = cm^3).

You can use these two rules to create a *conversion ratio* that will help you change one unit to another. Suppose you need to convert 2.3 kg to grams. First, write the relationship between kilograms and grams:

$$1{,}000 \text{ g} = 1 \text{ kg}$$

Next, divide both sides by what you wish to convert *from* (kilograms in this example):

$$\frac{1{,}000 \text{ g}}{1} = \frac{1 \text{ kg}}{1 \text{ kg}}$$

One kilogram divided by 1 kg equals 1, just as 10 divided by 10 equals 1. Therefore, the right side of the relationship becomes 1 and the equation is:

$$\frac{1{,}000 \text{ g}}{1 \text{ kg}} = 1$$

The 1 is usually understood, that is, not stated, and the operation is called *canceling*. Canceling leaves you with the fraction 1,000 g/1 kg, which is a conversion ratio that can be used to convert from kilograms to grams. You simply multiply the conversion ratio by the numerical value and unit you wish to convert:

$$= 2.3 \text{ kg} \times \frac{1{,}000 \text{ g}}{1 \text{ kg}}$$
$$= \frac{2.3 \times 1{,}000}{1} \frac{\text{kg} \times \text{g}}{\text{kg}}$$
$$= \boxed{2{,}300 \text{ g}}$$

The kilogram units cancel. Showing the whole operation with units only, you can see how you end up with the correct unit of grams:

$$\text{kg} \times \frac{\text{g}}{\text{kg}} = \frac{\text{kg·g}}{\text{kg}} = \text{g}$$

Since you did obtain the correct unit, you know that you used the correct conversion ratio. If you had blundered and used an inverted conversion ratio, you would obtain

$$2.3 \text{ kg} \times \frac{1 \text{ kg}}{1{,}000 \text{ g}} = .0023 \frac{\text{kg}^2}{\text{g}}$$

which yields the meaningless, incorrect units of kg^2/g. Carrying out the mathematical operations on the numbers and the units will always tell you whether or not you used the correct conversion ratio.

EXAMPLE A.4

A distance is reported as 100.0 km, and you want to know how far this is in miles.

Solution

First, you need to obtain a *conversion factor* from a textbook or reference book, which usually lists the conversion factors by properties in a table. Such a table will show two conversion factors for kilometers and miles: (1) 1 km = 0.621 mi and (2) 1 mi = 1.609 km. You select the factor that is in the same form as your problem; for example, your problem is 100.0 km = ? mi. The conversion factor in this form is 1 km = 0.621 mi.

Second, you convert this conversion factor into a *conversion ratio* by dividing the factor by what you wish to convert *from:*

conversion factor:	1 km = 0.621
divide factor by what you want to convert from:	$\frac{1 \text{ km}}{1 \text{ km}} = \frac{0.621 \text{ mi}}{1 \text{ km}}$
resulting conversion rate:	$\frac{0.621 \text{ mi}}{\text{km}}$

Note that if you had used the 1 mi = 1.609 km factor, the resulting units would be meaningless. The conversion ratio is now multiplied by the numerical value *and unit* you wish to convert:

$$100.0 \text{ km} \times \frac{0.621 \text{ mi}}{\text{km}}$$
$$(100.0)(0.621) \frac{\text{km·mi}}{\text{km}}$$
$$62.1 \text{ mi}$$

EXAMPLE A.5

A service station sells gasoline by the liter, and you fill your tank with 72 liters. How many gallons is this? (Answer: 19 gal)

Scientific Notation

Most of the properties of things that you might measure in your everyday world can be expressed with a small range of numerical values together with some standard unit of measure. The range of numerical values for most everyday things can be dealt with by using units (1s), tens (10s), hundreds (100s), or perhaps thousands (1,000s). But the actual universe contains some objects of incredibly large size that require some very big numbers to describe. The sun, for example, has a mass of about

1,970,000,000,000,000,000,000,000,000,000 kg. On the other hand, very small numbers are needed to measure the size and parts of an atom. The radius of a hydrogen atom, for example, is about 0.00000000005 m. Such extremely large and small numbers are cumbersome and awkward since there are so many zeros to keep track of, even if you are successful in carefully counting all the zeros. A method does exist to deal with extremely large or small numbers in a more condensed form. The method is called *scientific notation,* but it is also sometimes called *powers of ten* or *exponential notation,* since it is based on exponents of 10. Whatever it is called, the method is a compact way of dealing with numbers that not only helps you keep track of zeros but provides a simplified way to make calculations as well.

In algebra you save a lot of time (as well as paper) by writing $(a \times a \times a \times a \times a)$ as a^5. The small number written to the right and above a letter or number is a superscript called an *exponent.* The exponent means that the letter or number is to be multiplied by itself that many times, for example, a^5 means a multiplied by itself five times, or $a \times a \times a \times a \times a$. As you can see, it is much easier to write the exponential form of this operation than it is to write it out in the long form. Scientific notation uses an exponent to indicate the power of the base 10. The exponent tells how many times the base, 10, is multiplied by itself. For example,

$$10,000 = 10^4$$
$$1,000 = 10^3$$
$$100 = 10^2$$
$$10 = 10^1$$
$$1 = 10^0$$
$$0.1 = 10^{-1}$$
$$0.01 = 10^{-2}$$
$$0.001 = 10^{-3}$$
$$0.0001 = 10^{-4}$$

This table could be extended indefinitely, but this somewhat shorter version will give you an idea of how the method works. The symbol 10^4 is read as "ten to the fourth power" and means $10 \times 10 \times 10 \times 10$. Ten times itself four times is 10,000, so 10^4 is the scientific notation for 10,000. It is also equal to the number of zeros between the 1 and the decimal point, that is, to write the longer form of 10^4 you simply write 1, then move the decimal point four places to the *right;* 10 to the fourth power is 10,000.

The power of ten table also shows that numbers smaller than 1 have negative exponents. A negative exponent means a reciprocal:

$$10^{-1} = \frac{1}{10} = 0.1$$

$$10^{-2} = \frac{1}{100} = 0.01$$

$$10^3 = \frac{1}{1,000} = 0.001$$

To write the longer form of 10^{-4}, you simply write 1 then move the decimal point four places to the *left;* 10 to the negative fourth power is 0.0001.

Scientific notation usually, but not always, is expressed as the product of two numbers: (1) a number between 1 and 10 that is called the *coefficient* and (2) a power of ten that is called the *exponent.* For example, the mass of the sun that was given in long form earlier is expressed in scientific notation as

$$1.97 \times 10^{30} \text{ kg}$$

and the radius of a hydrogen atom is

$$5.0 \times 10^{-11} \text{ m}$$

In these expressions, the coefficients are 1.97 and 5.0, and the power of ten notations are the exponents. Note that in both of these examples, the exponent tells you where to place the decimal point if you wish to write the number all the way out in the long form. Sometimes scientific notation is written without a coefficient, showing only the exponent. In these cases the coefficient of 1.0 is understood, that is, not stated. If you try to enter a scientific notation in your calculator, however, you will need to enter the understood 1.0, or the calculator will not be able to function correctly. Note also that 1.97×10^{30} kg and the expressions 0.197×10^{31} kg and 19.7×10^{29} kg are all correct expressions of the mass of the sun. By convention, however, you will use the form that has one digit to the left of the decimal.

EXAMPLE A.6

What is 26,000,000 in scientific notation?

Solution

Count how many times you must shift the decimal point until one digit remains to the left of the decimal point. For numbers larger than the digit 1, the number of shifts tells you how much the exponent is increased, so the answer is

$$26 \times 10^7$$

which means the coefficient 2.6 is multiplied by 10 seven times.

EXAMPLE A.7

What is 0.000732 in scientific notation? (Answer: 7.32×10^{-4})

It was stated earlier that scientific notation provides a compact way of dealing with very large or very small numbers, but it provides a simplified way to make calculations as well. There are a few mathematical rules that will describe how the use of scientific notation simplifies these calculations.

To *multiply* two scientific notation numbers, the coefficients are multiplied as usual, and the exponents are *added* algebraically. For example, to multiply (2×10^2) by (3×10^3), first separate the coefficients from the exponents,

$$(2 \times 3) \times (10^2 \times 10^3),$$

then multiply the coefficients and add the exponents,

$$6 \times 10^{(2 + 3)} = 6 \times 10^5$$

Adding the exponents is possible because $10^2 \times 10^3$ means the same thing as $(10 \times 10) \times (10 \times 10 \times 10)$, which equals $(100) \times (1,000)$, or $100,000$, which is expressed as 10^{-5} in scientific notation. Note that two negative exponents add algebraically, for example $10^{-2} \times 10^{-3} = 10^{[(-2) + (-3)]} = 10^{-5}$. A negative and a positive exponent also add algebraically, as in $10^5 \times 10^{-3} = 10^{[(+5) + (-3)]} = 10^2$.

If the result of a calculation involving two scientific notation numbers does not have the conventional one digit to the left of the decimal, move the decimal point so it does, changing the exponent according to which way and how much the decimal point is moved. Note that the exponent increases by one number for each decimal point moved to the left. Likewise, the exponent decreases by one number for each decimal point moved to the right. For example, $938. \times 10^3$ becomes 9.38×10^5 when the decimal point is moved two places to the left.

To *divide* two scientific notation numbers, the coefficients are divided as usual and the exponents are *subtracted*. For example, to divide (6×10^6) by (3×10^2), first separate the coefficients from the exponents,

$$(6 \div 3) \times (10^6 \div 10^2)$$

then divide the coefficients and subtract the exponents,

$$2 \times 10^{(6 - 2)} = 2 \times 10^4$$

Note that when you subtract a negative exponent, for example, $10^{[(3) - (-2)]}$, you change the sign and add, $10^{(3 + 2)} = 10^5$.

EXAMPLE **A.8**

Solve the following problem concerning scientific notation:

$$\frac{(2 \times 10^4) \times (8 \times 10^{-6})}{8 \times 10^4}$$

Solution

First, separate the coefficients from the exponents,

$$\frac{2 \times 8}{8} \times \frac{10^4 \times 10^{-6}}{10^4}$$

then multiply and divide the coefficients and add and subtract the exponents as the problem requires,

$$2 \times 10^{\{[(4) + (-6)] - [4]\}}$$

Solving the remaining additions and subtractions of the coefficients gives

$$2 \times 10^{-6}$$

The Simple Line Graph

An equation describes a relationship between variables, and a graph helps you picture this relationship. A line graph pictures how changes in one variable correspond with changes in a second variable, that is, how the two variables change together. Usually one variable can be easily manipulated. The other variable is caused to change in value by the manipulation of the first variable. The **manipulated variable** is known by various names (*independent, input,* or *cause variable*), and the **responding variable** is known by various related names (*dependent, output,* or *effect variable*). The manipulated variable is usually placed on the horizontal axis, or *x*-axis, of the graph, so you could also identify it as the *x-variable*. The responding variable is placed on the vertical axis, or *y-axis*. This variable is identified as the *y-variable*.

Figure A.2 shows the mass of different volumes of water at room temperature. Volume is placed on the *x*-axis because the volume of water is easily manipulated, and the mass values change as a consequence of changing the values of volume. Note that both variables are named and that the measuring unit for each is identified on the graph.

Figure A.2 also shows a number *scale* on each axis that represents changes in the values of each variable. The scales are usually, but not always, linear. A **linear scale** has equal intervals that represent equal increases in the value of the variable. Thus a certain distance on the *x*-axis to the right represents a certain

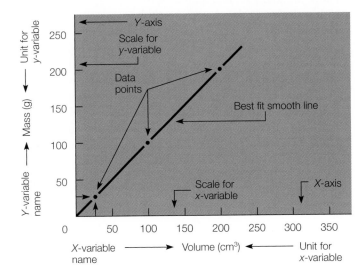

FIGURE A.2

The parts of a graph. On this graph, volume is placed on the *x*-axis and mass on the *y*-axis.

increase in the value of the *x*-variable. Likewise, certain distances up the *y*-axis represent certain increases in the value of the *y*-variable. The **origin** is the only point where both the *x*- and *y*-variables have a value of zero at the same time.

Figure A.2 shows three **data points.** A data point represents measurements of two related variables that were made at the same time. For example, a volume of 25 cm^3 of water was found to have a mass of 25 g. Locate 25 cm^3 on the *x*-axis and imagine a line moving straight up from this point on the scale. Locate 25 g on the *y*-axis and imagine a line moving straight out from this point on the scale. Where the lines meet is the data point for the 25 cm^3 and 25 g measurements. A data point is usually indicated with a small dot or x (dots are used in the graph in Figure A.2).

A "best fit" smooth line is drawn through all the data points as close to them as possible. If it is not possible to draw the straight line *through* all the data points, then a straight line is drawn that has the same number of data points on both sides of the line. Such a line will represent a "best approximation" of the relationship between the two variables. The origin is also used as a data point in this example because a volume of zero will have a mass of zero.

The smooth line tells you how the two variables get larger together. With the same *x*- and *y*-axis scale, a 45° line means that they are increasing in an exact direct proportion. A more flat or more upright line means that one variable is increasing faster than the other. The more you work with graphs, the easier it will become for you to analyze what the "picture" means. There are more exact ways to extract information from a graph, and one of these techniques is discussed next.

One way to determine the relationship between two variables that are graphed with a straight line is to calculate the **slope.** The slope is a *ratio* between the changes in one variable and the changes in the other. The ratio is between the change in the value of the *x*-variable and the change in the value of the *y*-variable. Recall that the symbol Δ (Greek letter delta) means "change in," so the symbol Δx means the "change in *x*." The first step in calculating the slope is to find out how much the *x*-variable is changing (Δx) in relation to how much the *y*-variable is changing (Δy). You can find this relationship by first drawing a dashed line to the right of the straight *line* (not the data points), so that the *x*-variable has increased by some convenient unit (Figure A.3). Where you start or end this dashed line will not matter since the ratio between the variables will be the same everywhere on the graph line. The Δx is determined by subtracting the initial value of the *x*-variable on the dashed line (x_i) from the final value of the *x*-variable on the dashed line x_f, or $\Delta x = x_f - x_i$. In Figure A.3, the dashed line has an x_f of 200 cm^3 and an x_i of 100 cm^3, so Δx is 200 cm^3 − 100 cm^3, or 100 cm^3. Note that Δx has both a number value and a unit.

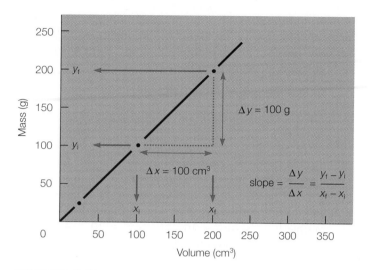

FIGURE A.3

The slope is a ratio between the changes in the *y*-variable and the changes in the *x*-variable, or $\Delta y/\Delta x$.

Now you need to find Δy. The example in Figure A.3 shows a dashed line drawn back up to the graph line from the *x*-variable dashed line. The value of Δy is $y_f - y_i$. In the example, $\Delta y = 200$ g − 100 g, or 100 g.

The slope of a straight graph line is the ratio of Δy to Δx, or

$$\text{slope} = \frac{\Delta y}{\Delta x}$$

In the example,

$$\text{slope} = \frac{100 \text{ g}}{100 \text{ cm}^3}$$

$$= 1 \frac{\text{g}}{\text{cm}^3} \text{ or } 1 \text{ g/cm}^3$$

Thus the slope is 1 g/cm^3, and this tells you how the variables change together. Since g/cm^3 is also the unit of density, you know that you have just calculated the density of water from a graph.

Note that the slope can be calculated only for two variables that are increasing together, that is, for variables that are in direct proportion and have a line that moves upward and to the right. If variables change in any other way, mathematical operations must be performed to change the variables *into* this relationship. Examples of such necessary changes include taking the inverse of one variable, squaring one variable, taking the inverse square, and so forth.

Appendix B
Solubilities Chart

	Acetate	Bromide	Carbonate	Chloride	Fluoride	Hydroxide	Iodide	Nitrate	Oxide	Phosphate	Sulfate	Sulfide
Aluminum	S	S	—	S	s	i	S	S	i	i	S	d
Ammonium	S	S	S	S	S	S	S	S	—	S	S	S
Barium	S	S	i	S	s	S	S	S	S	i	i	d
Calcium	S	S	i	S	i	s	S	S	s	i	s	d
Copper (I)	—	s	i	s	i	—	i	—	i	—	d	i
Copper (II)	S	S	i	S	S	i	S	S	i	i	S	i
Iron (II)	S	S	i	S	s	i	S	S	i	i	S	i
Iron (III)	S	S	i	S	s	i	S	S	i	i	S	d
Lead	S	s	i	s	i	i	s	S	i	i	i	i
Magnesium	S	S	i	S	i	i	S	S	i	i	S	d
Mercury (I)	s	i	i	i	d	d	i	S	i	i	i	i
Mercury (II)	S	s	i	S	d	i	i	S	i	i	i	i
Potassium	S	S	S	S	S	S	S	S	S	S	S	i
Silver	s	i	i	i	S	—	i	S	i	i	i	i
Sodium	S	S	S	S	S	S	S	S	d	S	S	S
Strontium	S	S	s	S	i	s	S	S	—	i	i	i
Zinc	S	S	i	S	S	i	S	S	i	i	S	i

S–soluble
i–insoluble
s–slightly soluble
d–decomposes

Appendix C
Relative Humidity Table

Dry-Bulb Temperature (°C)	Difference between Wet-Bulb and Dry-Bulb Temperatures (°C)																			
	1	2	3	4	5	6	7	8	9	10	11	12	13	14	15	16	17	18	19	20
0	81	64	46	29	13															
1	83	66	49	33	17															
2	84	68	52	37	22	7														
3	84	70	55	40	26	12														
4	86	71	57	43	29	16														
5	86	72	58	45	33	20	7													
6	86	73	60	48	35	24	11													
7	87	74	62	50	38	26	15													
8	87	75	63	51	40	29	19	8												
9	88	76	64	53	42	32	22	12												
10	88	77	66	55	44	34	24	15	6											
11	89	78	67	56	46	36	27	18	9											
12	89	78	68	58	48	39	29	21	12											
13	89	79	69	59	50	41	32	23	15	7										
14	90	79	70	60	51	42	34	26	18	10										
15	90	80	71	61	53	44	36	27	20	13	6									
16	90	81	71	63	54	46	38	30	23	15	8									
17	90	81	72	64	55	47	40	32	25	18	11									
18	91	82	73	65	57	49	41	34	27	20	14	7								
19	91	82	74	65	58	50	43	36	29	22	16	10								
20	91	83	74	66	59	51	44	37	31	24	18	12	6							
21	91	83	75	67	60	53	46	39	32	26	20	14	9							
22	92	83	76	68	61	54	47	40	34	28	22	17	11	6						
23	92	84	76	69	62	55	48	42	36	30	24	19	13	8						
24	92	84	77	69	62	56	49	43	37	31	26	20	15	10	5					
25	92	84	77	70	63	57	50	44	39	33	28	22	17	12	8					
26	92	85	78	71	64	58	51	46	40	34	29	24	19	14	10	5				
27	92	85	78	71	65	58	52	47	41	36	31	26	21	16	12	7				
28	93	85	78	72	65	59	53	48	42	37	32	27	22	18	13	9	5			
29	93	86	79	72	66	60	54	49	43	38	33	28	24	19	15	11	7			
30	93	86	79	73	67	61	55	50	44	39	35	30	25	21	17	13	9	5		
31	93	86	80	73	67	61	56	51	45	40	36	31	27	22	18	14	11	7		
32	93	86	80	74	68	62	57	51	46	41	37	32	28	24	20	16	12	9	5	
33	93	87	80	74	68	63	57	52	47	42	38	33	29	25	21	17	14	10	7	
34	93	87	81	75	69	63	58	53	48	43	39	35	30	28	23	19	15	12	8	5
35	94	87	81	75	69	64	59	54	49	44	40	36	32	28	24	20	17	13	10	7

Appendix D
Solutions for Group A Parallel Exercises

NOTE: Solutions that involve calculations of measurements are rounded up or down to conform to the rules for significant figures described in Appendix A.

EXERCISES **Chapter 1**

1.1. Answers will vary but should have the relationship of 100 cm in 1 m, for example, 178 cm = 1.78 m.

1.2. Since mass density is given by the relationship $\rho = m/V$, then

$$\rho = \frac{m}{V} = \frac{272 \text{ g}}{20.0 \text{ cm}^3}$$

$$= \frac{272}{20.0} \frac{\text{g}}{\text{cm}^3}$$

$$= \boxed{13.6 \frac{\text{g}}{\text{cm}^3}}$$

1.3. The volume of a sample of lead is given and the problem asks for the mass. From the relationship of $\rho = m/V$, solving for the mass (m) tells you that the mass density (ρ) times the volume (V), or $m = \rho V$. The mass density of lead, 11.4 g/cm³, can be obtained from Table 1.4, so

$$\rho = \frac{m}{V}$$

$$V\rho = \frac{mV}{V}$$

$$m = \rho V$$

$$m = \left(11.4 \frac{\text{g}}{\text{cm}^3}\right)(10.0 \text{ cm}^3)$$

$$11.4 \times 10.0 \frac{\text{g}}{\text{cm}^3} \times \text{cm}^3$$

$$114 \frac{\text{g} \cdot \text{cm}^3}{\text{cm}^3}$$

$$= \boxed{114 \text{ g}}$$

1.4. Solving the relationship $\rho = m/V$ for volume gives $V = m/\rho$, and

$$\rho = \frac{m}{V}$$

$$V\rho = \frac{mV}{V}$$

$$\frac{V\rho}{\rho} = \frac{m}{\rho}$$

$$V = \frac{m}{\rho}$$

$$V = \frac{600 \text{ g}}{3.00 \dfrac{\text{g}}{\text{cm}^3}}$$

$$= \frac{600}{3.00} \frac{\text{g}}{1} \times \frac{\text{cm}^3}{\text{g}}$$

$$= 200 \frac{\text{g} \cdot \text{cm}^3}{\text{g}}$$

$$= \boxed{200 \text{ cm}^3}$$

1.5. A 50.0 cm³ sample with a mass of 34.0 grams has a density of

$$\rho = \frac{m}{V} = \frac{34.0 \text{ g}}{50.0 \text{ cm}^3}$$

$$= \frac{34.0}{50.0} \frac{\text{g}}{\text{cm}^3}$$

$$= \boxed{0.680 \frac{\text{g}}{\text{cm}^3}}$$

According to Table 1.4, 0.680 g/cm³ is the mass density of gasoline, so the substance must be gasoline.

1.6. The problem asks for a mass and gives a volume, so you need a relationship between mass and volume. Table 1.4 gives the mass density of water as 1.00 g/cm³, which is a density that is easily remembered. The volume is given in liters (L), which should first be converted to cm³ because this is the unit in which density is expressed. The relationship of $\rho = m/V$ solved for mass is ρV, so the solution is

$$\rho = \frac{m}{V} \quad \therefore \quad m = \rho V$$

$$m = \left(1.00\ \frac{g}{cm^3}\right)(40,000\ cm^3)$$

$$= 1.00 \times 40,000\ \frac{g}{cm^3} \times cm^3$$

$$= 40,000\ \frac{g \cdot cm^3}{cm^3}$$

$$= 40,000\ g$$

$$= \boxed{40\ kg}$$

1.7. From Table 1.4, the mass density of aluminum is given as 2.70 g/cm³. Converting 2.1 kg to the same units as the density gives 2,100 g. Solving $\rho = m/V$ for the volume gives

$$V = \frac{m}{\rho} = \frac{2,100\ g}{2.70\ \frac{g}{cm^3}}$$

$$= \frac{2,100}{2.70}\ \frac{g}{1} \times \frac{cm^3}{g}$$

$$= 777.78\ \frac{g \cdot cm^3}{g}$$

$$= \boxed{780\ cm^3}$$

1.8. The length of one side of the box is 0.1 m. Reasoning: Since the density of water is 1.00 g/cm³, then the volume of 1,000 g of water is 1,000 cm³. A cubic box with a volume of 1,000 cm³ is 10 cm (since 10 × 10 × 10 = 1,000). Converting 10 cm to m units, the cube is 0.1 m on each edge.

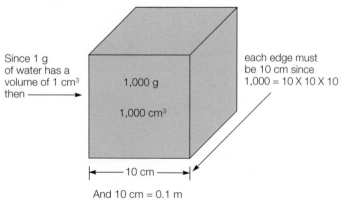

Since 1 g of water has a volume of 1 cm³ then ⟶

1,000 g

1,000 cm³

each edge must be 10 cm since 1,000 = 10 × 10 × 10

⟵ 10 cm ⟶

And 10 cm = 0.1 m

FIGURE A.4

Visualize the reasoning in 1.8.

1.9. The relationship between mass, volume, and density is $\rho = m/V$. The problem gives a volume, but not a mass. The mass, however, can be assumed to remain constant during the compression of the bread so the mass can be obtained from the original volume and density, or

$$\rho = \frac{m}{V} \quad \therefore \quad m = \rho V$$

$$m = \left(0.2\ \frac{g}{cm^3}\right)(3,000\ cm^3)$$

$$= 0.2 \times 3,000\ \frac{g}{cm^3} \times cm^3$$

$$= 600\ \frac{g \cdot cm^3}{cm^3}$$

$$= 600\ g$$

A mass of 600 g and the new volume of 1,500 cm³ means that the new density of the crushed bread is

$$\rho = \frac{m}{V}$$

$$= \frac{600\ g}{1,500\ cm^3}$$

$$= \frac{600}{1,500}\ \frac{g}{cm^3}$$

$$= \boxed{0.4\ \frac{g}{cm^3}}$$

1.10. According to Table 1.4, lead has a density of 11.4 g/cm³. Therefore, a 1.00 cm³ sample of lead would have a mass of

$$\rho = \frac{m}{V} \quad \therefore \quad m = \rho V$$

$$m = \left(11.4\ \frac{g}{cm^3}\right)(1.00\ cm^3)$$

$$= 11.4 \times 1.00\ \frac{g}{cm^3} \times cm^3$$

$$= 11.4\ \frac{g \cdot cm^3}{cm^3}$$

$$= 11.4\ g$$

Also according to Table 1.4, copper has a density of 8.96 g/cm³. To balance a mass of 11.4 g of lead, a volume of this much copper would be required:

$$\rho = \frac{m}{V} \quad \therefore \quad V = \frac{m}{\rho}$$

$$V = \frac{11.4\ g}{8.96\ \frac{g}{cm^3}}$$

$$= \frac{11.4}{8.96}\ \frac{g}{1} \times \frac{cm^3}{g}$$

$$= 1.27\ \frac{g \cdot cm^3}{g}$$

$$= \boxed{1.27\ cm^3}$$

2.1. Listing these quantities given in this problem, with their symbols, we have

$$d = 22 \text{ km}$$

$$t = 15 \text{ min}$$

$$\bar{v} = ?$$

The usual units for a speed problem are km/h or m/s, and the problem specifies that the answer should be in km/h. We see that 15 minutes is 15/60, or 1/4, or 0.25 of an hour. We will now make a new list of the quantities with the appropriate units:

$$d = 22 \text{ km}$$

$$t = 0.25 \text{ h}$$

$$\bar{v} = ?$$

These quantities are related in the average speed equation, which is already solved for the unknown average velocity:

$$\bar{v} = \frac{d}{t}$$

Substituting the known quantities, we have

$$\bar{v} = \frac{22 \text{ km}}{0.25 \text{ h}}$$

$$= \boxed{88 \frac{\text{km}}{\text{h}}}$$

2.2. Listing the quantities with their symbols:

$$\bar{v} = 3.0 \times 10^8 \text{ m/s}$$

$$t = 20.0 \text{ min}$$

$$d = ?$$

We see that the velocity units are meters per second, but the time units are minutes. We need to convert minutes to seconds, and:

$$\bar{v} = 3.0 \times 10^8 \text{ m/s}$$

$$t = 1.20 \times 10^3 \text{ s}$$

$$d = ?$$

These relationships can be found in the average speed equation, which can be solved for the unknown:

$$\bar{v} = \frac{d}{t} \qquad \therefore \qquad d = \bar{v}t$$

$$d = \left(3.0 \times 10^8 \frac{\text{m}}{\text{s}} \right) (1.20 \times 10^3 \text{ s})$$

$$= (3.0)(1.20) \times 10^{8+3} \frac{\text{m}}{\text{s}} \times \text{s}$$

$$= 3.6 \times 10^{11} \text{ m}$$

$$= \boxed{3.6 \times 10^8 \text{ km}}$$

2.3. Listing the quantities with their symbols, we can see the problem involves the quantities found in the definition of average speed:

$$\bar{v} = 350.0 \text{ m/s}$$

$$t = 5.00 \text{ s}$$

$$d = ?$$

$$v = \frac{d}{t} \qquad \therefore \qquad d = vt$$

$$d = \left(350.0 \frac{\text{m}}{\text{s}} \right) (5.00 \text{ s})$$

$$= (350.0)(5.00) \frac{\text{m}}{\text{s}} \times \text{s}$$

$$= \boxed{1,750 \text{ m}}$$

2.4. Note that the two speeds given (100.0 km/h and 50.0 km/h) are *average* speeds for two different legs of a trip. They are not initial and final speeds of an accelerating object, so you cannot add them together and divide by 2. The average speed for the total (entire) trip can be found from the definition of average speed, that is, average speed is the *total* distance covered divided by the *total* time elapsed. So, we start by finding the distance covered for each of the two legs of the trip:

$$\bar{v} = \frac{d}{t} \qquad \therefore \qquad d = \bar{v}t$$

$$\text{Leg 1 distance} = \left(100.0 \frac{\text{km}}{\text{h}} \right) (2.00 \text{ h})$$

$$= 200.0 \text{ km}$$

$$\text{Leg 2 distance} = \left(50.0 \frac{\text{km}}{\text{h}} \right) (1.00 \text{ h})$$

$$= 50.0 \text{ km}$$

Total distance (leg 1 plus leg 2) = 250.0 km
Total time = 3.00 h

$$\bar{v} = \frac{d}{t} = \frac{250.0 \text{ km}}{3.00 \text{ h}} = \boxed{83.3 \text{ km/h}}$$

2.5. The initial velocity, final velocity, and time are known and the problem asked for the acceleration. Listing these quantities with their symbols, we have

$$v_\text{i} = 0$$

$$v_\text{f} = 15.0 \text{ m/s}$$

$$t = 10.0 \text{ s}$$

$$a = ?$$

These are the quantities involved in the acceleration equation, which is already solved for the unknown:

$$a = \frac{v_\text{f} - v_\text{i}}{t}$$

$$a = \frac{15.0 \text{ m/s} - 0 \text{ m/s}}{10.0 \text{ s}}$$

$$= \frac{15.0}{10.0} \frac{m}{s} \times \frac{1}{s}$$

$$= \boxed{1.50 \frac{m}{s^2}}$$

2.6. The initial velocity, final velocity, and acceleration are known and the problem asked for the time. Listing these quantities with their symbols, we have

$$v_i = 20.0 \text{ m/s}$$

$$v_f = 25.0 \text{ m/s}$$

$$a = 3.0 \text{ m/s}^2$$

$$t = ?$$

These are the quantities involved in the acceleration equation, which must first be solved for the unknown time:

$$a = \frac{v_f - v_i}{t} \qquad \therefore \qquad t = \frac{v_f - v_i}{a}$$

$$t = \frac{25.0 \frac{m}{s} - 20.0 \frac{m}{s}}{3.0 \frac{m}{s^2}}$$

$$= \frac{5.00 \frac{m}{s}}{3.0 \frac{m}{s^2}}$$

$$= \frac{5.00}{3.0} \frac{m}{s} \times \frac{s \cdot s}{m}$$

$$= \boxed{1.7 \text{ s}}$$

2.7. The relationship between average velocity ($\bar{v}$), distance (d), and time (t) can be solved for time:

$$\bar{v} = \frac{d}{t}$$

$$\bar{v}t = d$$

$$t = \frac{d}{\bar{v}}$$

$$t = \frac{5,280 \text{ ft}}{2,360 \frac{ft}{s}}$$

$$= \frac{5,280}{2,360} \frac{ft}{1} \times \frac{s}{ft}$$

$$= 2.24 \frac{ft \cdot s}{ft}$$

$$= \boxed{2.24 \text{ s}}$$

2.8. The relationship between average velocity ($\bar{v}$), distance (d), and time (t) can be solved for distance:

$$\bar{v} = \frac{d}{t} \qquad \therefore \qquad d = \bar{v}t$$

$$d = \left(40.0 \frac{m}{s} \right)(0.4625 \text{ s})$$

$$= 40.0 \times 0.4625 \frac{m \cdot s}{s}$$

$$= \boxed{18.5 \text{ m}}$$

2.9. "How many minutes . . .," is a question about time and the distance is given. Since the distance is given in km and the speed in m/s, a unit conversion is needed. The easiest thing to do is to convert km to m. There are 1,000 m in a km, and

$$(1.50 \times 10^8 \text{ km}) \times (1 \times 10^3 \text{ m/km}) = 1.50 \times 10^{11} \text{ m}$$

The relationship between average velocity ($\bar{v}$), distance (d), and time (t) can be solved for time:

$$\bar{v} = \frac{d}{t} \qquad \therefore \qquad t = \frac{d}{\bar{v}}$$

$$t = \frac{1.50 \times 10^{11} \text{ m}}{3.00 \times 10^8 \frac{m}{s}}$$

$$= \frac{1.50}{3.00} \times 10^{11-8} \frac{m}{1} \times \frac{s}{m}$$

$$= 0.500 \times 10^3 \frac{m \cdot s}{m}$$

$$= 5.00 \times 10^2 \text{ s}$$

$$\frac{500 \text{ s}}{60 \frac{s}{min}} = \frac{500}{60} \frac{s}{1} \times \frac{min}{s}$$

$$= 8.33 \frac{s \cdot min}{s}$$

$$= \boxed{8.33 \text{ min}}$$

[Information on how to use scientific notation (also called powers of ten or exponential notation) is located in the Mathematical Review of appendix A.]
All significant figures are retained here because the units are defined exactly, without uncertainty.

2.10. The initial velocity (v_i) is given as 100.0 m/s, the final velocity (v_f) is given as 51.0 m/s, and the time is given as 5.00 s. Acceleration, including a deceleration or negative acceleration, is found from a change of velocity during a given time. Thus,

$$a = \frac{v_f - v_i}{t}$$

$$= \frac{\left(51.0\,\frac{m}{s}\right) - \left(100.0\,\frac{m}{s}\right)}{5.00\,s}$$

$$= \frac{-49.0\,\frac{m}{s}}{5.00\,s}$$

$$= -9.80\,\frac{m}{s} \times \frac{1}{s}$$

$$= \boxed{-9.80\,\frac{m}{s^2}}$$

(The negative sign means a negative acceleration, or deceleration.)

2.11. A ball thrown straight up decelerates to a velocity of zero, then accelerates back to the surface, just as a dropped ball would do from the height reached. Thus, the time required to decelerate upward is the same as the time required to accelerate downward. The ball returns to the surface with the same velocity with which it was thrown (neglecting friction). Therefore:

$$a = \frac{v_f - v_i}{t}$$

$$at = v_f - v_i$$

$$v_f = at + v_i$$

$$= \left(9.8\,\frac{m}{s^2}\right)(3.0\,s)$$

$$= (9.8)(3.0)\,\frac{m}{s^2} \times s$$

$$= 29\,\frac{m \cdot s}{s \cdot s}$$

$$= \boxed{29\,m/s}$$

2.12. These three questions are easily answered by using the three sets of relationships, or equations, that were presented in this chapter:

(a) $v_f = at + v_i$, and when v_i is zero,

$$v_f = at$$

$$v_f = \left(9.8\,\frac{m}{s^2}\right)(4.00\,s)$$

$$= 9.8 \times 4.00\,\frac{m}{s^2} \times s$$

$$= 39\,\frac{m \cdot s}{s \cdot s}$$

$$= \boxed{39\,m/s}$$

(b) $\bar{v} = \dfrac{v_f + v_i}{2} = \dfrac{39\,m/s + 0}{2} = 19.5\,m/s$

(c) $\bar{v} = \dfrac{d}{t}$ $\therefore$ $d = \bar{v}t = \left(19.5\,\dfrac{m}{s}\right)(4.00\,s)$

$$= 19.5 \times 4.00\,\frac{m}{s} \times s$$

$$= 78\,\frac{m \cdot s}{s}$$

$$= \boxed{78\,m}$$

2.13. Note that this problem can be solved with a series of three steps as in the previous problem. It can also be solved by the equation that combines all the relationships into one step. Either method is acceptable, but the following example of a one-step solution reduces the possibilities of error since fewer calculations are involved:

$$d = \frac{1}{2}gt^2 = \frac{1}{2}\left(9.8\,\frac{m}{s^2}\right)(5.00\,s)^2$$

$$= \frac{1}{2}\left(9.8\,\frac{m}{s^2}\right)(25.0\,s^2)$$

$$= \left(\frac{1}{2}\right)(9.8)(25.0)\,\frac{m}{s^2} \times s^2$$

$$= 4.90 \times 25.0\,\frac{m \cdot s^2}{s^2}$$

$$= 122.5\,m$$

$$= \boxed{120\,m}$$

2.14. Listing the known and unknown quantities:

$$F = 100\,N$$

$$m = 5\,kg$$

$$a = ?$$

These are the quantities of Newton's second law of motion, $F = ma$, and

$$F = ma \qquad \therefore \qquad a = \frac{F}{m}$$

$$= \frac{100\,\frac{kg \cdot m}{s^2}}{5\,kg}$$

$$= \frac{100}{5}\,\frac{kg \cdot m}{s^2} \times \frac{1}{kg}$$

$$= \boxed{20\,\frac{m}{s^2}}$$

2.15. Listing the known and unknown quantities:

$$m = 100\,kg$$

$$v = 6\,m/s$$

$$p = ?$$

These are the quantities found in the equation for momentum, $p = mv$, which is already solved for momentum (p). Thus,

$$p = mv$$

$$= (100 \text{ kg})\left(6\frac{\text{m}}{\text{s}}\right)$$

$$= \boxed{600\frac{\text{kg·m}}{\text{s}}}$$

2.16 Listing the known and unknown quantities:

$$w = 13{,}720 \text{ N}$$

$$v = 91 \text{ km/h}$$

$$p = ?$$

The equation for momentum is $p = mv$, which is already solved for momentum (p). The weight unit must be first converted to a mass unit:

$$w = mg \qquad \therefore \qquad m = \frac{w}{g}$$

$$= \frac{13{,}720\,\dfrac{\text{kg·m}}{\text{s}^2}}{9.8\,\dfrac{\text{m}}{\text{s}^2}}$$

$$= \frac{13{,}720}{9.8}\,\frac{\text{kg·m}}{\text{s}^2} \times \frac{\text{s}^2}{\text{m}}$$

$$= 1{,}400 \text{ kg}$$

The km/h unit should next be converted to m/s. Using the conversion factor from inside the front cover:

$$\frac{0.2778\,\dfrac{\text{m}}{\text{s}}}{1\,\dfrac{\text{km}}{\text{h}}} \times 91\,\frac{\text{km}}{\text{h}}$$

$$0.2778 \times 91\,\frac{\text{m}}{\text{s}} \times \frac{\text{h}}{\text{km}} \times \frac{\text{km}}{\text{h}}$$

$$25\,\frac{\text{m}}{\text{s}}$$

Now, listing the converted known and unknown quantities:

$$m = 1{,}400 \text{ kg}$$

$$v = 25 \text{ m/s}$$

$$p = ?$$

and solving for momentum (p),

$$p = mv$$

$$= (1{,}400 \text{ kg})\left(25\frac{\text{m}}{\text{s}}\right)$$

$$= \boxed{35{,}000\frac{\text{kg·m}}{\text{s}}}$$

2.17. Listing the known and unknown quantities:

$$\text{Bullet} \rightarrow m = 0.015 \text{ kg} \qquad \text{Rifle} \rightarrow m = 6 \text{ kg}$$

$$\text{Bullet} \rightarrow v = 200 \text{ m/s} \qquad \text{Rifle} \rightarrow v = ? \text{ m/s}$$

Note the mass of the bullet was converted to kg. This is a conservation of momentum question, where the bullet and rifle can be considered as a system of interacting objects:

$$\text{Bullet momentum} = -\text{rifle momentum}$$

$$(mv)_\text{b} = -(mv)_\text{r}$$

$$(mv)_\text{b} - (mv)_\text{r} = 0$$

$$(0.015 \text{ kg})\left(200\frac{\text{m}}{\text{s}}\right) - (6 \text{ kg})v_\text{r} = 0$$

$$\left(3 \text{ kg} \cdot \frac{\text{m}}{\text{s}}\right) - (6 \text{ kg·}v_\text{r}) = 0$$

$$\left(3 \text{ kg} \cdot \frac{\text{m}}{\text{s}}\right) = (6 \text{ kg·}v_\text{r})$$

$$v_\text{r} = \frac{3 \text{ kg} \cdot \dfrac{\text{m}}{\text{s}}}{6 \text{ kg}}$$

$$= \frac{3}{6}\,\frac{\text{kg}}{1} \times \frac{1}{\text{kg}} \times \frac{\text{m}}{\text{s}}$$

$$= \boxed{0.5\frac{\text{m}}{\text{s}}}$$

The rifle recoils with a velocity of 0.5 m/s.

2.18. Listing the known and unknown quantities:

$$\text{Astronaut} \rightarrow w = 2{,}156 \text{ N} \qquad \text{Wrench} \rightarrow m = 5.0 \text{ kg}$$

$$\text{Astronaut} \rightarrow v = ? \text{ m/s} \qquad \text{Wrench} \rightarrow v = 5.0 \text{ m/s}$$

Note that the astronaut's weight is given, but we need mass for the conservation of momentum equation. Mass can be found because the weight on earth was given, where we know $g = 9.8$ m/s^2. Thus the mass is

$$w = mg \qquad \therefore \qquad m = \frac{w}{g}$$

$$= \frac{2{,}156\,\dfrac{\text{kg·m}}{\text{s}^2}}{9.8\,\dfrac{\text{m}}{\text{s}^2}}$$

$$= \frac{2{,}156}{9.8}\,\frac{\text{kg·m}}{\text{s}^2} \times \frac{\text{s}^2}{\text{m}}$$

$$= 220 \text{ kg}$$

So the converted known and unknown quantities are:

$$\text{Astronaut} \rightarrow m = 220 \text{ kg} \qquad \text{Wrench} \rightarrow m = 5.0 \text{ kg}$$

$$\text{Astronaut} \rightarrow v = ? \text{ m/s} \qquad \text{Wrench} \rightarrow v = 5.0 \text{ m/s}$$

This is a conservation of momentum question, where the astronaut and wrench can be considered as a system of interacting objects:

$$\text{Wrench momentum} = -\text{astronaut momentum}$$

$$(mv)_w = -(mv)_a$$

$$(mv)_w - (mv)_a = 0$$

$$(5.0 \text{ kg})\left(5.0 \frac{m}{s}\right) - (220 \text{ kg})v_a = 0$$

$$\left(25 \text{ kg} \cdot \frac{m}{s}\right) - (220 \text{ kg} \cdot v_a) = 0$$

$$v_a = \frac{25 \text{ kg} \cdot \dfrac{m}{s}}{220 \text{ kg}}$$

$$= \frac{25}{220} \frac{\text{kg}}{1} \times \frac{1}{\text{kg}} \times \frac{m}{s}$$

$$= \boxed{0.11 \frac{m}{s}}$$

The astronaut moves away with a velocity of 0.11 m/s.

2.19. (a) Weight (w) is a downward force from the acceleration of gravity (g) on the mass (m) of an object. This relationship is the same as Newton's second law of motion, $F = ma$, and

$$w = mg = (1.25 \text{ kg})\left(9.8 \frac{m}{s^2}\right)$$

$$= (1.25)(9.8) \text{ kg} \times \frac{m}{s^2}$$

$$= 12.25 \frac{\text{kg} \cdot m}{s^2}$$

$$= \boxed{12 \text{ N}}$$

(b) First, recall that a force (F) is measured in newtons (N) and a newton has units of $N = \dfrac{\text{kg} \cdot m}{s^2}$. Second, the relationship between force (F), mass (m), and acceleration (a) is given by Newton's second law of motion, force = mass times acceleration, or $F = ma$. Thus,

$$F = ma \qquad \therefore \qquad a = \frac{F}{m} = \frac{10.0 \dfrac{\text{kg} \cdot m}{s^2}}{1.25 \text{ kg}}$$

$$= \frac{10.0}{1.25} \frac{\text{kg} \cdot m}{s^2} \times \frac{1}{\text{kg}}$$

$$= 8.00 \frac{\text{kg} \cdot m}{\text{kg} \cdot s^2}$$

$$= \boxed{8.00 \frac{m}{s^2}}$$

(Note how the units were treated mathematically in this solution and why it is necessary to show the units for a newton of force. The resulting unit in the answer

is a unit of acceleration, which provides a check that the problem was solved correctly.

2.20.
$$F = ma = (1.25 \text{ kg})\left(5.00 \frac{m}{s^2}\right)$$

$$= (1.25)(5.00) \text{ kg} \times \frac{m}{s^2}$$

$$= 6.25 \frac{\text{kg} \cdot m}{s^2}$$

$$= \boxed{6.25 \text{ N}}$$

(Note that the solution is correctly reported in *newton* units of force rather than kg·m/s².)

2.21. The bicycle tire exerts a backward force on the road, and the equal and opposite reaction force of the road on the bicycle produces the forward motion. (The motion is always in the direction of the applied force.) Therefore,

$$F = ma = (70.0 \text{ kg})\left(2.0 \frac{m}{s^2}\right)$$

$$= (70.0)(2.0) \text{ kg} \times \frac{m}{s^2}$$

$$= 140 \frac{\text{kg} \cdot m}{s^2}$$

$$= \boxed{140 \text{ N}}$$

2.22. The question requires finding a force in the metric system, which is measured in newtons of force. Since newtons of force are defined in kg, m, and s, unit conversions are necessary, and these should be done first.

$$1 \frac{\text{km}}{h} = \frac{1,000 \text{ m}}{3,600 \text{ s}} = 0.2778 \frac{m}{s}$$

Dividing both sides of this conversion factor by what you are converting *from* gives the conversion ratio of

$$\frac{0.2778 \dfrac{m}{s}}{\dfrac{\text{km}}{h}}$$

Multiplying this conversion ratio times the two velocities in km/h will convert them to m/s as follows:

$$\left(0.2778 \frac{\dfrac{m}{s}}{\dfrac{\text{km}}{h}}\right)\left(80.0 \frac{\text{km}}{h}\right)$$

$$= (0.2778)(80.0) \frac{m}{s} \times \frac{h}{\text{km}} \times \frac{\text{km}}{h}$$

$$= 22.2 \frac{m}{s}$$

$$\left(0.2778 \frac{\dfrac{m}{s}}{\dfrac{\text{km}}{h}}\right)\left(44.0 \frac{\text{km}}{h}\right)$$

$$= (0.2778)(44.0)\,\frac{m}{s} \times \frac{h}{km} \times \frac{km}{h}$$

$$= 12.2\,\frac{m}{s}$$

Now you are ready to find the appropriate relationship between the quantities involved. This involves two separate equations: Newton's second law of motion and the relationship of quantities involved in acceleration. These may be combined as follows:

$$F = ma \text{ and } a = \frac{v_f - v_i}{t} \qquad \therefore \qquad F = m\left(\frac{v_f - v_i}{t}\right)$$

Now you are ready to substitute quantities for the symbols and perform the necessary mathematical operations:

$$= (1,500 \text{ kg})\left(\frac{22.2 \text{ m/s} - 12.2 \text{ m/s}}{10.0 \text{ s}}\right)$$

$$= (1,500 \text{ kg})\left(\frac{10.0 \text{ m/s}}{10.0 \text{ s}}\right)$$

$$= 1,500 \times 1.00\,\frac{kg \cdot \frac{m}{s}}{s}$$

$$= 1,500\,\frac{kg \cdot m}{s} \times \frac{1}{s}$$

$$= 1,500\,\frac{kg \cdot m}{s \cdot s}$$

$$= 1,500\,\frac{kg \cdot m}{s^2}$$

$$= 1,500 \text{ N} = \boxed{1.5 \times 10^3 \text{ N}}$$

2.23. A unit conversion is needed as in the previous problem:

$$\left(90.0\,\frac{km}{h}\right)\left(0.2778\,\frac{\frac{m}{s}}{\frac{km}{h}}\right) = 25.0 \text{ m/s}$$

(a) $F = ma \qquad \therefore \qquad m = \frac{F}{a}$ and $a = \frac{v_f - v_i}{t}$, so

$$m = \frac{F}{\frac{v_f - v_i}{t}} = \frac{5,000.0\,\frac{kg \cdot m}{s^2}}{\frac{25.0 \text{ m/s} - 0}{5.0 \text{ s}}}$$

$$= \frac{5,000.0\,\frac{kg \cdot m}{s^2}}{5.0\,\frac{m}{s^2}}$$

$$= \frac{5,000.0}{5.0}\,\frac{kg \cdot m}{s^2} \times \frac{s^2}{m}$$

$$= 1,000\,\frac{kg \cdot m \cdot s^2}{m \cdot s^2}$$

$$= \boxed{1.0 \times 10^3 \text{ kg}}$$

(b) $w = mg$

$$= (1.0 \times 10^3 \text{ kg})\left(9.8\,\frac{m}{s^2}\right)$$

$$= (1.0 \times 10^3)(9.8) \text{ kg} \times \frac{m}{s^2}$$

$$= 9.8 \times 10^3\,\frac{kg \cdot m}{s^2}$$

$$= \boxed{9.8 \times 10^3 \text{ N}}$$

2.24. $w = mg$

$$= (70.0 \text{ kg})\left(9.8\,\frac{m}{s^2}\right)$$

$$= 70.0 \times 9.8 \text{ kg} \times \frac{m}{s^2}$$

$$= 686\,\frac{kg \cdot m}{s^2}$$

$$= \boxed{690 \text{ N}}$$

2.25. $F = \frac{mv^2}{r}$

$$= \frac{(0.20 \text{ kg})\left(3.0\,\frac{m}{s}\right)^2}{1.5 \text{ m}}$$

$$= \frac{(0.20 \text{ kg})\left(9.0\,\frac{m^2}{s^2}\right)}{1.5 \text{ m}}$$

$$= \frac{0.20 \times 9.0}{1.5}\,\frac{kg \cdot m^2}{s^2} \times \frac{1}{m}$$

$$= 1.2\,\frac{kg \cdot m \cdot m}{s^2 \cdot m}$$

$$= \boxed{1.2 \text{ N}}$$

2.26. **(a)** Newton's laws of motion consider the resistance to a change of motion, or mass, and not weight. The astronaut's mass is

$$w = mg \qquad \therefore \qquad m = \frac{w}{g} = \frac{1,960.0\,\frac{kg \cdot m}{s^2}}{9.8\,\frac{m}{s^2}}$$

$$= \frac{1,960.0}{9.8}\,\frac{kg \cdot m}{s^2} \times \frac{s^2}{m} = 200 \text{ kg}$$

(b) From Newton's second law of motion, you can see that the 100 N rocket gives the 200 kg astronaut an acceleration of:

$$F = ma \qquad \therefore \qquad a = \frac{F}{m} = \frac{100\,\frac{kg \cdot m}{s^2}}{200 \text{ kg}}$$

$$= \frac{100 \text{ kg} \cdot m}{200 \text{ s}^2} \times \frac{1}{kg} = 0.5 \text{ m/s}^2$$

(c) An acceleration of 0.5 m/s² for 2.0 s will result in a final velocity of

$$a = \frac{v_f - v_i}{t} \qquad \therefore \qquad v_f = at + v_i$$

$$= (0.5 \text{ m/s}^2)(2.0 \text{ s}) + 0 \text{ m/s}$$

$$= \boxed{1 \text{ m/s}}$$

EXERCISES Chapter 3

3.1. Listing the known and unknown quantities:

$$F = 200 \text{ N}$$

$$d = 3 \text{ m}$$

$$W = ?$$

These are the quantities found in the equation for work, $W = Fd$, which is already solved for work (W). Thus,

$$W = Fd$$

$$= \left(200 \frac{\text{kg·m}}{\text{s}^2}\right)(3 \text{ m})$$

$$= (200)(3) \text{ N·m}$$

$$= \boxed{600 \text{ J}}$$

3.2. Listing the known and unknown quantities:

$$F = 440 \text{ N}$$

$$d = 5.0 \text{ m}$$

$$w = 880 \text{ N}$$

$$W = ?$$

These are the quantities found in the equation for work, $W = Fd$, which is already solved for work (W). As you can see in the equation, the force exerted and the distance the box was moved are the quantities used in determining the work accomplished. The weight of the box is a different variable, and one that is not used in this equation. Thus,

$$W = Fd$$

$$= \left(440 \frac{\text{kg·m}}{\text{s}^2}\right)(5.0 \text{ m})$$

$$= 2,200 \text{ N·m}$$

$$= \boxed{2,200 \text{ J}}$$

3.3. Note that 10.0 kg is a mass quantity, and not a weight quantity. Weight is found from $w = mg$, a form of Newton's second law of motion. Thus the force that must

be exerted to lift the backpack is its weight, or (10.0 kg) × (9.8 m/s²) which is 98 N. Therefore, a force of 98 N was exerted on the backpack through a distance of 1.5 m, and

$$W = Fd$$

$$= \left(98 \frac{\text{kg·m}}{\text{s}^2}\right)(1.5 \text{ m})$$

$$= 147 \text{ N·m}$$

$$= \boxed{150 \text{ J}}$$

3.4. Weight is defined as the force of gravity acting on an object, and the greater the force of gravity the harder it is to lift the object. The force is proportional to the mass of the object, as the equation $w = mg$ tells you. Thus the force you exert when lifting is $F = w = mg$, so the work you do on an object you lift must be $W = mgh$.

You know the mass of the box and you know the work accomplished. You also know the value of the acceleration due to gravity, g, so the list of known and unknown quantities is:

$$m = 102 \text{ kg}$$

$$g = 9.8 \text{ m/s}^2$$

$$W = 5,000 \text{ J}$$

$$h = ?$$

The equation $W = mgh$ is solved for work, so the first thing to do is to solve it for h, the unknown height in this problem (note that height is also a distance):

$$W = mgh \qquad \therefore \qquad h = \frac{W}{mg}$$

$$= \frac{5,000 \frac{\text{kg·m}}{\text{s}^2} \times \text{m}}{(102 \text{ kg})\left(9.8 \frac{\text{m}}{\text{s}^2}\right)}$$

$$= \frac{5,000.0}{102 \times 9.8} \frac{\text{kg·m}}{\text{s}^2} \times \frac{\text{m}}{1} \times \frac{1}{\text{kg}} \times \frac{\text{s}^2}{\text{m}}$$

$$= \frac{5,000}{999.6} \text{ m}$$

$$= \boxed{5 \text{ m}}$$

3.5. A student running up the stairs has to lift herself, so her weight is the required force needed. Thus the force exerted is $F = w = mg$, and the work done is $W = mgh$. You know the mass of the student, the height, and the time. You also know the value of the acceleration due to gravity, g, so the list of known and unknown quantities is:

$$m = 60.0 \text{ kg}$$

$$g = 9.8 \text{ m/s}^2$$

$$h = 5.00 \text{ m}$$

$$t = 3.94 \text{ s}$$

$$P = ?$$

The equation $P = \dfrac{mgh}{t}$ is already solved for power, so:

$$P = \frac{mgh}{t}$$

$$= \frac{(60.0 \text{ kg}) \left(9.8 \dfrac{m}{s^2}\right)(5.00 \text{ m})}{3.92 \text{ s}}$$

$$= \frac{(60.0)(9.8)(5.00)}{(3.94)} \frac{\left(\dfrac{kg \cdot m}{s^2}\right) \times m}{s}$$

$$= \frac{2,940}{3.92} \frac{N \cdot m}{s}$$

$$= 750 \frac{J}{s}$$

$$= \boxed{750 \text{ W}}$$

3.6.

(a) $\dfrac{1.00 \text{ hp}}{746 \text{ W}} \times 1,400 \text{ W}$

$\dfrac{1,400}{746} \dfrac{hp \cdot W}{W}$

$\boxed{1.9 \text{ hp}}$

(b) $\dfrac{746 \text{ W}}{1 \text{ hp}} \times 3.5 \text{ hp}$

$746 \times 3.5 \dfrac{W \cdot hp}{hp}$

$2,611 \text{ W}$

$\boxed{2,600 \text{ hp}}$

3.7. Listing the known and unknown quantities:

$$m = 2,000 \text{ kg}$$
$$v = 72 \text{ km/h}$$
$$KE = ?$$

These are the quantities found in the equation for kinetic energy, $KE = 1/2mv^2$, which is already solved. However, note that the velocity is in units of km/h, which must be changed to m/s before doing anything else (it must be m/s because all energy and work units are in units of the joule [J]. A joule is a newton-meter, and a newton is a kg·m/s²). Using the conversion factor from inside the front cover of your text,

$$\frac{0.2778 \dfrac{m}{s}}{1.0 \dfrac{km}{h}} \times 72 \frac{km}{h}$$

$$(0.2778)(72) \frac{m}{s} \times \frac{h}{km} \times \frac{km}{h}$$

$$20 \frac{m}{s}$$

and

$$KE = \frac{1}{2}mv^2$$

$$= \frac{1}{2}(2,000 \text{ kg})\left(20 \frac{m}{s}\right)^2$$

$$= \frac{1}{2}(2,000 \text{ kg})\left(400 \frac{m^2}{s^2}\right)$$

$$= \frac{1}{2} \times 2,000 \times 400 \ \frac{kg \cdot m^2}{s^2}$$

$$= 400,000 \frac{kg \cdot m}{s^2} \times m$$

$$= 40,000 \text{ N·m}$$

$$= \boxed{4 \times 10^5 \text{ J}}$$

Scientific notation is used here to simplify a large number and to show one significant figure.

3.8. Recall the relationship between work and energy—that you do work on an object when you throw it, giving it kinetic energy, and the kinetic energy it has will do work on something else when stopping. Because of the relationship between work and energy you can calculate (1) the work you do, (2) the kinetic energy a moving object has as a result of your work, and (3) the work it will do when coming to a stop, and all three answers should be the same. Thus you do not have a force or a distance to calculate the work needed to stop a moving car, but you can simply calculate the kinetic energy of the car. Both answers should be the same.

Before you start, note that the velocity is in units of km/hr, which must be changed to m/s before doing anything else (it must be m/s because all energy and work units are in units of the joule [J]. A joule is a newton-meter, and a newton is a kg·m/s²). Using the conversion factor from inside the front cover,

$$\frac{0.2778 \dfrac{m}{s}}{1.0 \dfrac{km}{h}} \times 54.0 \frac{km}{h}$$

$$0.2778 \times 54.0 \ \frac{m}{s} \times \frac{h}{km} \times \frac{km}{h}$$

$$15.0 \frac{m}{s}$$

and

$$KE = \frac{1}{2}mv^2$$

$$= \frac{1}{2}(1,000.0 \text{ kg})\left(15.0 \frac{m}{s}\right)^2$$

$$= \frac{1}{2}(1,000.0 \text{ kg})\left(225 \frac{m^2}{s^2}\right)$$

$$= \frac{1}{2} \times 1,000.0 \times 225 \frac{kg \cdot m^2}{s^2}$$

$$= 112{,}500 \ \frac{\text{kg·m}}{\text{s}^2} \times \text{m}$$

$$= 112{,}500 \ \text{N·m}$$

$$= \boxed{1.13 \times 10^5 \ \text{J}}$$

Scientific notation is used here to simplify a large number and to easily show three significant figures. The answer could likewise be expressed as 113 kJ.

3.9.   **(a)** $W = Fd$

$$= (10 \ \text{lb})(5 \ \text{ft})$$

$$= (10)(5) \ \text{ft} \times \text{lb}$$

$$= 50 \ \text{ft·lb}$$

(b) The distance of the bookcase from some horizontal reference level did not change, so the gravitational potential energy does not change.

3.10. The force (F) needed to lift the book is equal to the weight (w) of the book, or $F = w$. Since $w = mg$, then $F = mg$. Work is defined as the product of a force moved through a distance, or $W = Fd$. The work done in lifting the book is therefore $W = mgd$, and:

(a) $W = mgd$

$$= (2.0 \ \text{kg})(9.8 \ \text{m/s}^2)(2.00 \ \text{m})$$

$$= (2.0)(9.8)(2.00) \ \frac{\text{kg·m}}{\text{s}^2} \times \text{m}$$

$$= 39.2 \ \frac{\text{kg·m}^2}{\text{s}^2}$$

$$= 39.2 \ \text{J} = \boxed{39 \ \text{J}}$$

(b) $PE = mgh = \boxed{39 \ \text{J}}$

(c) $PE_{\text{lost}} = KE_{\text{gained}} = mgh = \boxed{39 \ \text{J}}$

(or)

$$v = \sqrt{2gh} = \sqrt{(2)(9.8 \ \text{m/s}^2)(2.00 \ \text{m})}$$

$$= \sqrt{39.2 \ \text{m}^2/\text{s}^2}$$

$$= 6.26 \ \text{m/s}$$

$$KE = \frac{1}{2}mv^2 = \left(\frac{1}{2}\right)(2.0 \ \text{kg})(6.26 \ \text{m/s})^2$$

$$= \left(\frac{1}{2}\right)(2.0 \ \text{kg})(39.2 \ \text{m}^2/\text{s}^2)$$

$$= (1.0)(39.2) \ \frac{\text{kg·m}^2}{\text{s}^2}$$

$$= \boxed{39 \ \text{J}}$$

3.11. Note that the gram unit must be converted to kg to be consistent with the definition of a newton-meter, or joule unit of energy:

$$KE = \frac{1}{2}mv^2 = \left(\frac{1}{2}\right)(0.15 \ \text{kg})(30.0 \ \text{m/s})^2$$

$$= \left(\frac{1}{2}\right)(0.15 \ \text{kg})(900 \ \text{m}^2/\text{s}^2)$$

$$= \left(\frac{1}{2}\right)(0.15)(900) \ \frac{\text{kg·m}^2}{\text{s}^2}$$

$$= 67.5 \ \text{J} = \boxed{68 \ \text{J}}$$

3.12. The km/h unit must first be converted to m/s before finding the kinetic energy. Note also that the work done to put an object in motion is equal to the energy of motion, or kinetic energy that it has as a result of the work. The work needed to bring the object to a stop is also equal to the kinetic energy of the moving object:

Unit conversion:

$$1 \ \frac{\text{km}}{\text{h}} = 0.2778 \ \frac{\frac{\text{m}}{\text{s}}}{\frac{\text{km}}{\text{h}}} = \left(90.0 \ \frac{\text{km}}{\text{h}}\right)\left(0.2778 \ \frac{\frac{\text{m}}{\text{s}}}{\frac{\text{km}}{\text{h}}}\right) = 25.0 \ \text{m/s}$$

(a) $KE = \frac{1}{2}mv^2 = \frac{1}{2}(1{,}000.0 \ \text{kg})\left(25.0 \ \frac{\text{m}}{\text{s}}\right)^2$

$$= \frac{1}{2}(1{,}000.0 \ \text{kg})\left(625 \ \frac{\text{m}^2}{\text{s}^2}\right)$$

$$= \frac{1}{2}(1{,}000.0)(625) \ \frac{\text{kg·m}^2}{\text{s}^2}$$

$$= 312.5 \ \text{kJ} = \boxed{313 \ \text{kJ}}$$

(b) $W = Fd = KE = \boxed{313 \ \text{kJ}}$

(c) $KE = W = Fd = \boxed{313 \ \text{kJ}}$

3.13. $KE = \frac{1}{2}mv^2$

$$= \frac{1}{2}(60.0 \ \text{kg})\left(2.0 \ \frac{\text{m}}{\text{s}}\right)^2$$

$$= \frac{1}{2}(60.0 \ \text{kg})\left(4.0 \ \frac{\text{m}^2}{\text{s}^2}\right)$$

$$= 30.0 \times 4.0 \ \text{kg} \times \left(\frac{\text{m}^2}{\text{s}^2}\right)$$

$$= \boxed{120 \ \text{J}}$$

$$KE = \frac{1}{2}mv^2$$

$$= \frac{1}{2}(60.0 \ \text{kg})\left(4.0 \ \frac{\text{m}}{\text{s}}\right)^2$$

$$= \frac{1}{2}(60.0 \text{ kg})\left(16 \frac{\text{m}^2}{\text{s}^2}\right)$$

$$= 30.0 \times 16 \text{ kg} \times \left(\frac{\text{m}^2}{\text{s}^2}\right)$$

$$= \boxed{480 \text{ J}}$$

Thus, doubling the speed results in a fourfold increase in kinetic energy.

3.14.

$$KE = \frac{1}{2}mv^2$$

$$= \frac{1}{2}(70.0 \text{ kg})(6.00 \text{ m/s})^2$$

$$= (35.0 \text{ kg})(36.0 \text{ m}^2/\text{s}^2)$$

$$= 35.0 \times 36.0 \text{ kg} \times \frac{\text{m}^2}{\text{s}^2}$$

$$= \boxed{1,260 \text{ J}}$$

$$KE = \frac{1}{2}mv^2$$

$$= \frac{1}{2}(140.0 \text{ kg})(6.00 \text{ m/s})^2$$

$$= (70.0 \text{ kg})(36.0 \text{ m}^2/\text{s}^2)$$

$$= 70.0 \times 36.0 \text{ kg} \times \frac{\text{m}^2}{\text{s}^2}$$

$$= \boxed{2,520 \text{ J}}$$

Thus, doubling the mass results in a doubling of the kinetic energy.

3.15. (a) The force needed is equal to the weight of the student. The English unit of a pound is a force unit, so

$$W = Fd$$

$$= (170.0 \text{ lb})(25.0 \text{ ft})$$

$$= \boxed{4,250 \text{ ft·lb}}$$

(b) Work (W) is defined as a force (F) moved through a distance (d), or $W = Fd$. Power (P) is defined as work (W) per unit of time (t), or $P = W/t$. Therefore,

$$P = \frac{Fd}{t}$$

$$= \frac{(170.0 \text{ lb})(25.0 \text{ ft})}{10.0 \text{ s}}$$

$$= \frac{(170.0)(25.0)}{10.0} \frac{\text{ft·lb}}{\text{s}}$$

$$= 425 \frac{\text{ft·lb}}{\text{s}}$$

One hp is defined as $550 \dfrac{\text{ft·lb}}{\text{s}}$ and

$$\frac{425 \text{ ft·lb/s}}{550 \dfrac{\text{ft·lb/s}}{\text{hp}}} = \boxed{0.77 \text{ hp}}$$

Note that the student's power rating (425 ft·lb/s) is less than the power rating defined as 1 horsepower (550 ft·lb/s). Thus, the student's horsepower must be *less* than 1 horsepower. A simple analysis such as this will let you know if you inverted the ratio or not.

3.16. (a) The force (F) needed to lift the elevator is equal to the weight of the elevator. Since the work (W) is equal to Fd and power (P) is equal to W/t, then

$$P = \frac{Fd}{t} \qquad \therefore \qquad t = \frac{Fd}{P}$$

$$= \frac{[2,000.0 \text{ lb}][20.0 \text{ ft}]}{\left(550 \dfrac{\text{ft·lb}}{\text{s}}\right)[20.0 \text{ hp}]}$$

$$= \frac{40,000}{11,000} \frac{\text{ft·lb}}{\dfrac{\text{ft·lb}}{\text{s}}} \times \frac{1}{\cancel{\text{hp}}} \times \cancel{\text{hp}}$$

$$= \frac{40,000}{11,000} \frac{\text{ft·lb}}{1} \times \frac{\text{s}}{\text{ft·lb}}$$

$$= 3.64 \frac{\cancel{\text{ft·lb}}\cdot\text{s}}{\cancel{\text{ft·lb}}}$$

$$= \boxed{3.6 \text{ s}}$$

(b)

$$\bar{v} = \frac{d}{t}$$

$$= \frac{20.0 \text{ ft}}{3.6 \text{ s}}$$

$$= \boxed{5.6 \text{ ft/s}}$$

3.17. Since $PE_{\text{lost}} = KE_{\text{gained}}$ then $mgh = \dfrac{1}{2}mv^2$. Solving for v,

$$v = \sqrt{2gh} = \sqrt{(2)(32.0 \text{ ft/s}^2)(9.8 \text{ ft})}$$

$$= \sqrt{(2)(32.0)(9.8) \text{ ft}^2/\text{s}^2}$$

$$= \sqrt{627 \text{ ft}^2/\text{s}^2}$$

$$= 25 \text{ ft/s}$$

3.18.
$$KE = \frac{1}{2}mv^2 \qquad \therefore \qquad v = \sqrt{\frac{2KE}{m}}$$

$$= \sqrt{\frac{(2)\left(200,000 \dfrac{\text{kg·m}^2}{\text{s}^2}\right)}{1,000.0 \text{ kg}}}$$

$$= \frac{400{,}000}{1{,}000.0} \frac{\text{kg·m}^2}{\text{s}^2} \times \frac{1}{\text{kg}}$$

$$= \sqrt{\frac{400{,}000}{1{,}000.0} \frac{\text{kg·m}^2}{\text{kg·s}^2}}$$

$$= \sqrt{400 \ \text{m}^2/\text{s}^2}$$

$$= \boxed{20 \ \text{m/s}}$$

3.19. The maximum velocity occurs at the lowest point with a gain of kinetic energy equivalent to the loss of potential energy in falling 3.0 in (which is 0.25 ft), so

$$KE_{\text{gained}} = PE_{\text{lost}}$$

$$\frac{1}{2}mv^2 = mgh$$

$$v = \sqrt{2gh}$$

$$= \sqrt{(2)(32 \ \text{ft/s}^2)(0.25 \ \text{ft})}$$

$$= \sqrt{(2)(32)(0.25)\text{ft/s}^2 \times \text{ft}}$$

$$= \sqrt{16 \ \text{ft}^2/\text{s}^2}$$

$$= \boxed{4.0 \ \text{ft/s}}$$

3.20. **(a)** $W = Fd$ and the force F that is needed to lift the load upward is mg, so $W = mgh$. Power is W/t, so

$$P = \frac{mgh}{t}$$

$$= \frac{(250.0 \ \text{kg})(9.8 \ \text{m/s}^2)(80.0 \ \text{m})}{39.2 \ \text{s}}$$

$$= \frac{(250.0)(9.8)(80.0)}{39.2} \frac{\text{kg}}{1} \times \frac{\text{m}}{\text{s}^2} \times \frac{\text{m}}{1} \times \frac{1}{\text{s}}$$

$$= \frac{196{,}000}{39.2} \frac{\text{kg·m}^2}{\text{s}^2} \times \frac{1}{\text{s}}$$

$$= 5{,}000 \ \frac{\text{J}}{\text{s}}$$

$$= \boxed{5.0 \ \text{kW}}$$

(b) There are 746 watts per horsepower, so

$$\frac{5{,}000 \ \text{W}}{746 \ \dfrac{\text{W}}{\text{hp}}} = \frac{5{,}000}{746} \frac{\text{W}}{1} \times \frac{\text{hp}}{\text{W}}$$

$$= 6.70 \ \frac{\text{W·hp}}{\text{W}}$$

$$= \boxed{6.70 \ \text{hp}}$$

4.1. Listing the known and unknown quantities:

body temperature $\quad T_F = 98.6°$

$$T_C = ?$$

These are the quantities found in the equation for conversion of Fahrenheit to Celsius, $T_C = \dfrac{5}{9}(T_F - 32°)$, where T_F is the temperature in Fahrenheit and T_C is the temperature in Celsius. This equation describes a relationship between the two temperature scales and is used to convert a Fahrenheit temperature to Celsius. The equation is already solved for the Celsius temperature, T_C. Thus,

$$T_C = \frac{5}{9}(T_F - 32°)$$

$$= \frac{5}{9}(98.6° - 32°)$$

$$= \frac{333°}{9}$$

$$= \boxed{37° \ \text{C}}$$

4.2. $\quad Q = mc\Delta T$

$$= (221 \ \text{g})\left(0.093 \ \frac{\text{cal}}{\text{gC°}}\right)(38.0°\text{C} - 20.0°\text{C})$$

$$= (221)(0.093)(18.0) \ \text{g} \times \frac{\text{cal}}{\text{gC°}} \times °\text{C}$$

$$= 370 \ \frac{\text{g·cal·}°\text{C}}{\text{gC°}}$$

$$= \boxed{370 \ \text{cal}}$$

4.3. First, you need to know the energy of the moving bike and rider. Since the speed is given as 36.0 km/h, convert to m/s by multiplying times 0.2778 m/s per km/h:

$$\left(36.0 \ \frac{\text{km}}{\text{h}}\right)\left(0.2778 \ \frac{\text{m/s}}{\text{km/h}}\right)$$

$$= (36.0)(0.2778) \ \frac{\text{km}}{\text{h}} \times \frac{\text{h}}{\text{km}} \times \frac{\text{m}}{\text{s}}$$

$$= 10.0 \ \text{m/s}$$

Then,

$$KE = \frac{1}{2}mv^2$$

$$= \frac{1}{2}(100.0 \ \text{kg})(10.0 \ \text{m/s})^2$$

$$= \frac{1}{2}(100.0 \ \text{kg})(100 \ \text{m}^2/\text{s}^2)$$

$$= \frac{1}{2}(100.0)(100) \ \frac{\text{kg·m}^2}{\text{s}^2}$$

$$= 5{,}000 \ \text{J}$$

Second, this energy is converted to the calorie heat unit through the mechanical equivalent of heat relationship, that 1.0 kcal = 4,184 J, or that 1.0 cal = 4.184 J. Thus,

$$\frac{5,000 \text{ J}}{4,184 \text{ J/kcal}}$$

$$1.195 \frac{\text{J}}{1} \times \frac{\text{kcal}}{\text{J}}$$

$$\boxed{1.20 \text{ kcal}}$$

4.4. First, you need to find the energy of the falling bag. Since the potential energy lost equals the kinetic energy gained, the energy of the bag just as it hits the ground can be found from

$$PE = mgh$$

$$= (15.53 \text{ kg})(9.8 \text{ m/s}^2)(5.50 \text{ m})$$

$$= (15.53)(9.8)(5.50) \ \frac{\text{kg} \cdot \text{m}}{\text{s}^2} \times \text{m}$$

$$= 837 \text{ J}$$

In calories, this energy is equivalent to

$$\frac{837 \text{ J}}{4,184 \text{ J/kcal}} = 0.200 \text{ kcal}$$

Second, the temperature change can be calculated from the equation giving the relationship between a quantity of heat (Q), mass (m), specific heat of the substance (c), and the change of temperature:

$$Q = mc\Delta T \quad \therefore \quad \Delta T = \frac{Q}{mc}$$

$$= \frac{0.200 \text{ kcal}}{(15.53 \text{ kg})\left(0.200 \dfrac{\text{kcal}}{\text{kgC}^\circ}\right)}$$

$$= \frac{0.200}{(15.53)(0.200)} \ \frac{\text{kcal}}{1} \times \frac{1}{\text{kg}} \times \frac{\text{kgC}^\circ}{\text{kcal}}$$

$$= 0.0644 \ \frac{\text{kcal} \cdot \text{kgC}^\circ}{\text{kcal} \cdot \text{kg}}$$

$$= \boxed{6.44 \times 10^{-2} \, {}^\circ\text{C}}$$

4.5. The Calorie used by dietitians is a kilocalorie; thus 250.0 Cal is 250.0 kcal. The mechanical energy equivalent is 1 kcal = 4,184 J, so (250.0 kcal)(4,184 J/kcal) = 1,046,250 J.

Since $W = Fd$ and the force needed is equal to the weight (mg) of the person, $W = mgh = (75.0 \text{ kg})(9.8 \text{ m/s}^2)$ (10.0 m) = 7,350 J for each stairway climb.

A total of 1,046,250 J of energy from the French fries would require 1,046,250 J/7,350 J per climb, or 142.3 trips up the stairs.

4.6. For unit consistency,

$$T_C = \frac{5}{9}(T_F - 32^\circ) = \frac{5}{9}(68^\circ - 32^\circ) = \frac{5}{9}(36^\circ) = 20^\circ\text{C}$$

$$= \frac{5}{9}(32^\circ - 32^\circ) = \frac{5}{9}(0^\circ) = 0^\circ\text{C}$$

Glass Bowl:

$$Q = mc\Delta T$$

$$= (0.5 \text{ kg})\left(0.2 \frac{\text{kcal}}{\text{kg}^\circ\text{C}}\right)(20^\circ\text{C})$$

$$= (0.5)(0.2)(20) \ \frac{\text{kg}}{1} \times \frac{\text{kcal}}{\text{kgC}^\circ} \times \frac{^\circ\text{C}}{1}$$

$$= \boxed{2 \text{ kcal}}$$

Iron Pan:

$$Q = mc\Delta T$$

$$= (0.5 \text{ kg})\left(0.11 \frac{\text{kcal}}{\text{kgC}^\circ}\right)(20^\circ\text{C})$$

$$= (0.5)(0.11)(20) \ \text{kg} \times \frac{\text{kcal}}{\text{kgC}^\circ} \times {}^\circ\text{C}$$

$$= \boxed{1 \text{ kcal}}$$

4.7. Note that a specific heat expressed in cal/g has the same numerical value as a specific heat expressed in kcal/kg because you can cancel the k units. You could convert 896 cal to 0.896 kcal, but one of the two conversion methods is needed for consistency with other units in the problem.

$$Q = mc\Delta T \quad \therefore \quad m = \frac{Q}{c\Delta T}$$

$$= \frac{896 \text{ cal}}{\left(0.056 \dfrac{\text{cal}}{\text{gC}^\circ}\right)(80.0^\circ\text{C})}$$

$$= \frac{896}{(0.056)(80.0)} \ \frac{\text{cal}}{1} \times \frac{\text{gC}^\circ}{\text{cal}} \times \frac{1}{\text{C}}$$

$$= 200 \text{ g}$$

$$= \boxed{0.20 \text{ kg}}$$

4.8. Since a watt is defined as a joule/s, finding the total energy in joules will tell the time:

$$Q = mc\Delta T$$

$$= (250.0 \text{ g})\left(1.00 \frac{\text{cal}}{\text{gC}^\circ}\right)(60.0^\circ\text{C})$$

$$= (250.0)(1.00)(60.0) \ \text{g} \times \frac{\text{cal}}{\text{gC}^\circ} \times {}^\circ\text{C}$$

$$= 1.50 \times 10^4 \text{ cal}$$

This energy in joules is

$$(1.50 \times 10^4 \text{ cal})\left(4.184 \frac{\text{J}}{\text{cal}}\right) = 62,800 \text{ J}$$

A 300-watt heater uses energy at a rate of $300 \dfrac{\text{J}}{\text{s}}$, so $\dfrac{62,800 \text{ J}}{300 \text{ J/s}}$

= 209 s is required, which is $\dfrac{209\ \text{s}}{60\ \dfrac{\text{s}}{\text{min}}}$ = 3.48 min, or

$$\boxed{\text{about } 3\tfrac{1}{2} \text{ min}}$$

4.9. $\quad Q = mc\Delta T \quad \therefore \quad c = \dfrac{Q}{m\Delta T}$

$$= \dfrac{60.0\ \text{cal}}{(100.0\ \text{g})(20.0°\text{C})}$$

$$= \dfrac{60.0}{(100.0)(20.0)}\ \dfrac{\text{cal}}{\text{gC}°}$$

$$= \boxed{0.0300\ \dfrac{\text{cal}}{\text{gC}°}}$$

4.10. Since the problem specified a solid changing to a liquid without a temperature change, you should recognize that this is a question about a phase change only. The phase change from solid to liquid (or liquid to solid) is concerned with the latent heat of fusion. For water, the latent heat of fusion is given as 80.0 cal/g, and

$m = 250.0\ \text{g}$ $\qquad Q = mL_f$

$L_{f\,(\text{water})} = 80.0\ \text{cal/g}$

$\qquad Q = ?$ $\qquad\qquad = (250.0\ \text{g})\left(80.0\ \dfrac{\text{cal}}{\text{g}}\right)$

$$= 250.0 \times 80.0\ \dfrac{\text{g·cal}}{\cancel{\text{g}}}$$

$$= 20,000\ \text{cal} = \boxed{20.0\ \text{kcal}}$$

4.11. To change water at 80.0°C to steam at 100.0°C requires two separate quantities of heat that can be called Q_1 and Q_2. The quantity Q_1 is the amount of heat needed to warm the water from 80.0°C to the boiling point, which is 100.0°C at sea level pressure ($\Delta T = 20.0$°C). The relationship between the variable involved is $Q = mc\Delta T$. The quantity Q_2 is the amount of heat needed to take 100.0° water through the phase change to steam (water vapor) at 100.0°C. The phase change from a liquid to a gas (or gas to liquid) is concerned with the latent heat of vaporization. For water, the latent heat of vaporization is given as 540.0 cal/g.

$m = 250.0\ \text{g}$ $\qquad Q_1 = mc\Delta T$

$L_{v\,(\text{water})} = 540.0\ \text{cal/g}$

$\qquad Q = ?$ $\qquad\quad = (250.0\ \text{g})\left(1.00\ \dfrac{\text{cal}}{\text{gC}°}\right)(20.0°\text{C})$

$$= (250.0)(1.00)(20.0)\ \text{g} \times \dfrac{\text{cal}}{\text{gC}°} \times °\text{C}$$

$$= 5,000\ \dfrac{\text{g·cal·}°\cancel{\text{C}}}{\cancel{\text{gC}}°}$$

$$= 5,000\ \text{cal}$$

$$= 5.00\ \text{kcal}$$

$Q_2 = mL_v$

$$= (250.0\ \text{g})\left(540.0\ \dfrac{\text{cal}}{\text{g}}\right)$$

$$= 250.0 \times 540.0\ \dfrac{\text{g·cal}}{\cancel{\text{g}}}$$

$$= 135,000\ \text{cal}$$

$$= 135.0\ \text{kcal}$$

$Q_{\text{Total}} = Q_1 + Q_2$

$$= 5.00\ \text{kcal} + 135.0\ \text{kcal}$$

$$= \boxed{140.0\ \text{kcal}}$$

4.12. To change 20.0°C water to steam at 125.0°C requires three separate quantities of heat. First, the quantity Q_1 is the amount of heat needed to warm the water from 20.0°C to 100.0°C ($\Delta T = 80.0$°C). The quantity Q_2 is the amount of heat needed to take 100.0°C water to steam at 100.0°C. Finally, the quantity Q_3 is the amount of heat needed to warm the steam from 100.0° to 125.0°C. According to Table 4.4, the c for steam is 0.480 cal/g°C.

$m = 100.0\ \text{g}$ $\qquad Q_1 = mc\Delta T$

$\Delta T_{\text{water}} = 80.0$°C

$\qquad\qquad\qquad = (100.0\ \text{g})\left(1.00\ \dfrac{\text{cal}}{\text{gC}°}\right)(80.0°\text{C})$

$\Delta T_{\text{steam}} = 25.0$°C

$L_{v(\text{water})} = 540.0\ \text{cal/g}$ $\quad = (100.0)(1.00)(80.0)\ \text{g} \times \dfrac{\text{cal}}{\text{gC}°} \times °\text{C}$

$c_{\text{steam}} = 0.480\ \text{cal/gC}°$

$$= 8,000\ \dfrac{\text{g·cal·}°\cancel{\text{C}}}{\text{g}\cancel{\text{C}}°}$$

$$= 8,000\ \text{cal}$$

$$= 8.00\ \text{kcal}$$

$Q_2 = mL_v$

$$= (100.0\ \text{g})\left(540.0\ \dfrac{\text{cal}}{\text{g}}\right)$$

$$= 100.0 \times 540.0\ \dfrac{\text{g·cal}}{\cancel{\text{g}}}$$

$$= 54,000\ \text{cal}$$

$$= 54.00\ \text{kcal}$$

$Q_3 = mc\Delta T$

$$= (100.0\ \text{g})\left(0.480\ \dfrac{\text{cal}}{\text{gC}°}\right)(25.0°\text{C})$$

$$= (100.0)(0.480)(25.0)\ \text{g} \times \dfrac{\text{cal}}{\text{gC}°} \times °\text{C}$$

$$= 1,200\ \dfrac{\text{g·cal·}°\cancel{\text{C}}}{\text{g}\cancel{\text{C}}°}$$

$$= 1,200\ \text{cal}$$

$$= 1.20\ \text{kcal}$$

$Q_{\text{total}} = Q_1 + Q_2 + Q_3$

$$= 8.00\ \text{kcal} + 54.00\ \text{kcal} + 1.20\ \text{kcal}$$

$$= \boxed{63.20\ \text{kcal}}$$

4.13. (a) **Step 1:** Cool water from 18.0°C to 0°C.

$$Q_1 = mc\Delta T$$

$$= (400.0 \text{ g})\left(1.00 \frac{\text{cal}}{\text{gC°}}\right)(18.0°\text{C})$$

$$= (400.0)(1.00)(18.0) \text{ g} \times \frac{\text{cal}}{\text{gC°}} \times °\text{C}$$

$$= 7{,}200 \frac{\text{g·cal·}°\cancel{C}}{\text{g}\cancel{C°}}$$

$$= 7{,}200 \text{ cal}$$

$$= 7.20 \text{ kcal}$$

Step 2: Find the energy needed for the phase change of water at 0°C to ice at 0°C.

$$Q_2 = mL_f$$

$$= (400.0 \text{ g})\left(80.0 \frac{\text{cal}}{\text{g}}\right)$$

$$= 400.0 \times 80.0 \frac{\text{g·cal}}{\cancel{g}}$$

$$= 32{,}000 \text{ cal}$$

$$= 32.0 \text{ kcal}$$

Step 3: Cool the ice from 0°C to ice at −5.00°C.

$$Q_3 = mc\Delta T$$

$$= (400.0 \text{ g})\left(0.500 \frac{\text{cal}}{\text{gC°}}\right)(5.00°\text{C})$$

$$= 400.0 \times 0.500 \times 5.00 \text{ g} \times \frac{\text{cal}}{\text{gC°}} \times °\text{C}$$

$$= 1{,}000 \frac{\text{g·cal·}°\cancel{C}}{\text{g·}\cancel{C°}}$$

$$= 1{,}000 \text{ cal}$$

$$= 1.00 \text{ kcal}$$

$$Q_{\text{total}} = Q_1 + Q_2 + Q_3$$

$$= 7.20 \text{ kcal} + 32.0 \text{ kcal} + 1.00 \text{ kcal}$$

$$= \boxed{40.2 \text{ kcal}}$$

(b) $Q = mL_v \quad \therefore \quad m = \dfrac{Q}{L_v}$

$$= \frac{40{,}200 \text{ cal}}{40.0 \dfrac{\text{cal}}{\text{g}}}$$

$$= \frac{40{,}200}{40.0} \frac{\text{cal}}{1} \times \frac{\text{g}}{\text{cal}}$$

$$= 1{,}005 \frac{\text{cal·g}}{\cancel{\text{cal}}}$$

$$= \boxed{1.01 \times 10^3 \text{ g}}$$

4.14. $W = J(Q_H - Q_L)$

$$= 4{,}184 \frac{\text{J}}{\text{kcal}}(0.3000 \text{ kcal} - 0.2000 \text{ kcal})$$

$$= 4{,}184 \frac{\text{J}}{\text{kcal}}(0.1000 \text{ kcal})$$

$$= 4{,}184 \times 0.1000 \frac{\text{J·kcal}}{\text{kcal}}$$

$$= \boxed{418.4 \text{ J}}$$

4.15. $W = J(Q_H - Q_L)$

$$= 4{,}184 \frac{\text{J}}{\text{kcal}}(55.0 \text{ kcal} - 40.0 \text{ kcal})$$

$$= 4{,}184 \frac{\text{J}}{\text{kcal}}(15.0 \text{ kcal})$$

$$= 4{,}184 \times 15.0 \frac{\text{J·kcal}}{\text{kcal}}$$

$$= 62{,}760 \text{ J}$$

$$= \boxed{62.8 \text{ kJ}}$$

EXERCISES **Chapter 5**

5.1. (a) $f = 3$ Hz $\quad T = \dfrac{1}{f}$

$\lambda = 2$ cm

$T = ?$ $\qquad = \dfrac{1}{3\dfrac{1}{\text{s}}}$

$$= \frac{1}{3}\frac{\text{s}}{1}$$

$$= \boxed{0.3 \text{ s}}$$

(b) $f = 3$ Hz $\quad v = \lambda f$

$\lambda = 2$ cm

$T = ?$ $\qquad = (2 \text{ cm})\left(3\dfrac{1}{\text{s}}\right)$

$$= 2 \times 3 \text{ cm} \times \frac{1}{\text{s}}$$

$$= \boxed{6 \frac{\text{cm}}{\text{s}}}$$

5.2. **Step 1:**

$t = 20.0°\text{C}$ $\quad v_{T_{P(m/s)}} = v_0 + \left(\dfrac{0.600 \text{ m/s}}{°\text{C}}\right)(T_P)$

$f = 20{,}000$ Hz

$v = ?$ $\qquad = 331\dfrac{\text{m}}{\text{s}} + \left(\dfrac{0.600 \text{ m/s}}{°\text{C}}\right)(20.0°\text{C})$

$\lambda = ?$

$$= 331\frac{\text{m}}{\text{s}} + 12.0\frac{\text{m}}{\text{s}}$$

$$= 343\frac{\text{m}}{\text{s}}$$

Step 2:

$$v = \lambda f \quad \therefore \quad \lambda = \frac{v}{f}$$

$$= \frac{343 \, \frac{m}{s}}{20{,}000 \, \frac{1}{s}}$$

$$= \frac{343}{20{,}000} \, \frac{m}{s} \times \frac{s}{1}$$

$$= 0.01715 \, m$$

$$= \boxed{0.02 \, m}$$

5.3. $f_1 = 440 \, Hz \qquad f_b = f_2 - f_1$

$\quad\;\; f_2 = 446 \, Hz \qquad\quad = (446 \, Hz) - (440 \, Hz)$

$\quad\;\; f_b = ? \qquad\qquad\quad = \boxed{6 \, Hz}$

Note that the smaller frequency is subtracted from the larger one to avoid negative beats.

5.4. **Step 1:** Assume room temperature (20.0°C) to obtain the velocity:

$f = 2.00 \times 10^7 \, Hz \quad v_{T_{P(m/s)}} = v_0 + \left(\frac{0.600 \, \frac{m}{s}}{°C} \right) (T_P)$

$\lambda = ?$

$v = ?$

$$= 331 \, \frac{m}{s} + \left(\frac{0.600 \, \frac{m}{s}}{°C} \right) (20.0°C)$$

$$= 331 \, \frac{m}{s} + 12.0 \, \frac{m}{s}$$

$$= 343 \, \frac{m}{s}$$

Step 2:

$$v = \lambda f \quad \therefore \quad \lambda = \frac{v}{f}$$

$$= \frac{343 \, \frac{m}{s}}{2.00 \times 10^7 \, \frac{1}{s}}$$

$$= \frac{343}{2.00 \times 10^7} \, \frac{m}{s} \times \frac{s}{1}$$

$$= 1.715 \times 10^{-5} \, m$$

$$= \boxed{1.7 \times 10^{-5} \, m}$$

5.5. **Step 1:** Assume room temperature (20.0°C) to obtain the velocity (yes, you can have room temperature outside a room):

$d = 150 \, m$

$t = ?$

$v = ?$

$$v_{T_{P(m/s)}} = v_0 + \left(\frac{0.600 \, \frac{m}{s}}{°C} \right) (T_P)$$

$$= 331 \, \frac{m}{s} + \left(\frac{0.600 \, \frac{m}{s}}{°C} \right) (20.0°C)$$

$$= 331 \, \frac{m}{s} + 12.0 \, \frac{m}{s}$$

$$= 343 \, \frac{m}{s}$$

Step 2:

$$v = \frac{d}{t} \quad \therefore \quad t = \frac{d}{v}$$

$$= \frac{150.0 \, m}{343 \, \frac{m}{s}}$$

$$= \frac{150.0}{343} \, \frac{m}{1} \times \frac{s}{m}$$

$$= 0.4373177 \, s$$

$$= \boxed{0.437 \, s}$$

5.6. **Step 1:** Find the velocity of sound in ft/s at a temperature of 20.0°C

$t = 0.500 \, s \qquad v_{T_{P(ft/s)}} = v_0 + \left(\frac{2.00 \, \frac{ft}{s}}{°C} \right) (T_P)$

$T = 20.0°C$

$d = ?$

$v = ? \qquad\qquad\quad = 1{,}087 \, \frac{ft}{s} + \left(\frac{2.00 \, \frac{ft}{s}}{°C} \right) (20.0°C)$

$$= 1{,}087 \, \frac{ft}{s} + 40.0 \, \frac{ft}{s}$$

$$= 1{,}127 \, \frac{ft}{s}$$

Step 2:

$$v = \frac{d}{t} \quad \therefore \quad d = vt$$

$$= \left(1{,}127 \, \frac{ft}{s} \right) (0.500 \, s)$$

$$= 1{,}127 \times 0.500 \, \frac{ft}{s} \times s$$

$$= 563.5 \, ft$$

$$= 564 \, ft$$

Step 3: The distance to the building is half the distance the sound traveled, so

$$\frac{564}{2} = \boxed{282 \, ft}$$

5.7. **Step 1:**

$t = 1.75$ s $\qquad v = \dfrac{d}{t} \quad \therefore \quad d = vt$

$v = 1{,}530$ m/s

$d = ?$ $\qquad\qquad = \left(1{,}530\,\dfrac{\text{m}}{\text{s}}\right)(1.75\text{ s})$

$\qquad\qquad\qquad = 1{,}530 \times 1.75\,\dfrac{\text{m}}{\text{s}} \times \text{s}$

$\qquad\qquad\qquad = 2{,}677.5$ m

Step 2: The sonar signal traveled from the ship to the bottom, then back to the ship, so the distance to the bottom is half of the distance traveled:

$\dfrac{2{,}677.5\text{ m}}{2} = 1{,}338.75$ m

$\qquad\qquad = \boxed{1{,}340\text{ m}}$

5.8.

$f = 660$ Hz $\qquad v = \lambda f$

$\lambda = 9.0$ m $\qquad\qquad = (9.0\text{ m})\left(660\,\dfrac{1}{\text{s}}\right)$

$v = ?$

$\qquad\qquad\qquad = 9.0 \times 660\text{ m} \times \dfrac{1}{\text{s}}$

$\qquad\qquad\qquad = 5{,}940\,\dfrac{\text{m}}{\text{s}}$

$\qquad\qquad\qquad = \boxed{5{,}900\,\dfrac{\text{m}}{\text{s}}}$

5.9. **Step 1:**

$t = 2.5$ s $\qquad v_{T_{\text{P(m/s)}}} = v_0 + \left(\dfrac{0.600\,\frac{\text{m}}{\text{s}}}{{}^{\circ}\text{C}}\right)(T_\text{P})$

$T = 20.0^{\circ}$C

$d = ?$

$v = ?$ $\qquad = 331\,\dfrac{\text{m}}{\text{s}} + \left(\dfrac{0.600\,\frac{\text{m}}{\text{s}}}{{}^{\circ}\text{C}}\right)(20.0^{\circ}\text{C})$

$\qquad\qquad = 331\,\dfrac{\text{m}}{\text{s}} + 12.0\,\dfrac{\text{m}}{\text{s}}$

$\qquad\qquad = 343\,\dfrac{\text{m}}{\text{s}}$

Step 2:

$v = \dfrac{d}{t} \quad \therefore \quad d = vt$

$\qquad\qquad = \left(343\,\dfrac{\text{m}}{\text{s}}\right)(2.5\text{ s})$

$\qquad\qquad = 343 \times 2.5\,\dfrac{\text{m}}{\text{s}} \times \text{s}$

$\qquad\qquad = 857.5$ m

$\qquad\qquad = \boxed{860\text{ m}}$

5.10. According to table 5.1, sound moves through air at 0°C with a velocity of 331 m/s and through steel with a velocity of 5,940 m/s. Therefore,

(a) $v = 331$ m/s $\qquad v = \dfrac{d}{t} \quad \therefore \quad d = vt$

$t = 8.00$ s

$d = ?$ $\qquad\qquad = \left(331\,\dfrac{\text{m}}{\text{s}}\right)(8.00\text{ s})$

$\qquad\qquad\qquad = 331 \times 8.00\,\dfrac{\text{m}}{\text{s}} \times \text{s}$

$\qquad\qquad\qquad = 2{,}648$ m

$\qquad\qquad\qquad = \boxed{2.65\text{ km}}$

(b) $v = 5{,}940$ m/s $\qquad v = \dfrac{d}{t} \quad \therefore \quad d = vt$

$t = 8.00$ s

$d = ?$ $\qquad\qquad = \left(5{,}940\,\dfrac{\text{m}}{\text{s}}\right)(8.00\text{ s})$

$\qquad\qquad\qquad = 5{,}940 \times 8.00\,\dfrac{\text{m}}{\text{s}} \times \text{s}$

$\qquad\qquad\qquad = 47{,}520$ m

$\qquad\qquad\qquad = \boxed{47.5\text{ km}}$

5.11. $\qquad v = f\lambda$

$\qquad\qquad = \left(10\,\dfrac{1}{\text{s}}\right)(0.50\text{ m})$

$\qquad\qquad = 5\,\dfrac{\text{m}}{\text{s}}$

5.12. The distance between two *consecutive* condensations (or rarefactions) is one wavelength, so $\lambda = 3.00$ m and

$\qquad v = f\lambda$

$\qquad\qquad = \left(112.0\,\dfrac{1}{\text{s}}\right)(3.00\text{ m})$

$\qquad\qquad = 336\,\dfrac{\text{m}}{\text{s}}$

5.13. **(a)** One complete wave every 4.0 s means that $T = 4.0$ s.

(b) $\qquad f = \dfrac{1}{T}$

$\qquad\qquad = \dfrac{1}{4.0\text{ s}}$

$\qquad\qquad = \dfrac{1}{4.0}\,\dfrac{1}{\text{s}}$

$\qquad\qquad = 0.25\,\dfrac{1}{\text{s}}$

$\qquad\qquad = \boxed{0.25\text{ Hz}}$

5.14. The distance from one condensation to the next is one wavelength, so

$$v = f\lambda \quad \therefore \quad \lambda = \frac{v}{f}$$

$$= \frac{330 \, \dfrac{m}{s}}{260 \, \dfrac{1}{s}}$$

$$= \frac{330}{260} \, \frac{m}{s} \times \frac{s}{1}$$

$$= \boxed{1.3 \text{ m}}$$

5.15. **(a)** $v = f\lambda = \left(256 \, \dfrac{1}{s}\right)(1.34 \text{ m})$ $= \boxed{343 \text{ m/s}}$

(b) $= \left(440.0 \, \dfrac{1}{s}\right)(0.780 \text{ m})$ $= \boxed{343 \text{ m/s}}$

(c) $= \left(750.0 \, \dfrac{1}{s}\right)(0.457 \text{ m})$ $= \boxed{343 \text{ m/s}}$

(d) $= \left(2,500.0 \, \dfrac{1}{s}\right)(0.1372 \text{ m})$ $= \boxed{343 \text{ m/s}}$

5.16. The speed of sound at 0.0°C is 1,087 ft/s, and

(a) $v_{T_F} = v_0 + \left[\dfrac{2.00 \text{ ft/s}}{°C}\right][T_P]$

$$= 1,087 \text{ ft/s} + \left[\frac{2.00 \text{ ft/s}}{°C}\right][0.0°C]$$

$$= 1,087 + (2.00)(0.0) \text{ ft/s} + \frac{ft/s}{°C} \times °C$$

$$= 1,087 \text{ ft/s} + 0.0 \text{ ft/s}$$

$$= \boxed{1,087 \text{ ft/s}}$$

(b) $v_{20°} = 1,087 \text{ ft/s} + \left[\dfrac{2.00 \text{ ft/s}}{°C}\right][20.0°C]$

$$= 1,087 \text{ ft/s} + 40.0 \text{ ft/s}$$

$$= \boxed{1,127 \text{ ft/s}}$$

(c) $v_{40°} = 1,087 \text{ ft/s} + \left[\dfrac{2.00 \text{ ft/s}}{°C}\right][40.0°C]$

$$= 1,087 \text{ ft/s} + 80.0 \text{ ft/s}$$

$$= \boxed{1,167 \text{ ft/s}}$$

(d) $v_{80°} = 1,087 \text{ ft/s} + \left[\dfrac{2.00 \text{ ft/s}}{°C}\right][80.0°C]$

$$= 1,087 \text{ ft/s} + 160 \text{ ft/s}$$

$$= \boxed{1,247 \text{ ft/s}}$$

5.17. For consistency with the units of the equation given, 43.7°F is first converted to 6.50°C. The velocity of sound in this air is:

$$v_{T_F} = v_0 + \left[\frac{2.00 \text{ ft/s}}{°C}\right][T_P]$$

$$= 1,087 \text{ ft/s} + \left[\frac{2.00 \text{ ft/s}}{°C}\right][6.50°C]$$

$$= 1,087 \text{ ft/s} + 13.0 \text{ ft/s}$$

$$= 1,100 \text{ ft/s}$$

The distance that a sound with this velocity travels in the given time is

$$v = \frac{d}{t} \quad \therefore \quad d = vt$$

$$= (1,100 \text{ ft/s})(4.80 \text{ s})$$

$$= (1,100)(4.80) \, \frac{ft \cdot s}{s}$$

$$= 5,280 \text{ ft}$$

$$\frac{5,280 \text{ ft}}{2}$$

$$= 2,640 \text{ ft}$$

$$= \boxed{2,600 \text{ ft}}$$

Since the sound traveled from the rifle to the cliff and then back, the cliff must be about one-half mile away.

5.18. This problem requires three steps, (1) conversion of the °F temperature value to °C, (2) calculating the velocity of sound in air at this temperature, and (3) calculating the distance from the calculated velocity and the given time:

$$v_{T_F} = v_0 + \left[\frac{2.00 \text{ ft/s}}{°C}\right][T_P]$$

$$= 1,087 \text{ ft/s} + \left[\frac{2.00 \text{ ft/s}}{°C}\right][26.67°C]$$

$$= 1,087 \text{ ft/s} + 53.0 \text{ ft/s} = 1,140 \text{ ft/s}$$

$$v = \frac{d}{t} \quad \therefore \quad d = vt$$

$$= (1,140 \text{ ft/s})(4.63 \text{ s})$$

$$= \boxed{5,280 \text{ ft (one mile)}}$$

5.19. **(a)** $v = f\lambda \quad \therefore \quad \lambda = \dfrac{v}{f}$

$$= \frac{1,125 \, \dfrac{ft}{s}}{440 \, \dfrac{1}{s}}$$

$$= \frac{1,125}{440} \, \frac{ft}{s} \times \frac{s}{1}$$

$$= 2.56 \, \frac{ft \cdot s}{s}$$

$$= \boxed{2.6 \text{ ft}}$$

(b) $v = f\lambda$ $\quad\therefore\quad$ $\lambda = \dfrac{v}{f}$

$$= \dfrac{5{,}020}{440}\,\dfrac{\text{ft}}{\text{s}} \times \dfrac{\text{s}}{1}$$

$$= 11.4\text{ ft} = \boxed{11\text{ ft}}$$

EXERCISES Chapter 6

6.1. First, recall that a negative charge means an excess of electrons. Second, the relationship between the total charge (q), the number of electrons (n), and the charge of a single electron (e) is $q = ne$. The fundamental charge of a single ($n = 1$) electron (e) is 1.60×10^{-19} C. Thus

$$q = ne \quad\therefore\quad n = \dfrac{q}{e}$$

$$= \dfrac{1.00 \times 10^{-14}\text{ C}}{1.60 \times 10^{-19}\,\dfrac{\text{C}}{\text{electron}}}$$

$$= \dfrac{1.00 \times 10^{-14}}{1.60 \times 10^{-19}}\,\dfrac{\text{C}}{1} \times \dfrac{\text{electron}}{\text{C}}$$

$$= 6.25 \times 10^{4}\,\dfrac{\text{C·electron}}{\text{C}}$$

$$= \boxed{6.25 \times 10^{4}\text{ electron}}$$

6.2. **(a)** Both balloons have negative charges so the force is repulsive, pushing the balloons away from each other.

(b) The magnitude of the force can be found from Coulomb's law:

$$F = \dfrac{kq_1q_2}{d^2}$$

$$= \dfrac{(9.00 \times 10^9\text{ N·m}^2/\text{C}^2)(3.00 \times 10^{-14}\text{ C})(2.00 \times 10^{-12}\text{ C})}{(2.00 \times 10^{-2}\text{m})^2}$$

$$= \dfrac{(9.00 \times 10^9)(3.00 \times 10^{-14})(2.00 \times 10^{-12})}{4.00 \times 10^{-4}}\,\dfrac{\dfrac{\text{N·m}^2}{\text{C}^2} \times \text{C} \times \text{C}}{\text{m}^2}$$

$$= \dfrac{5.40 \times 10^{-16}}{4.00 \times 10^{-4}}\,\dfrac{\text{N·m}^2}{\text{C}^2} \times \text{C}^2 \times \dfrac{1}{\text{m}^2}$$

$$= \boxed{1.35 \times 10^{-12}\text{ N}}$$

6.3. $\dfrac{\text{potential}}{\text{difference}} = \dfrac{\text{work}}{\text{charge}}$

or

$$V = \dfrac{W}{q}$$

$$= \dfrac{7.50\text{ J}}{5.00\text{ C}}$$

$$= 1.50\,\dfrac{\text{J}}{\text{C}}$$

$$= \boxed{1.50\text{ V}}$$

6.4. $\dfrac{\text{electric}}{\text{current}} = \dfrac{\text{charge}}{\text{time}}$

or

$$I = \dfrac{q}{t}$$

$$= \dfrac{6.00\text{ C}}{2.00\text{ s}}$$

$$= 3.00\,\dfrac{\text{C}}{\text{s}}$$

$$= \boxed{3.00\text{ A}}$$

6.5. A current of 1.00 amp is defined as 1.00 coulomb/s. Since the fundamental charge of the electron is 1.60×10^{-19} C/electron,

$$\dfrac{1.00\,\dfrac{\text{C}}{\text{s}}}{1.60 \times 10^{-19}\,\dfrac{\text{C}}{\text{electron}}}$$

$$= 6.25 \times 10^{18}\,\dfrac{\text{C}}{\text{s}} \times \dfrac{\text{electron}}{\text{C}}$$

$$= \boxed{6.25 \times 10^{18}\,\dfrac{\text{electrons}}{\text{s}}}$$

6.6. $R = \dfrac{V}{I}$

$$= \dfrac{120.0\text{ V}}{4.00\text{ A}}$$

$$= 30.0\,\dfrac{\text{V}}{\text{A}}$$

$$= \boxed{30.0\ \Omega}$$

6.7. $R = \dfrac{V}{I}$ $\quad\therefore\quad$ $I = \dfrac{V}{R}$

$$= \dfrac{120.0\text{ V}}{60.0\,\dfrac{\text{V}}{\text{A}}}$$

$$= \dfrac{120.0}{60.0}\,\text{V} \times \dfrac{\text{A}}{\text{V}}$$

$$= \boxed{2.00\text{ A}}$$

6.8. **(a)** $R = \dfrac{V}{I}$ $\quad\therefore\quad$ $V = IR$

$$= (1.20\text{ A})\left(10.0\,\dfrac{\text{V}}{\text{A}}\right)$$

$$= \boxed{12.0\text{ V}}$$

(b) Power $=$ (current)(potential difference)

or

$$P = IV$$

$$= \left(1.20\,\dfrac{\text{C}}{\text{s}}\right)\left(12.0\,\dfrac{\text{J}}{\text{C}}\right)$$

$$= (1.20)(12.0) \frac{\cancel{C}}{s} \times \frac{J}{\cancel{C}}$$

$$= 14.4 \frac{J}{s}$$

$$= \boxed{14.4 \text{ W}}$$

6.9. Note that there are two separate electrical units that are rates: (1) the amp (coulomb/s), and (2) the watt (joule/s). The question asked for a rate of using energy. Energy is measured in joules, so you are looking for the power of the radio in watts. To find watts ($P = IV$), you will need to calculate the current (I) since it is not given. The current can be obtained from the relationship of Ohm's law:

$$I = \frac{V}{R}$$

$$= \frac{3.00 \text{ V}}{15.0 \frac{V}{A}}$$

$$= 0.200 \text{ A}$$

$$P = IV$$

$$= (0.200 \text{ C/s})(3.00 \text{ J/C})$$

$$= \boxed{0.600 \text{ W}}$$

6.10. $\text{cost} = \dfrac{(\text{watts})(\text{time})(\text{rate})}{1,000 \frac{W}{kW}}$

$$= \frac{(1,200 \text{ W})(0.25 \text{ h})\left(\frac{\$0.10}{\text{kWh}}\right)}{1,000 \frac{W}{kW}}$$

$$= \frac{(1,200)(0.25)(0.10)}{1,000} \frac{W}{1} \times \frac{h}{1} \times \frac{\$}{\text{kWh}} \times \frac{kW}{W}$$

$$= \boxed{\$0.03} \text{ (3 cents)}$$

6.11. The relationship between power (P), current (I), and volts (V) will provide a solution. Since the relationship considers power in watts the first step is to convert horsepower to watts. One horsepower is equivalent to 746 watts, so:

$$(746 \text{ W/hp})(2.00 \text{ hp}) = 1,492 \text{ W}$$

$$P = IV \quad \therefore \quad I = \frac{P}{V}$$

$$= \frac{1,492 \frac{J}{s}}{12.0 \frac{J}{C}}$$

$$= \frac{1,492}{12.0} \frac{\cancel{J}}{s} \times \frac{C}{\cancel{J}}$$

$$= 124.3 \frac{C}{s}$$

$$= \boxed{124 \text{ A}}$$

6.12. **(a)** The rate of using energy is joule/s, or the watt. Since 1.00 hp = 746 W,

inside motor: $(746 \text{ W/hp})(1/3 \text{ hp}) = 249 \text{ W}$

outside motor: $(746 \text{ W/hp})(1/3 \text{ hp}) = 249 \text{ W}$

compressor motor: $(746 \text{ W/hp})(3.70 \text{ hp}) = 2,760 \text{ W}$

$249 \text{ W} + 249 \text{ W} + 2,760 \text{ W} = \boxed{3,258 \text{ W}}$

(b) $\dfrac{(3,258 \text{ W})(1.00 \text{ h})\left(\frac{\$0.10}{\text{kWh}}\right)}{1,000 \frac{W}{kW}} = \$0.33 \text{ per hour}$

(c) $(\$0.33/\text{h})(12 \text{ h/day})(30 \text{ day/mo}) = \boxed{\$118.80}$

6.13. The solution is to find how much current each device draws and then to see if the total current is less or greater than the breaker rating:

$$\text{Toaster: } I = \frac{V}{R} = \frac{120 \text{ V}}{15 \text{ V/A}} = 8.0 \text{ A}$$

$$\text{Motor: } (0.20 \text{ hp})(746 \text{ W/hp}) = 150 \text{ W}$$

$$I = \frac{P}{V} = \frac{150 \text{ J/s}}{102 \text{ J/C}} = 1.3 \text{ A}$$

Three 100 W bulbs: $3 \times 100 \text{ W} = 300 \text{ W}$

$$I = \frac{P}{V} = \frac{300 \text{ J/s}}{120 \text{ J/C}} = 2.5 \text{ A}$$

$$\text{Iron: } I = \frac{P}{V} = \frac{600 \text{ J/s}}{120 \text{ J/C}} = 5.0 \text{ A}$$

The sum of the currents is 8.0 A + 1.3 A + 2.5 A + 5.0 A = 16.8 A, so the total current is greater than 15.0 amp and the circuit breaker will trip.

6.14. **(a)** $V_p = 1,200 \text{ V}$

$N_p = 1 \text{ loop}$

$N_s = 200 \text{ loops}$

$V_s = ?$

$$\frac{V_p}{N_p} = \frac{V_s}{N_s} \quad \therefore \quad V_s = \frac{V_p N_s}{N_p}$$

$$V_s = \frac{(1,200 \text{ V})(200 \text{ loop})}{1 \text{ loop}}$$

$$= \boxed{240,000 \text{ V}}$$

(b) $I_P = 40 \text{ A} \qquad V_p I_p = V_s I_s \quad \therefore \quad I_s = \dfrac{V_p I_p}{V_s}$

$$I_s = ? \qquad I_s = \frac{1,200 \text{ V} \times 40 \text{ A}}{240,000 \text{ V}}$$

$$= \frac{1,200 \times 40}{240,000} \frac{\text{V·A}}{\cancel{V}}$$

$$= \boxed{0.2 \text{ A}}$$

6.15. (a) $V_s = 12$ V $\qquad \dfrac{V_P}{N_P} = \dfrac{V_s}{N_s} \quad \therefore \quad \dfrac{N_P}{N_s} = \dfrac{V_P}{V_s}$

$\quad I_s = 0.5$ A

$\quad V_P = 120$ V $\qquad \dfrac{N_P}{N_s} = \dfrac{120 \, \text{V}}{12 \, \text{V}} = \dfrac{10}{1}$

$\quad \dfrac{N_P}{N_s} = ?$ $\qquad\qquad\qquad$ or

$\qquad\qquad$ | 10 primary to 1 secondary |

(b) $I_P = ?$ $\qquad V_P I_P = V_s I_s \quad \therefore \quad I_P = \dfrac{V_s I_s}{V_P}$

$\qquad\qquad I_P = \dfrac{(12 \, \text{V})(0.5 \, \text{A})}{120 \, \text{V}}$

$\qquad\qquad\quad = \dfrac{12 \times 0.5}{120} \dfrac{\text{V·A}}{\text{V}}$

$\qquad\qquad\quad = \boxed{0.05 \text{ A}}$

(c) $P_s = ?$ $\qquad P_s = I_s V_s$

$\qquad\qquad\quad = (0.5 \, \text{A})(12 \, \text{V})$

$\qquad\qquad\quad = 0.5 \times 12 \dfrac{\text{C}}{\text{s}} \times \dfrac{\text{J}}{\text{C}}$

$\qquad\qquad\quad = 6 \dfrac{\text{J}}{\text{s}}$

$\qquad\qquad\quad = \boxed{6 \text{ W}}$

6.16. (a) $V_P = 120$ V $\qquad \dfrac{V_P}{N_P} = \dfrac{V_s}{N_s} \quad \therefore \quad V_s = \dfrac{V_P N_s}{N_P}$

$\quad N_P = 50$ loops

$\quad N_s = 150$ loops $\qquad V_s = \dfrac{120 \, \text{V} \times 150 \, \text{loops}}{50 \, \text{loops}}$

$\quad I_P = 5.0$ A

$\quad V_s = ?$ $\qquad\qquad\quad = \dfrac{120 \times 150}{50} \dfrac{\text{V·loops}}{\text{loops}}$

$\qquad\qquad\qquad\quad = \boxed{360 \text{ V}}$

(b) $I_s = ?$ $\qquad V_P I_P = V_s I_s \quad \therefore \quad I_s = \dfrac{V_P I_P}{V_s}$

$\qquad\qquad I_s = \dfrac{(120 \, \text{V})(5.0 \, \text{A})}{360 \, \text{V}}$

$\qquad\qquad\quad = \dfrac{120 \times 5.0}{360} \dfrac{\text{V·A}}{\text{V}}$

$\qquad\qquad\quad = \boxed{1.7 \text{ A}}$

(c) $P_s = ?$ $\qquad P_s = I_s V_s$

$\qquad\qquad\quad = \left(1.7 \dfrac{\text{C}}{\text{s}}\right)\left(360 \dfrac{\text{J}}{\text{C}}\right)$

$\qquad\qquad\quad = 1.7 \times 360 \dfrac{\text{C}}{\text{s}} \times \dfrac{\text{J}}{\text{C}}$

$\qquad\qquad\quad = 612 \dfrac{\text{J}}{\text{s}}$

$\qquad\qquad\quad = \boxed{600 \text{ W}}$

7.1. The relationship between the speed of light in a transparent material (v), the speed of light in a vacuum ($c = 3.00 \times 10^8$ m/s) and the index of refraction (n) is $n = c/v$. According to Table 7.1, the index of refraction for water is $n = 1.33$ and for ice is $n = 1.31$.

(a) $\qquad\qquad c = 3.00 \times 10^8$ m/s

$\qquad\qquad\quad n = 1.33$

$\qquad\qquad\quad v = ?$

$\qquad\qquad\quad n = \dfrac{c}{v} \quad \therefore \quad v = \dfrac{c}{n}$

$\qquad\qquad\quad v = \dfrac{3.00 \times 10^8 \text{ m/s}}{1.33}$

$\qquad\qquad\qquad = \boxed{2.26 \times 10^8 \text{ m/s}}$

(b) $\qquad\qquad c = 3.00 \times 10^8$ m/s

$\qquad\qquad\quad n = 1.31$

$\qquad\qquad\quad v = ?$

$\qquad\qquad\quad v = \dfrac{3.00 \times 10^8 \text{ m/s}}{1.31}$

$\qquad\qquad\qquad = \boxed{2.29 \times 10^8 \text{ m/s}}$

7.2. $\qquad\qquad d = 1.50 \times 10^8$ km

$\qquad\qquad\quad = 1.50 \times 10^{11}$ m

$\qquad\qquad\quad c = 3.00 \times 10^8$ m/s

$\qquad\qquad\quad t = ?$

$\qquad\qquad\quad v = \dfrac{d}{t} \quad \therefore \quad t = \dfrac{d}{v}$

$\qquad\qquad\quad t = \dfrac{1.50 \times 10^{11} \text{ m}}{3.00 \times 10^8 \, \frac{\text{m}}{\text{s}}}$

$\qquad\qquad\quad = \dfrac{1.50 \times 10^{11}}{3.00 \times 10^8} \text{ m} \times \dfrac{\text{s}}{\text{m}}$

$\qquad\qquad\quad = 5.00 \times 10^2 \dfrac{\text{m·s}}{\text{m}}$

$\qquad\qquad\quad = \dfrac{5.00 \times 10^2 \text{ s}}{60.0 \, \frac{\text{s}}{\text{min}}}$

$\qquad\qquad\quad = \dfrac{5.00 \times 10^2}{60.0} \text{ s} \times \dfrac{\text{min}}{\text{s}}$

$\qquad\qquad\quad = \boxed{8.33 \text{ min}}$

7.3. $\qquad\qquad d = 6.00 \times 10^9$ km

$\qquad\qquad\quad = 6.00 \times 10^{12}$ m

$\qquad\qquad\quad c = 3.00 \times 10^8$ m/s

$\qquad\qquad\quad t = ?$

$$v = \frac{d}{t} \quad \therefore \quad t = \frac{d}{v}$$

$$t = \frac{6.00 \times 10^{12}\,\text{m}}{3.00 \times 10^8\,\dfrac{\text{m}}{\text{s}}}$$

$$= \frac{6.00 \times 10^{12}}{3.00 \times 10^8}\,\cancel{\text{m}} \times \frac{\text{s}}{\cancel{\text{m}}}$$

$$= 2.00 \times 10^4\,\text{s}$$

$$= \frac{2.00 \times 10^4\,\text{s}}{3{,}600\,\dfrac{\text{s}}{\text{h}}}$$

$$= \frac{2.00 \times 10^4}{3.600 \times 10^3}\,\cancel{\text{s}} \times \frac{\text{h}}{\cancel{\text{s}}}$$

$$= \boxed{5.56\,\text{h}}$$

7.4. From equation 7.1, note that both angles are measured from the normal and that the angle of incidence (θ_i) equals the angle of reflection (θ_r), or

$$\theta_i = \theta_r \quad \therefore \quad \boxed{\theta_i = 10°}$$

7.5.
$$v = 2.20 \times 10^8\,\text{m/s}$$

$$c = 3.00 \times 10^8\,\text{m/s}$$

$$n = ?$$

$$n = \frac{c}{v}$$

$$= \frac{3.00 \times 10^8\,\dfrac{\text{m}}{\text{s}}}{2.20 \times 10^8\,\dfrac{\text{m}}{\text{s}}}$$

$$= 1.36$$

According to Table 7.1, the substance with an index of refraction of 1.36 is $\boxed{\text{ethyl alcohol.}}$

7.6. **(a)** From equation 7.3:

$$\lambda = 6.00 \times 10^{-7}\,\text{m} \quad c = \lambda f \quad \therefore \quad f = \frac{c}{\lambda}$$

$$c = 3.00 \times 10^8\,\text{m/s}$$

$$f = ?$$

$$f = \frac{3.00 \times 10^8\,\dfrac{\text{m}}{\text{s}}}{6.00 \times 10^{-7}\,\text{m}}$$

$$= \frac{3.00 \times 10^8}{6.00 \times 10^{-7}}\,\dfrac{\cancel{\text{m}}}{\text{s}} \times \frac{1}{\cancel{\text{m}}}$$

$$= 5.00 \times 10^{14}\,\frac{1}{\text{s}}$$

$$= \boxed{5.00 \times 10^{14}\,\text{Hz}}$$

(b) From equation 7.5:

$$f = 5.00 \times 10^{14}\,\text{Hz}$$

$$h = 6.63 \times 10^{-34}\,\text{J·s}$$

$$E = ?$$

$$E = hf$$

$$= (6.63 \times 10^{-34}\,\text{J·s})\left(5.00 \times 10^{14}\,\frac{1}{\text{s}}\right)$$

$$= (6.63 \times 10^{-34})(5.00 \times 10^{14})\,\text{J·}\cancel{\text{s}} \times \frac{1}{\cancel{\text{s}}}$$

$$= \boxed{3.32 \times 10^{-19}\,\text{J}}$$

7.7. First, you can find the energy of one photon of the peak intensity wavelength (5.60×10^{-7} m) by using equation 7.3 to find the frequency, then equation 7.5 to find the energy:

Step 1: $c = \lambda f \quad \therefore \quad f = \dfrac{c}{\lambda}$

$$= \frac{3.00 \times 10^8\,\dfrac{\text{m}}{\text{s}}}{5.60 \times 10^{-7}\,\text{m}}$$

$$= 5.36 \times 10^{14}\,\text{Hz}$$

Step 2: $E = hf$

$$= (6.63 \times 10^{-34}\,\text{J·s})(5.36 \times 10^{14}\,\text{Hz})$$

$$= 3.55 \times 10^{-19}\,\text{J}$$

Step 3: Since one photon carries an energy of 3.55×10^{-19} J and the overall intensity is 1,000.0 W, each square meter must receive an average of

$$\frac{1{,}000.0\,\dfrac{\text{J}}{\text{s}}}{3.55 \times 10^{-19}\,\dfrac{\text{J}}{\text{photon}}}$$

$$\frac{1.000 \times 10^3}{3.55 \times 10^{-19}}\,\frac{\cancel{\text{J}}}{\text{s}} \times \frac{\text{photon}}{\cancel{\text{J}}}$$

$$\boxed{2.82 \times 10^{21}\,\frac{\text{photon}}{\text{s}}}$$

7.8. **(a)** $f = 4.90 \times 10^{14}\,\text{Hz} \quad c = \lambda f \quad \therefore \quad \lambda = \dfrac{c}{f}$

$$c = 3.00 \times 10^8\,\text{m/s}$$

$$\lambda = ?$$

$$\lambda = \frac{3.00 \times 10^8\,\dfrac{\text{m}}{\text{s}}}{4.90 \times 10^{14}\,\dfrac{1}{\text{s}}}$$

$$= \frac{3.00 \times 10^8}{4.90 \times 10^{14}}\,\frac{\text{m}}{\cancel{\text{s}}} \times \frac{\cancel{\text{s}}}{1}$$

$$= \boxed{6.12 \times 10^{-7}\,\text{m}}$$

(b) According to Table 7.2, this is the frequency and wavelength of orange light.

7.9. $f = 5.00 \times 10^{20}$ Hz

$h = 6.63 \times 10^{-34}$ J·s

$E = ?$

$$E = hf$$

$$= (6.63 \times 10^{-34} \text{ J·s}) \left(5.00 \times 10^{20} \frac{1}{s} \right)$$

$$= (6.63 \times 10^{-34})(5.00 \times 10^{20}) \text{ J·s} \times \frac{1}{\text{s}}$$

$$= \boxed{3.32 \times 10^{-13} \text{ J}}$$

7.10. $\lambda = 1.00$ mm

$\quad = 0.001$ m

$f = ?$

$c = 3.00 \times 10^8$ m/s

$h = 6.63 \times 10^{-34}$ J·s

$E = ?$

Step 1: $c = \lambda f \quad \therefore \quad f = \dfrac{v}{\lambda}$

$$f = \frac{3.00 \times 10^8 \frac{m}{s}}{1.00 \times 10^{-3} \text{ m}}$$

$$= \frac{3.00 \times 10^8}{1.00 \times 10^{-3}} \frac{\text{m}}{\text{s}} \times \frac{1}{\text{m}}$$

$$= 3.00 \times 10^{11} \text{Hz}$$

Step 2: $E = hf$

$$= (6.63 \times 10^{-34} \text{ J·s}) \left(3.00 \times 10^{11} \frac{1}{s} \right)$$

$$= (6.63 \times 10^{-34})(3.00 \times 10^{11}) \text{ J·s} \times \frac{1}{\text{s}}$$

$$= \boxed{1.99 \times 10^{-22} \text{ J}}$$

7.11. The index of refraction is found from $n = \dfrac{c}{v}$, where n is the index of refraction of a transparent material, c is the speed of light in a vaccum, and v is the speed of light in the material. The index of refraction of glass is found in Table 7.1 ($n = 1.50$).

$n = 1.50$

$c = 3.00 \times 10^8$ m/s

$v = ?$

$n = \dfrac{c}{v} \quad \therefore \quad v = \dfrac{c}{n}$

$$= \frac{3.00 \times 10^8 \frac{m}{s}}{1.50}$$

$$= \boxed{2.00 \times 10^8 \frac{m}{s}}$$

7.12. Listing the known and unknown quantities:

Wavelength	$\lambda = 5.60 \times 10^{-7}$ m
Speed of light	$c = 3.00 \times 10^{-8}$ m/s
Frequency	$f = ?$

The relationship between the wavelength (λ), frequency (f), and speed of light in a vacuum (c), is found in equation 7.3, $c = \lambda f$.

$$c = \lambda f \quad \therefore \quad f = \frac{c}{\lambda}$$

$$= \frac{3.00 \times 10^8 \frac{m}{s}}{5.60 \times 10^{-7} \text{ m}}$$

$$= \frac{3.00 \times 10^8}{5.60 \times 10^{-7}} \frac{\text{m}}{\text{s}} \times \frac{1}{\text{m}}$$

$$= 5.40 \times 10^{14} \frac{1}{s}$$

$$= \boxed{5.40 \times 10^{14} \text{ Hz}}$$

7.13. Listing the known and unknown quantities:

Frequency	$f = 5.00 \times 10^{14}$ Hz
Planck's constant	$h = 6.63 \times 10^{-34}$ J·s
Energy	$E = ?$

The relationship between the frequency (f) and energy (E) of a photon is found in equation 7.4, $E = hf$.

$$E = hf$$

$$= (6.63 \times 10^{-34} \text{ J·s}) \left(5.00 \times 10^{14} \frac{1}{s} \right)$$

$$= (6.63 \times 10^{-34})(5.00 \times 10^{14}) \text{ J·s} \times \frac{1}{s}$$

$$= 3.32 \times 10^{-19} \frac{\text{J·s}}{\text{s}}$$

$$= \boxed{3.32 \times 10^{-19} \text{ J}}$$

7.14. Listing the known and unknown quantities:

Frequency	$f = 6.50 \times 10^{14}$ Hz
Planck's constant	$h = 6.63 \times 10^{-34}$ J·s
Energy	$E = ?$

The relationship between the frequency (f) and energy (E) of a photon is found in equation 7.4, $E = hf$.

$$E = hf$$

$$= (6.63 \times 10^{-34} \text{ J·s}) \left(6.50 \times 10^{14} \frac{1}{s} \right)$$

$$= (6.63 \times 10^{-34})(6.50 \times 10^{14}) \text{ J·s} \times \frac{1}{s}$$

$$= 4.31 \times 10^{-19} \frac{J \cdot s}{s}$$

$$= \boxed{4.31 \times 10^{-19} \, J}$$

7.15   Listing the known and unknown quantities:

Wavelength $\lambda = 5.60 \times 10^{-7}$ m

Planck's constant $h = 6.63 \times 10^{-34}$ J·s

Speed of light $c = 3.00 \times 10^{8}$ m/s

Frequency $f = ?$

First, you can find the energy of one photon of the peak intensity wavelength (5.60×10^{-7} m) by using the relationship between the wavelength (λ), frequency (f), and speed of light in a vacuum (c), $c = \lambda f$, then use the relationship between the frequency (f) and energy (E) of a photon, $E = hf$, to find the energy:

Step 1: $c = \lambda f$ $\therefore$ $f = \dfrac{c}{\lambda}$

$$= \frac{3.00 \times 10^{8} \, \frac{m}{s}}{5.60 \times 10^{-7} \, m}$$

$$= \frac{3.00 \times 10^{8}}{5.60 \times 10^{-7}} \, \frac{m}{s} \times \frac{1}{m}$$

$$= 5.36 \times 10^{14} \, \frac{1}{s}$$

$$= 5.36 \times 10^{14} \, Hz$$

Step 2: $E = hf$

$$= (6.63 \times 10^{-34} \, J \cdot s)\left(5.36 \times 10^{14} \, \frac{1}{s}\right)$$

$$= (6.63 \times 10^{-34})(5.36 \times 10^{14}) \, J \cdot s \times \frac{1}{s}$$

$$= 3.55 \times 10^{-19} \, \frac{J \cdot s}{s}$$

$$= 3.55 \times 10^{-19} \, J$$

Step 3: Since one photon carries an energy of 3.55×10^{-19} J and the overall intensity is 1,000.0 W, for each square meter there must be an average of

$$\frac{1,000.0 \, \frac{J}{s}}{3.55 \times 10^{-19} \, \frac{J}{photon}}$$

$$\frac{1.000 \times 10^{-3}}{3.55 \times 10^{-19}} \, \frac{J}{s} \times \frac{photon}{J}$$

$$\boxed{2.82 \times 10^{21} \, \frac{photon}{s}}$$

8.1. $m = 1.68 \times 10^{-27}$ kg

$v = 3.22 \times 10^{3}$ m/s

$h = 6.63 \times 10^{-34}$ J·s

$\lambda = ?$

$$\lambda = \frac{h}{mv}$$

$$= \frac{6.63 \times 10^{-34} \, J \cdot s}{(1.68 \times 10^{-27} \, kg)\left(3.22 \times 10^{3} \, \frac{m}{s}\right)}$$

$$= \frac{6.63 \times 10^{-34}}{(1.68 \times 10^{-27})(3.22 \times 10^{3})} \, \frac{J \cdot s}{kg \times \frac{m}{s}}$$

$$= \frac{6.63 \times 10^{-34}}{5.41 \times 10^{-24}} \, \frac{\frac{kg \cdot m^{2}}{s \cdot s} \times s}{kg \times \frac{m}{s}}$$

$$= 1.23 \times 10^{-10} \, \frac{kg \cdot m \cdot m}{s} \times \frac{1}{kg} \times \frac{s}{m}$$

$$= \boxed{1.23 \times 10^{-10} \, m}$$

8.2.   **(a)** $n = 6$

$E_{L} = -13.6$ eV

$E_{6} = ?$

$$E_{n} = \frac{E_{L}}{n^{2}}$$

$$E_{6} = \frac{-13.6 \, eV}{6^{2}}$$

$$= \frac{-13.6 \, eV}{36}$$

$$= \boxed{-1.378 \, eV}$$

(b) $= (-0.378 \, eV)\left(1.60 \times 10^{-19} \, \frac{J}{eV}\right)$

$$= (-0.378)(1.60 \times 10^{-19}) \, eV \times \frac{J}{eV}$$

$$= \boxed{-6.05 \times 10^{-20} \, J}$$

8.3.   **(a)** Energy is related to the frequency and Planck's constant in equation 8.1, $E = hf$. From equation 8.4,

$$hf = E_{H} - E_{L} \quad \therefore \quad E = E_{H} - E_{L}$$

For $n = 6$, $E_{H} = 6.05 \times 10^{-20}$ J

For $n = 2$, $E_{L} = 5.44 \times 10^{-19}$ J

$$E = ? \, J$$

$$E = E_H - E_L$$
$$= (-6.05 \times 10^{-20}\,\text{J}) - (-5.44 \times 10^{-19}\,\text{J})$$
$$= \boxed{4.84 \times 10^{-19}\,\text{J}}$$

(b) $E_H = -0.377\,\text{eV}^*$

$E_L = -3.40\,\text{eV}^*$

$E = ?\,\text{eV}$

$E = E_H - E_L$

$= (-0.377\,\text{eV}) - (-3.40\,\text{eV})$

$= \boxed{3.02\,\text{eV}}$

*From figure 8.11

8.4. For $n = 6$, $E_H = -6.05 \times 10^{-20}\,\text{J}$

For $n = 2$, $E_L = -5.44 \times 10^{-19}\,\text{J}$

$h = 6.63 \times 10^{-34}\,\text{J·s}$

$f = ?$

$$hf = E_H - E_L \quad \therefore \quad f = \frac{E_H - E_L}{h}$$

$$f = \frac{(-6.05 \times 10^{-20}\,\text{J}) - (-5.44 \times 10^{-19}\,\text{J})}{6.63 \times 10^{-34}\,\text{J·s}}$$

$$= \frac{4.84 \times 10^{-19}}{6.63 \times 10^{-34}}\,\frac{\cancel{\text{J}}}{\cancel{\text{J}}\text{·s}}$$

$$= 7.29 \times 10^{14}\,\frac{1}{\text{s}}$$

$$= \boxed{7.29 \times 10^{14}\,\text{Hz}}$$

8.5. $(n = 1) = -13.6\,\text{eV} \qquad E_n = \dfrac{E_1}{n^2}$

$E = ? \qquad\qquad = \dfrac{-13.6\,\text{eV}}{1^2}$

$\qquad\qquad\qquad = -13.6\,\text{eV}$

Since the energy of the electron is $-13.6\,\text{eV}$, it will require $13.6\,\text{eV}$ (or $2.17 \times 10^{-18}\,\text{J}$) to remove the electron.

8.6. $q/m = -1.76 \times 10^{11}\,\text{C/kg}$

$q = -1.60 \times 10^{-19}\,\text{C}$

$m = ?$

$$\text{mass} = \frac{\text{charge}}{\text{charge/mass}}$$

$$= \frac{-1.60 \times 10^{-19}\,\text{C}}{-1.76 \times 10^{11}\,\dfrac{\text{C}}{\text{kg}}}$$

$$= \frac{-1.60 \times 10^{-19}}{-1.76 \times 10^{11}}\,\cancel{\text{C}} \times \frac{\text{kg}}{\cancel{\text{C}}}$$

$$= \boxed{9.09 \times 10^{-31}\,\text{kg}}$$

8.7. $\lambda = -1.67 \times 10^{-10}\,\text{m}$

$m = 9.11 \times 10^{-31}\,\text{kg}$

$v = ?$

$$\lambda = \frac{h}{mv} \qquad \therefore \qquad v = \frac{h}{m\lambda}$$

$$v = \frac{6.63 \times 10^{-34}\,\text{J·s}}{(9.11 \times 10^{-31}\,\text{kg})(1.67 \times 10^{-10}\,\text{m})}$$

$$= \frac{6.63 \times 10^{-34}}{(9.11 \times 10^{-31})(1.67 \times 10^{-10})}\,\frac{\text{J·s}}{\text{kg·m}}$$

$$= \frac{6.63 \times 10^{-34}}{1.52 \times 10^{-40}}\,\frac{\dfrac{\text{kg·m}^2}{\text{s·}\cancel{\text{s}}} \times \cancel{\text{s}}}{\text{kg·m}}$$

$$= 4.36 \times 10^{6}\,\frac{\cancel{\text{kg}}\text{·m·}\cancel{\text{m}}}{\text{s}} \times \frac{1}{\cancel{\text{kg}}} \times \frac{1}{\cancel{\text{m}}}$$

$$= \boxed{4.36 \times 10^{6}\,\frac{\text{m}}{\text{s}}}$$

8.8. **(a)** Boron: $1s^2 2s^2 2p^1$
(b) Aluminum: $1s^2 2s^2 2p^6 3s^2 3p^1$
(c) Potassium: $1s^2 2s^2 2p^6 3s^2 3p^6 4s^1$

8.9. **(a)** Boron is atomic number 5 and there are 5 electrons.
(b) Aluminum is atomic number 13 and there are 13 electrons.
(c) Potassium is atomic number 19 and there are 19 electrons.

8.10. **(a)** Argon: $1s^2 2s^2 2p^6 3s^2 3p^6$
(b) Zinc: $1s^2 2s^2 2p^6 3s^2 3p^6 4s^2 3d^{10}$
(c) Bromine: $1s^2 2s^2 2p^6 3s^2 3p^6 4s^2 3d^{10} 4p^5$

8.11. Atomic weight is the weighted average of the isotopes as they occur in nature. Thus

Lithium-6: $6.01512\,\text{u} \times 0.0742 = 0.446\,\text{u}$
Lithium-7: $7.016\,\text{u} \times 0.9258 = 6.4054\,\text{u}$

Lithium-6 contributes 0.446 u of the weighted average and lithium-7 contributes 6.4954 u. The atomic weight of lithium is therefore

$$0.446\,\text{u}$$
$$\underline{+6.4954\,\text{u}}$$
$$6.941\,\text{u}$$

8.12. Recall that the subscript is the atomic number, which identifies the number of protons. In a neutral atom, the number of protons equals the number of electrons, so the atomic number tells you the number of electrons, too. The superscript is the mass number, which identifies the number of neutrons and the number of protons in the nucleus. The number of neutrons is therefore the mass number minus the atomic number.

	Protons	Neutrons	Electrons
(a)	6	6	6
(b)	1	0	1
(c)	18	22	18
(d)	1	1	1
(e)	79	118	79
(f)	92	143	92

8.13.

		Period	Family
(a)	Radon (Rn)	6	VIIIA
(b)	Sodium (Na)	3	IA
(c)	Copper (Cu)	4	IB
(d)	Neon (Ne)	2	VIIIA
(e)	Iodine (I)	5	VIIA
(f)	Lead (Pb)	6	IVA

8.14. Recall that the number of outer-shell electrons is the same as the family number for the representative elements:

(a) Li: 1   (d) Cl: 7
(b) N: 5   (e) Ra: 2
(c) F: 7   (f) Be: 2

8.15. The same information that was used in question 5 can be used to draw the dot notation (see Figure 8.20):

(a) $\overset{\cdot}{\underset{\cdot}{B}}\cdot$ (c) $\overset{\cdot}{Ca}\cdot$ (e) $\cdot\overset{\cdot\cdot}{\underset{\cdot\cdot}{O}}\colon$

(b) $\colon\overset{\cdot\cdot}{Br}\colon$ (d) $K\cdot$ (f) $\cdot\overset{\cdot\cdot}{\underset{\cdot}{S}}\colon$

8.16. The charge is found by identifying how many electrons are lost or gained in achieving the noble gas structure:

(a) Boron 3+
(b) Bromine 1−
(c) Calcium 2+
(d) Potassium 1+
(e) Oxygen 2−
(f) Nitrogen 3−

8.17. Metals have one, two, or three outer electrons and are located in the left two-thirds of the periodic table. Semiconductors are adjacent to the line that separates the metals and nonmetals. Look at the periodic table on the inside back cover and you will see:

(a) Krypton—nonmetal
(b) Cesium—metal
(c) Silicon—semiconductor
(d) Sulfur—nonmetal
(e) Molybdenum—metal
(f) Plutonium—metal

8.18. (a) Bromine gained an electron to acquire a 1− charge, so it must be in family VIIA (the members of this family have seven electrons and need one more to acquire the noble gas structure).
(b) Potassium must have lost one electron, so it is in IA.
(c) Aluminum lost three electrons, so it is in IIIA.
(d) Sulfur gained two electrons, so it is in VIA.
(e) Barium lost two electrons, so it is in IIA.
(f) Oxygen gained two electrons, so it is in VIA.

8.19. (a) $^{16}_{8}O$ (c) $^{3}_{1}H$
(b) $^{23}_{11}Na$ (d) $^{35}_{17}Cl$

9.1.

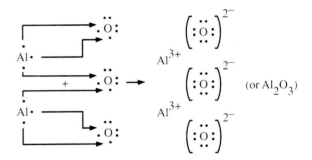

9.2. (a) Sulfur is in family VIA, so sulfur has six valence electrons and will need two more to achieve a stable outer structure like the noble gases. Two more outer shell electrons will give the sulfur atom a charge of 2−. Copper^{2+} will balance the 2− charge of sulfur, so the name is copper(II) sulfide. Note the "-ide" ending for compounds that have only two different elements.

(b) Oxygen is in family VIA, so oxygen has six valence electrons and will have a charge of 2−. Using the crossover technique in reverse, you can see that the charge on the oxygen is 2−, and the charge on the iron is 3−. Therefore the name is iron(III) oxide.

(c) From information in (a) and (b), you know that oxygen has a charge of 2−. The chromium ion must have the same charge to make a neutral compound as it must be, so the name is chromium(II) oxide. Again, note the "-ide" ending for a compound with two different elements.

(d) Sulfur has a charge of 2−, so the lead ion must have the same positive charge to make a neutral compound. The name is lead(II) sulfide.

9.3. The name of some common polyatomic ions are in Table 9.3. Using this table as a reference, the names are

(a) hydroxide
(b) sulfite
(c) hypochlorite
(d) nitrate
(e) carbonate
(f) perchlorate

9.4. The Roman numeral tells you the charge on the variable-charge elements. The charges for the polyatomic ions are found in Table 9.3. The charges for metallic elements can be found in Tables 9.1 and 9.2. Using these resources and the crossover technique, the formulas are as follows:

(a) $Fe(OH)_3$ (d) NH_4NO_3
(b) $Pb_3(PO_4)_2$ (e) $KHCO_3$
(c) $ZnCO_3$ (f) K_2SO_3

9.5. Table 9.7 has information about the meaning of prefixes and stem names used in naming covalent compounds. (a), for example, asks for the formula of carbon tetrachloride. Carbon has no prefixes, so there is one carbon atom, and it comes first in the formula because it comes first in the name. The "tetra-" prefix means four, so there are four chlorine atoms. The name ends in "-ide," so you know there are only two elements in the compound. The symbols can be obtained from the list of elements on the inside back cover of this text. Using all this information from the name, you can think out the formula for carbon tetrachloride. The same process is used for the other compounds and formulas:

(a) CCl_4 (d) SO_3
(b) H_2O (e) N_2O_5
(c) MnO_2 (f) As_2S_5

9.6. Again using information from Table 9.7, this question requires you to reverse the thinking procedure you learned in question 5.

(a) carbon monoxide
(b) carbon dioxide
(c) carbon disulfide
(d) dinitrogen monoxide
(e) tetraphosphorus trisulfide
(f) dinitrogen trioxide

9.7. The types of bonds formed are predicted by using the electronegativity scale in Table 9.5 and finding the absolute difference. On this basis:

(a) Difference = 1.7, which means ionic bond
(b) Difference = 0, which means covalent
(c) Difference = 0, which means covalent
(d) Difference = 0.4, which means covalent
(e) Difference = 3.0, which means ionic
(f) Difference = 1.6, which means polar covalent and almost ionic

10.1. (a) $MgCl_2$ is an ionic compound, so the formula has to be empirical.

(b) C_2H_2 is a covalent compound, so the formula might be molecular. Since it is not the simplest whole number ratio (which would be CH), then the formula is molecular.

(c) BaF_2 is ionic; the formula is empirical.

(d) C_8H_{18} is not the simplest whole number ratio of a covalent compound, so the formula is molecular.

(e) CH_4 is covalent, but the formula might or might not be molecular (?).

(f) S_8 is a nonmetal bonded to a nonmetal (itself); this is a molecular formula.

10.2. (a) $CuSO_4$

1 of Cu = 1 × 63.5 u = 63.5 u

1 of S = 1 × 32.1 u = 32.1 u

4 of O = 4 × 16.0 u = $\underline{64.0\ u}$
159.6 u

(b) CS_2

1 of C = 1 × 12.0 u = 12.0 u

2 of S = 2 × 32.0 u = $\underline{64.0\ u}$
76.0 u

(c) $CaSO_4$

1 of Ca = 1 × 40.1 u = 40.1 u

1 of S = 1 × 32.0 u = 32.0 u

4 of O = 4 × 16.0 u = $\underline{64.0\ u}$
136.1 u

(d) Na_2CO_3

2 of Na = 2 × 23.0 u = 46.0 u

1 of C = 1 × 12.0 u = 12.0 u

3 of O = 3 × 16.0 u = $\underline{48.0\ u}$
106.0 u

10.3. (a) FeS_2

For Fe: $\dfrac{(55.9\ u\ Fe)(1)}{119.9\ u\ FeS_2} \times 100\%\ FeS_2 = 46.6\%\ Fe$

For S: $\dfrac{(32.0\ u\ S)(2)}{119.9\ u\ FeS_2} \times 100\%\ FeS_2 = 53.4\%\ S$

or $(100\%\ FeS_2) - (46.6\%\ Fe) = 53.4\%\ S$

(b) H_3BO_3

For H: $\dfrac{(1.0\ u\ H)(3)}{61.8\ u\ H_3BO_3} \times 100\%\ H_3BO_3 = 4.85\%\ H$

For B: $\dfrac{(10.8\ u\ B)(1)}{61.8\ u\ H_3BO_3} \times 100\%\ H_3BO_3 = 17.5\%\ B$

For O: $\dfrac{(16\ u\ O)(3)}{61.8\ u\ H_3BO_3} \times 100\%\ H_3BO_3 = 77.7\%\ O$

(c) $NaHCO_3$

For Na: $\dfrac{(23.0 \text{ u Na})(1)}{84.0 \text{ u } \cancel{NaHCO_3}} \times 100\% \, \cancel{NaHCO_3} = 27.4\%$ Na

For H: $\dfrac{(1.0 \text{ u H})(1)}{84.0 \text{ u } \cancel{NaHCO_3}} \times 100\% \, \cancel{NaHCO_3} = 1.2\%$ H

For C: $\dfrac{(12.0 \text{ u C})(1)}{84.0 \text{ u } \cancel{NaHCO_3}} \times 100\% \, \cancel{NaHCO_3} = 14.3\%$ C

For O: $\dfrac{(16.0 \text{ u O})(3)}{84.0 \text{ u } \cancel{NaHCO_3}} \times 100\% \, \cancel{NaHCO_3} = 57.1\%$ O

(d) $C_9H_8O_4$

For C: $\dfrac{(12.0 \text{ u C})(9)}{180.0 \text{ u } \cancel{C_9H_8O_4}} \times 100\% \, \cancel{C_9H_8O_4} = 60.0\%$ C

For H: $\dfrac{(1.0 \text{ u H})(8)}{180.0 \text{ u } \cancel{C_9H_8O_4}} \times 100\% \, \cancel{C_9H_8O_4} = 4.4\%$ H

For O: $\dfrac{(16.0 \text{ u O})(4)}{180.0 \text{ u } \cancel{C_9H_8O_4}} \times 100\% \, \cancel{C_9H_8O_4} = 35.6\%$ O

10.4. **(a)** $2 SO_2 + O_2 \rightarrow 2 SO_3$

(b) $4 P + 5 O_2 \rightarrow 2 P_2O_5$

(c) $2 Al + 6 HCl \rightarrow 2 AlCl_3 + 3 H_2$

(d) $2 NaOH + H_2SO_4 \rightarrow Na_2SO_4 + 2 H_2O$

(e) $Fe_2O_3 + 3 CO \rightarrow 2 Fe + 3 CO_2$

(f) $3 Mg(OH)_2 + 2 H_3PO_4 \rightarrow Mg_3(PO_4)_2 + 6 H_2O$

10.5. **(a)** General form of $XY + AZ \rightarrow XZ + AY$ with precipitate formed: Ion exchange reaction.

(b) General form of $X + Y \rightarrow XY$: Combination reaction.

(c) General form of $XY \rightarrow X + Y + \dots$: Decomposition reaction.

(d) General form of $X + Y \rightarrow XY$: Combination reaction.

(e) General form of $XY + A \rightarrow AY + X$: Replacement reaction.

(f) General form of $X + Y \rightarrow XY$: Combination reaction.

10.6. **(a)** $C_5H_{12(g)} + 8 O_{2(g)} \rightarrow 5 CO_{2(g)} + 6 H_2O_{(g)}$

(b) $HCl_{(aq)} + NaOH_{(aq)} \rightarrow NaCl_{(aq)} + H_2O_{(l)}$

(c) $2 Al_{(s)} + Fe_2O_{3(s)} \rightarrow Al_2O_{3(s)} + 2 Fe_{(l)}$

(d) $Fe_{(s)} + CuSO_{4(aq)} \rightarrow FeSO_{4(aq)} + Cu_{(s)}$

(e) $MgCl_{(aq)} + Fe(NO_3)_{2(aq)} \rightarrow$ No reaction (all possible compounds are soluble and no gas or water was formed).

(f) $C_6H_{10}O_{5(s)} + 6 O_{2(g)} \rightarrow 6 CO_{2(g)} + 5 H_2O_{(g)}$

10.7. **(a)** $2 KClO_3 \xrightarrow{\Delta} 2 KCl_{(s)} + 3 O_2 \uparrow$

(b) $2 Al_2O_{3(l)} \xrightarrow{\text{elec}} 4 Al_{(s)} + 3 O_2 \uparrow$

(c) $CaCO_{3(s)} \xrightarrow{\Delta} CaO_{(s)} + CO_2 \uparrow$

10.8. **(a)** $2 Na_{(s)} + 2 H_2O_{(l)} \rightarrow 2 NaOH_{(aq)} + H_2 \uparrow$

(b) $Au_{(s)} + HCl_{(aq)} \rightarrow$ No reaction (gold is below hydrogen in the activity series).

(c) $Al_{(s)} + FeCl_{3(aq)} \rightarrow AlCl_{3(aq)} + Fe_{(s)}$

(d) $Zn_{(s)} + CuCl_{2(aq)} \rightarrow ZnCl_{2(aq)} + Cu_{(s)}$

10.9. **(a)** $NaOH_{(aq)} + HNO_{3(aq)} \rightarrow NaNO_{3(aq)} + H_2O_{(l)}$

(b) $CaCl_{2(aq)} + KNO_{3(aq)} \rightarrow$ No reaction

(c) $3 Ba(NO_3)_{2(aq)} + 2 Na_3PO_{4(aq)} \rightarrow 6 NaNO_{3(aq)} + Ba_3(PO_4)_2 \downarrow$

(d) $2 KOH_{(aq)} + ZnSO_{4(aq)} \rightarrow K_2SO_{4(aq)} + Zn(OH)_2 \downarrow$

10.10. One mole of oxygen combines with 2 moles of acetylene, so 0.5 mole of oxygen would be needed for 1 mole of acetylene. Therefore, 1 L of C_2H_2 requires 0.5 L of O_2.

EXERCISES # Chapter 11

11.1. $m_{\text{solute}} = 1.75 \text{ g}$

$m_{\text{solution}} = 50.0 \text{ g}$

% weight = ?

$\% \text{ solute} = \dfrac{m_{\text{solute}}}{m_{\text{solution}}} \times 100\% \text{ solution}$

$= \dfrac{1.75 \text{ g NaCl}}{50.0 \text{ g } \cancel{\text{solution}}} \times 100\% \, \cancel{\text{solution}}$

$= \boxed{3.50\% \text{ NaCl}}$

11.2. $m_{\text{solution}} = 103.5 \text{ g}$

$m_{\text{solute}} = 3.50 \text{ g}$

% weight = ?

$\% \text{ solute} = \dfrac{m_{\text{solute}}}{m_{\text{solution}}} \times 100\% \text{ solution}$

$= \dfrac{3.50 \text{ g NaCl}}{103.5 \text{ g } \cancel{\text{solution}}} \times 100\% \, \cancel{\text{solution}}$

$= \boxed{3.38\% \text{ NaCl}}$

11.3. Since ppm is defined as the weight unit of solute in 1,000,000 weight units of solution, the percent by weight can be calculated just like any other percent. The weight of the dissolved sodium and chlorine ions is the part, and the weight of the solution is the whole, so

$\% = \dfrac{\text{part}}{\text{whole}} \times 100\%$

$= \dfrac{30,113 \text{ g NaCl ions}}{1,000,000 \text{ g seawater}} \times 100\% \text{ seawater}$

$= \boxed{3.00\% \text{ NaCl ions}}$

11.4. $m_{\text{solution}} = 250 \text{ g}$

% solute = 3.0%

$m_{\text{solute}} = ?$

$\% \text{ solute} = \dfrac{m_{\text{solute}}}{m_{\text{solution}}} \times 100\% \text{ solution}$

$\therefore$

$m_{\text{solute}} = \dfrac{(m_{\text{solution}})(\% \text{ solute})}{100\% \text{ solution}}$

$$= \frac{(250 \text{ g})(3.0\%)}{100\%}$$

$$= \boxed{7.5 \text{ g}}$$

11.5.   % solution = 12% solution

$$V_{\text{solution}} = 200 \text{ mL}$$

$$V_{\text{solute}} = ?$$

$$\% \text{ solution} = \frac{V_{\text{solute}}}{V_{\text{solution}}} \times 100\% \text{ solution}$$

$$\therefore$$

$$V_{\text{solute}} = \frac{(\% \text{ solution})(V_{\text{solution}})}{100\% \text{ solution}}$$

$$= \frac{(12\% \text{ solution})(200 \text{ mL})}{100\% \text{ solution}}$$

$$= \boxed{24 \text{ mL alcohol}}$$

11.6.   % solution = 40%

$$V_{\text{solution}} = 50 \text{ mL}$$

$$V_{\text{solute}} = ?$$

$$\% \text{ solution} = \frac{V_{\text{solute}}}{V_{\text{solution}}} \times 100\% \text{ solution}$$

$$\therefore$$

$$V_{\text{solute}} = \frac{(\% \text{ solution})(V_{\text{solution}})}{100\% \text{ solution}}$$

$$= \frac{(40\% \text{ solution})(50 \text{ mL})}{100\% \text{ solution}}$$

$$= \boxed{20 \text{ mL alcohol}}$$

11.7.   **(a)** $\% \text{ concentration} = \dfrac{\text{ppm}}{1 \times 10^4}$

$$= \frac{5}{1 \times 10^4}$$

$$= \boxed{0.0005\% \text{ DDT}}$$

     (b) $\% \text{ part} = \dfrac{\text{part}}{\text{whole}} \times 100\% \text{ whole}$

$$\therefore$$

$$\text{whole} = \frac{(100\%)(\text{part})}{\% \text{ part}}$$

$$= \frac{(100\%)(17.0 \text{ g})}{0.0005\%} = \boxed{3,400,000 \text{ g or } 3,400 \text{ kg}}$$

11.8.

(a) $\overbrace{\text{HC}_2\text{H}_3\text{O}_{2(aq)}}^{\text{acid}} + \boxed{\text{H}_2\text{O}_{(l)}}_{\text{base}} \rightarrow \text{H}_3\text{O}^+_{(aq)} + \text{C}_2\text{H}_3\text{O}_2{}^-_{(aq)}$

(b) $\boxed{\text{C}_6\text{H}_6\text{NH}_{2(l)}}_{\text{base}} + \overbrace{\text{H}_2\text{O}_{(l)}}^{\text{acid}} \rightarrow \text{C}_6\text{H}_6\text{NH}_3{}^+_{(aq)} + \text{OH}^-_{(aq)}$

(c) $\overbrace{\text{HClO}_{4(aq)}}^{\text{acid}} + \boxed{\text{HC}_2\text{H}_3\text{O}_{2(aq)}}_{\text{base}} \rightarrow \text{H}_2\text{C}_2\text{H}_3\text{O}_2{}^+_{(aq)} + \text{ClO}_4{}^-_{(aq)}$

(d) $\boxed{\text{H}_2\text{O}_{(l)}}_{\text{base}} + \overbrace{\text{H}_2\text{O}_{(l)}}^{\text{acid}} \rightarrow \text{H}_3\text{O}^+_{(aq)} + \text{OH}^-_{(aq)}$

EXERCISES Chapter 12

12.1.   **(a)**

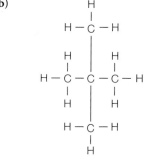

     (b)

     (c)   2,2-dimethylpropane

12.2.   *n*-hexane

3-methylpentane

2-methylpentane

2,2-dimethylbutane

$$H-\underset{\underset{H}{|}}{\overset{\overset{H}{|}}{C}}-\underset{\underset{CH_3}{|}}{\overset{\overset{CH_3}{|}}{C}}-\underset{\underset{H}{|}}{\overset{\overset{H}{|}}{C}}-\underset{\underset{H}{|}}{\overset{\overset{H}{|}}{C}}-H$$

12.3.

(a)
$$H-\underset{\underset{H}{|}}{\overset{\overset{H}{|}}{C}}-\underset{\underset{H}{|}}{\overset{\overset{H}{|}}{C}}-\underset{\underset{CH_3}{|}}{\overset{\overset{CH_3}{|}}{C}}-\underset{\underset{H}{|}}{\overset{\overset{CH_3}{|}}{C}}-\underset{\underset{H}{|}}{\overset{\overset{H}{|}}{C}}-\underset{\underset{H}{|}}{\overset{\overset{H}{|}}{C}}-\underset{\underset{H}{|}}{\overset{\overset{H}{|}}{C}}-\underset{\underset{H}{|}}{\overset{\overset{H}{|}}{C}}-H$$

(b)
$$H-\overset{\overset{H}{|}}{C}=\underset{\underset{H}{|}}{\overset{\overset{CH_3}{|}}{C}}-\underset{\underset{H}{|}}{\overset{\overset{H}{|}}{C}}-\underset{\underset{H}{|}}{\overset{\overset{H}{|}}{C}}-\underset{\underset{H}{|}}{\overset{\overset{H}{|}}{C}}-H$$

(c)
$$H-\underset{\underset{H}{|}}{\overset{\overset{H}{|}}{C}}-\underset{\underset{H}{|}}{\overset{\overset{H}{|}}{C}}-C\equiv C-\underset{\underset{CH_3}{|}}{\overset{\overset{CH_3}{|}}{C}}-\underset{\underset{H}{|}}{\overset{\overset{H}{|}}{C}}-\underset{\underset{H}{|}}{\overset{\overset{H}{|}}{C}}-H$$

12.4.　(a) 2-chloro-4-methylpentane

(b) 2-methyl-l-pentene

(c) 3-ethyl-4-methyl-2-pentene

12.5.　The 2,2,3-trimethylbutane is more highly branched, so it will have the higher octane rating.

2,2,3-trimethylbutane

$$H-\underset{\underset{H}{|}}{\overset{\overset{H}{|}}{C}}-\underset{\underset{CH_3}{|}}{\overset{\overset{CH_3}{|}}{C}}-\underset{\underset{H}{|}}{\overset{\overset{CH_3}{|}}{C}}-\underset{\underset{H}{|}}{\overset{\overset{H}{|}}{C}}-H$$

2,2,-dimethylpentane

$$H-\underset{\underset{H}{|}}{\overset{\overset{H}{|}}{C}}-\underset{\underset{CH_3}{|}}{\overset{\overset{CH_3}{|}}{C}}-\underset{\underset{H}{|}}{\overset{\overset{H}{|}}{C}}-\underset{\underset{H}{|}}{\overset{\overset{H}{|}}{C}}-\underset{\underset{H}{|}}{\overset{\overset{H}{|}}{C}}-H$$

12.6　(a) alcohol
(b) amide
(c) ether
(d) ester
(e) organic acid

13.1.　(a) cobalt-60: 27 protons, 33 neutrons

(b) potassium-40: 19 protons, 21 neutrons

(c) neon-24: 10 protons, 14 neutrons

(d) lead-208: 82 protons, 126 neutrons

13.2.　(a) $^{60}_{27}Co$　　(c) $^{24}_{10}Ne$

(b) $^{40}_{19}K$　　(d) $^{204}_{82}Pb$

13.3.　(a) cobalt-60: Radioactive because odd numbers of protons (27) and odd numbers of neutrons (33) are usually unstable.

(b) potassium-40: Radioactive, again having an odd number of protons (19) and an odd number of neutrons (21).

(c) neon-24: Stable, because even numbers of protons and neutrons are usually stable.

(d) lead-208: Stable, because even numbers of protons and neutrons *and* because 82 is a particularly stable number of nucleons.

13.4.　(a) $^{56}_{26}Fe \rightarrow \,^{0}_{-1}e + \,^{56}_{27}Co$

(b) $^{7}_{4}Be \rightarrow \,^{0}_{-1}e + \,^{7}_{5}B$

(c) $^{64}_{29}Cu \rightarrow \,^{0}_{-1}e + \,^{64}_{30}Zn$

(d) $^{24}_{11}Na \rightarrow \,^{0}_{-1}e + \,^{24}_{12}Mg$

(e) $^{214}_{82}Pb \rightarrow \,^{0}_{-1}e + \,^{214}_{83}Bi$

(f) $^{32}_{15}P \rightarrow \,^{0}_{-1}e + \,^{32}_{16}S$

13.5.　(a) $^{235}_{92}U \rightarrow \,^{4}_{2}He + \,^{231}_{90}Th$

(b) $^{226}_{88}Ra \rightarrow \,^{4}_{2}He + \,^{222}_{86}Rn$

(c) $^{239}_{94}Pu \rightarrow \,^{4}_{2}He + \,^{235}_{92}U$

(d) $^{214}_{83}Bi \rightarrow \,^{4}_{2}He + \,^{210}_{81}Tl$

(e) $^{230}_{90}Th \rightarrow \,^{4}_{2}He + \,^{226}_{88}Ra$

(f) $^{210}_{84}Po \rightarrow \,^{4}_{2}He + \,^{206}_{82}Pb$

13.6.　Thirty-two days is four half-lives. After the first half-life (8 days), 1/2 oz will remain. After the second half-life (8 + 8, or 16 days), 1/4 oz will remain. After the third half-life (8 + 8 + 8, or 24 days), 1/8 oz will remain. After the fourth half-life (8 + 8 + 8 + 8, or 32 days), 1/16 oz will remain, or 6.3×10^{-2} oz.

Glossary

A

absolute humidity a measure of the actual amount of water vapor in the air at a given time—for example, in grams per cubic meter

absolute magnitude a classification scheme to compensate for the distance differences to stars; calculations of the brightness that stars would appear to have if they were all at a defined, standard distance of 10 parsecs

absolute scale temperature scale set so that zero is at the theoretical lowest temperature possible, which would occur when all random motion of molecules has ceased

absolute zero the theoretical lowest temperature possible, which occurs when all random motion of molecules has ceased

abyssal plain the practically level plain of the ocean floor

acceleration a change in velocity per change in time; by definition, this change in velocity can result from a change in speed, a change in direction, or a combination of changes in speed and direction

accretion disk fat bulging disk of gas and dust from the remains of the gas cloud that forms around a protostar

achondrites homogeneously textured stony meteorites

acid any substance that is a proton donor when dissolved in water; generally considered a solution of hydronium ions in water that can neutralize a base, forming a salt and water

acid-base indicator a vegetable dye used to distinguish acid and base solutions by a color change

air mass a large, more or less uniform body of air with nearly the same temperature and moisture conditions throughout

air mass weather the weather experienced within a given air mass; characterized by slow, gradual changes from day to day

alcohol an organic compound with a general formula of ROH, where R is one of the hydrocarbon groups—for example, methyl or ethyl

aldehyde an organic molecule with the general formula RCHO, where R is one of the hydrocarbon groups—for example, methyl or ethyl

alkali metals members of family IA of the periodic table, having common properties of shiny, low-density metals that can be cut with a knife and that react violently with water to form an alkaline solution

alkaline earth metals members of family IIA of the periodic table, having common properties of soft, reactive metals that are less reactive than alkali metals

alkanes hydrocarbons with single covalent bonds between the carbon atoms

alkenes hydrocarbons with a double covalent carbon-carbon bond

alkyne hydrocarbon with a carbon-carbon triple bond

allotropic forms elements that can have several different structures with different physical properties—for example, graphite and diamond are two allotropic forms of carbon

alpha particle the nucleus of a helium atom (two protons and two neutrons) emitted as radiation from a decaying heavy nucleus; also known as an alpha ray

alpine glaciers glaciers that form at high elevations in mountainous regions

alternating current an electric current that first moves one direction, then the opposite direction with a regular frequency

amino acids organic functional groups that form polypeptides and proteins

amp unit of electric current; equivalent to C/s

ampere full name of the unit amp

amplitude the extent of displacement from the equilibrium condition; the size of a wave from the rest (equilibrium) position

angle of incidence angle of an incident (arriving) ray or particle to a surface; measured from a line perpendicular to the surface (the normal)

angle of reflection angle of a reflected ray or particle from a surface; measured from a line perpendicular to the surface (the normal)

angular momentum quantum number in the quantum mechanics model of the atom, one of four descriptions of the energy state of an electron wave; this quantum number describes the energy sublevels of electrons within the main energy levels of an atom

angular unconformity a boundary in rock where the bedding planes above and below the time interruption unconformity are not parallel, meaning probable tilting or folding followed by a significant period of erosion, which in turn was followed by a period of deposition

annular eclipse occurs when the penumbra reaches the surface of the earth; as seen from the earth, the sun forms a bright ring around the disk of the new moon

Antarctic circle parallel identifying the limit toward the equator where the sun appears above the horizon all day for six months during the summer; located at 66.5°S latitude

anticline an arch-shaped fold in layered bedrock

anticyclone a high-pressure center with winds flowing away from the center; associated with clear, fair weather

antinode region of maximum amplitude between adjacent nodes in a standing wave

apogee the point at which the Moon's elliptical orbit takes the Moon farthest from Earth

apparent local noon the instant when the Sun crosses the celestial meridian at any particular longitude

apparent local solar time the time found from the position of the sun in the sky; the shadow of the gnomon on a sundial

apparent magnitude a classification scheme for different levels of brightness of stars that you see; brightness values range from 1 to 6 with the number 1 (first magnitude) assigned to the brightest star and the number 6 (sixth magnitude) assigned to the faintest star that can be seen

apparent solar day the interval between two consecutive crossings of the celestial meridian by the sun

aquifer a layer of sand, gravel, or other highly permeable material beneath the surface that is saturated with water and is capable of producing water in a well or spring

Arctic circle parallel identifying the limit toward the equator where the sun appears above the horizon all day for one day up to six months during the summer; located at 66.5°N latitude

arid dry climate classification; receives less than 25 cm (10 in) precipitation per year

aromatic hydrocarbon organic compound with at least one benzene ring structure; cyclic hydrocarbons and their derivatives

artesian term describing the condition where confining pressure forces groundwater from a well to rise above the aquifer

asbestos The common name for any one of several incombustible fibrous minerals that will not melt or ignite and can be woven into a fireproof cloth or used directly in fireproof insulation; about six different commercial varieties of asbestos are used, one of which has been linked to cancer under heavy exposure

asteroids small rocky bodies left over from the formation of the solar system; most are accumulated in a zone between the orbits of Mars and Jupiter

asthenosphere a plastic, mobile layer of the earth's structure that extends around the earth below the lithosphere; ranges in thickness from a depth of 130 km to 160 km

astronomical unit the radius of Earth's orbit is defined as one astronomical unit (A.U.)

atmospheric stability the condition of the atmosphere related to the temperature of the air at increasing altitude compared to the temperature of a rising parcel of air at increasing altitude

atom the smallest unit of an element that can exist alone or in combination with other elements

atomic mass unit relative mass unit (u) of an isotope based on the standard of the carbon-12 isotope, which is defined as a mass of exactly 12.00 u; one atomic mass unit (1 u) is 1/12 the mass of a carbon-12 atom

atomic number the number of protons in the nucleus of an atom

atomic weight weighted average of the masses of stable isotopes of an element as they occur in nature, based on the abundance of each isotope of the element and the atomic mass of the isotope compared to C-12

autumnal equinox one of two times a year that daylight and night are of equal length; occurs on or about September 23 and identifies the beginning of the fall season

avalanche a mass movement of a wide variety of materials such as rocks, snow, trees, soils, and so forth in a single chaotic flow; also called debris avalanche

Avogadro's number the number of C-12 atoms in exactly 12.00 g of C; 6.02×10^{23} atoms or other chemical units; the number of chemical units in one mole of a substance

axis the imaginary line about which a planet or other object rotates

B

background radiation ionizing radiation (alpha, beta, gamma, etc.) from natural sources; between 100 and 500 millirems/yr of exposure to natural radioactivity from the environment

Balmer series a set of four line spectra, narrow lines of color emitted by hydrogen atom electrons as they drop from excited states to the ground state

band of stability a region of a graph of the number of neutrons versus the number of protons in nuclei; nuclei that have the neutron-to-proton ratios located in this band do not undergo radioactive decay

barometer an instrument that measures atmospheric pressure, used in weather forecasting and in determining elevation above sea level

base any substance that is a proton acceptor when dissolved in water; generally considered a solution that forms hydroxide ions in water that can neutralize an acid, forming a salt and water

basin a large, bowl-shaped fold in the land into which streams drain; also a small enclosed or partly enclosed body of water

batholith a large volume of magma that has cooled and solidified below the surface, forming a large mass of intrusive rock

beat rhythmic increases and decreases of volume from constructive and destructive interference between two sound waves of slightly different frequencies

beta particle high-energy electron emitted as ionizing radiation from a decaying nucleus; also known as a beta ray

big bang theory current model of galactic evolution in which the universe was created from an intense and brilliant explosion from a primeval fireball

binding energy the energy required to break a nucleus into its constituent protons and neutrons; also the energy equivalent released when a nucleus is formed

black hole the theoretical remaining core of a supernova that is so dense that even light cannot escape

blackbody radiation electromagnetic radiation emitted by an ideal material (the blackbody) that perfectly absorbs and perfectly emits radiation

body wave a seismic wave that travels through the earth's interior, spreading outward from a disturbance in all directions

Bohr model model of the structure of the atom that attempted to correct the deficiencies of the solar system model and account for the Balmer series

boiling point the temperature at which a phase change of liquid to gas takes place through boiling; the same temperature as the condensation point

boundary the division between two regions of differing physical properties

Bowen's reaction series crystallization series that occurs as a result of the different freezing point temperatures of various minerals present in magma

breaker a wave whose front has become so steep that the top part has broken forward of the wave, breaking into foam, especially against a shoreline

British thermal unit the amount of energy or heat needed to increase the temperature of 1 pound of water 1 degree Fahrenheit (abbreviated Btu)

C

calorie the amount of energy (or heat) needed to increase the temperature of 1 gram of water 1 degree Celsius

Calorie the dieter's "calorie"; equivalent to 1 kilocalorie

carbohydrates organic compounds that include sugars, starches, and cellulose; carbohydrates are used by plants and animals for structure, protection, and food

carbon film a type of fossil formed when the volatile and gaseous constituents of a buried organic structure are distilled away, leaving a carbon film as a record

carbonation in chemical weathering   a reaction that occurs naturally between carbonic acid (H_2CO_3) and rock minerals

cast   sediments deposited by groundwater in a mold, taking the shape and external features of the organism that was removed to form the mold, then gradually changing to sedimentary rock

cathode rays   negatively charged particles (electrons) that are emitted from a negative terminal in an evacuated glass tube

celestial equator   line of the equator of the earth directly above the earth; the equator of the earth projected on the celestial sphere

celestial meridian   an imaginary line in the sky directly above you that runs north through the north celestial pole, south through the south celestial pole, and back around the other side to make a big circle around the earth

celestial sphere   a coordinate system of lines used to locate objects in the sky by imagining a huge turning sphere surrounding the earth with the stars and other objects attached to the sphere; the latitude and longitude lines of the earth's surface are projected to the celestial sphere

cellulose   a polysaccharide abundant in plants that forms the fibers in cell walls that preserve the structure of plant materials

Celsius scale   referent scale that defines numerical values for measuring hotness or coldness, defined as degrees of temperature; based on the reference points of the freezing point of water and the boiling point of water at sea-level pressure, with 100 degrees between the two points

cementation   process by which spaces between buried sediment particles under compaction are filled with binding chemical deposits, binding the particles into a rigid, cohesive mass of a sedimentary rock

Cenozoic   one of four geologic eras; the time of recent life, meaning the fossils of this era are identical to the life found on the earth today

centigrade   alternate name for the Celsius scale

centrifugal force   an apparent outward force on an object following a circular path that is a consequence of the third law of motion

centripetal force   the force required to pull an object out of its natural straight-line path and into a circular path; centripetal means "center seeking"

Cepheid variable   a bright variable star that can be used to measure distance

chain reaction   a self-sustaining reaction where some of the products are able to produce more reactions of the same kind; in a nuclear chain reaction neutrons are the products that produce more nuclear reactions in a self-sustaining series

chemical bond   an attractive force that holds atoms together in a compound

chemical change   a change in which the identity of matter is altered and new substances are formed

chemical energy   a form of energy involved in chemical reactions associated with changes in internal potential energy; a kind of potential energy that is stored and later released during a chemical reaction

chemical equation   concise way of describing what happens in a chemical reaction

chemical equilibrium   occurs when two opposing reactions happen at the same time and at the same rate

chemical reaction   a change in matter where different chemical substances are created by forming or breaking chemical bonds

chemical sediments   ions from rock materials that have been removed from solution, for example, carbonate ions removed by crystallization or organisms to form calcium carbonate chemical sediments

chemical weathering   the breakdown of minerals in rocks by chemical reactions with water, gases of the atmosphere, or solutions

chemistry   the science concerned with the study of the composition, structure, and properties of substances and the transformations they undergo

Chinook   a warm wind that has been warmed by compression; also called Santa Ana

chondrites   subdivision of stony meteorites containing small, spherical lumps of silicate minerals or glass

chondrules   small, spherical lumps of silicate minerals or glass found in some meteorites

cinder cone volcano   a volcanic cone that formed from cinders, sharp-edged rock fragments that cooled from frothy blobs of lava as they were thrown into the air

cirque   a bowl-like depression in the side of a mountain, usually at the upper end of a mountain valley, formed by glacial erosion

clastic sediments   weathered rock fragments that are in various states of being

broken down from solid bedrock; boulders, gravel, and silt

climate   the general pattern of weather that occurs in a region over a number of years

coalescence process (meteorology)   the process by which large raindrops form from the merging and uniting of millions of tiny water droplets

cold front   the front that is formed as a cold air mass moves into warmer air

combination chemical reaction   a synthesis reaction in which two or more substances combine to form a single compound

comets   celestial objects originating from the outer edges of the solar system that move about the sun in highly elliptical orbits; solar heating and pressure from the solar wind form a tail on the comet that points away from the Sun

compaction   the process of pressure from a depth of overlying sediments squeezing the deeper sediments together and squeezing water out

composite volcano   a volcanic cone that formed from a buildup of alternating layers of cinders, ash, and lava flows

compound   a pure chemical substance that can be decomposed by a chemical change into simpler substances with a fixed mass ratio

compressive stress   a force that tends to compress the surface as the earth's plates move into each other

concentration   an arbitrary description of the relative amounts of solute and solvent in a solution; a larger amount of solute makes a concentrated solution, and a small amount of solute makes a dilute concentration

condensation (sound)   a compression of gas molecules; a pulse of increased density and pressure that moves through the air at the speed of sound

condensation (water vapor)   where more vapor or gas molecules are returning to the liquid state than are evaporating

condensation nuclei   tiny particles such as tiny dust, smoke, soot, and salt crystals that are suspended in the air on which water condenses

condensation point   the temperature at which a gas or vapor changes back to a liquid

conduction   the transfer of heat from a region of higher temperature to a region of lower temperature by increased kinetic energy moving from molecule to molecule

constructive interference   the condition in which two waves arriving at the same place, at the same time and in phase, add amplitudes to create a new wave

continental air mass   dry air masses that form over large land areas

continental climate   a climate influenced by air masses from large land areas; hot summers and cold winters

continental drift   a concept that continents shift positions on the earth's surface, moving across the surface rather than being fixed, stationary landmasses

continental glaciers   glaciers that cover a large area of a continent, e.g., Greenland and the Antarctic

continental shelf   a feature of the ocean floor; the flooded margins of the continents that form a zone of relatively shallow water adjacent to the continents

continental slope   a feature of the ocean floor; a steep slope forming the transition between the continental shelf and the deep ocean basin

control rods   rods inserted between fuel rods in a nuclear reactor to absorb neutrons and thus control the rate of the nuclear chain reaction

controlled experiment   an experiment that allows for a comparison of two events that are identical in all but one respect

convection   transfer of heat from a region of higher temperature to a region of lower temperature by the displacement of high-energy molecules—for example, the displacement of warmer, less dense air (higher kinetic energy) by cooler, more dense air (lower kinetic energy)

convection cell   complete convective circulation pattern; also, slowly turning regions in the plastic asthenosphere that might drive the motion of plate tectonics

convection zone (of a star)   part of the interior of a star according to a model; the region directly above the radiation zone where gases are heated by the radiation zone below and move upward by convection to the surface, where they emit energy in the form of visible light, ultraviolet radiation, and infrared radiation

conventional current   opposite to electron current—that is, considers an electric current to consist of a drift of positive charges that flow from the positive terminal to the negative terminal of a battery

convergent boundaries   boundaries that occur between two plates moving toward each other

core (of the earth)   the center part of the earth, which consists of a solid inner part and

liquid outer part, making up about 15 percent of the earth's total volume and about one-third of its mass

core (of a star)   dense, very hot region of a star where nuclear fusion reactions release gamma and X-ray radiation

Coriolis effect   the apparent deflection due to the rotation of the earth; it is to the right in the Northern Hemisphere

correlation   the determination of the equivalence in geologic age by comparing the rocks in two separate locations

coulomb   unit used to measure quantity of electric charge; equivalent to the charge resulting from the transfer of 6.24 billion billion particles such as the electron

Coulomb's law   relationship between charge, distance, and magnitude of the electrical force between two bodies

covalent bond   a chemical bond formed by the sharing of a pair of electrons

covalent compound   chemical compound held together by a covalent bond or bonds

creep   the slow downhill movement of soil down a steep slope

crest   the high mound of water that is part of a wave; also refers to the condensation, or high-pressure part, of a sound wave

critical angle   limit to the angle of incidence when all light rays are reflected internally

critical mass   mass of fissionable material needed to sustain a chain reaction

crude oil   petroleum pumped from the ground that has not yet been refined into usable products

crust   the outermost part of the earth's interior structure; the thin, solid layer of rock that rests on top of the Mohorovicic discontinuity

curie   unit of nuclear activity defined as 3.70×10^{10} nuclear disintegrations per second

cycle   a complete vibration

cyclone   a low-pressure center where the winds move into the low-pressure center and are forced upward; a low-pressure center with clouds, precipitation, and stormy conditions

D

data   measurement information used to describe something

data points   points that may be plotted on a graph to represent simultaneous measurements of two related variables

daylight saving time   setting clocks ahead one hour during the summer to more effectively utilize the longer days of summer, then setting the clocks back in the fall

decibel scale   a nonlinear scale of loudness based on the ratio of the intensity level of a sound to the intensity at the threshold of hearing

decomposition chemical reaction   a chemical reaction in which a compound is broken down into the elements that make up the compound, into simpler compounds, or into elements and simpler compounds

deep-focus earthquakes   earthquakes that occur in the lower part of the upper mantle, between 350 and 700 km below the surface of the earth

deflation   the widespread removal of base materials from the surface by the wind

degassing   process whereby gases and water vapor were released from rocks heated to melting during the early stages of the formation of a planet

delta   a somewhat triangular deposit at the mouth of a river formed where a stream flowing into a body of water slowed and lost its sediment-carrying ability

density   the compactness of matter described by a ratio of mass (or weight) per unit volume

density current   an ocean current that flows because of density differences in seawater

destructive interference   the condition in which two waves arriving at the same point at the same time out of phase add amplitudes to create zero total disturbance

dew   condensation of water vapor into droplets of liquid on surfaces

dew point temperature   the temperature at which condensation begins

diastrophism   all-inclusive term that means any and all possible movements of the earth's plates, including drift, isostatic adjustment, and any other process that deforms or changes the earth's surface by movement

diffuse reflection   light rays reflected in many random directions, as opposed to the parallel rays reflected from a perfectly smooth surface such as a mirror

dike   a tabular-shaped intrusive rock that formed when magma moved into joints or faults that cut across other rock bodies

direct current   an electrical current that always moves in one direction

direct proportion   when two variables increase or decrease together in the same ratio (at the same rate)

disaccharides   two monosaccharides joined together with the loss of a water molecule; examples of disaccharides are sucrose (table sugar), lactose, and maltose

dispersion   the effect of spreading colors of light into a spectrum with a material that has an index of refraction that varies with wavelength

divergent boundaries   boundaries that occur between two plates moving away from each other

divide   line separating two adjacent watersheds

dome   a large, upwardly bulging, symmetrical fold that resembles a dome

Doppler effect   an apparent shift in the frequency of sound or light due to relative motion between the source of the sound or light and the observer

double bond   covalent bond formed when two pairs of electrons are shared by two atoms

dune   a hill, low mound, or ridge of windblown sand or other sediments

E

earthflow   a mass movement of a variety of materials such as soil, rocks, and water with a thick, fluid-like flow

earthquake   a quaking, shaking, vibrating, or upheaval of the earth's surface

earthquake epicenter   point on the earth's surface directly above an earthquake focus

earthquake focus   place where seismic waves originate beneath the surface of the earth

echo   a reflected sound that can be distinguished from the original sound, which usually arrives 0.1 s or more after the original sound

eclipse   when the shadow of a celestial body falls on the surface of another celestial body

El Niño   changes in atmospheric pressure systems, ocean currents, water temperatures, and wind patterns that seem to be linked to worldwide changes in the weather

elastic rebound   the sudden snap of stressed rock into new positions; the recovery from elastic strain that results in an earthquake

elastic strain   an adjustment to stress in which materials recover their original shape after a stress is released

electric circuit   consists of a voltage source that maintains an electrical potential, a continuous conducting path for a current to follow, and a device where work is done by the electrical potential; a switch in the circuit

is used to complete or interrupt the conducting path

electric current   the flow of electric charge

electric field   force field produced by an electrical charge

electric field lines   a map of an electric field representing the direction of the force that a positive test charge would experience; the direction of an electric field shown by lines of force

electric generator   a mechanical device that uses wire loops rotating in a magnetic field to produce electromagnetic induction in order to generate electricity

electric potential energy   potential energy due to the position of a charge near other charges

electrical conductors   materials that have electrons that are free to move throughout the material; for example, metals

electrical energy   a form of energy from electromagnetic interactions; one of five forms of energy—mechanical, chemical, radiant, electrical, and nuclear

electrical force   a fundamental force that results from the interaction of electrical charge and is billions and billions of times stronger than the gravitational force; sometimes called the "electromagnetic force" because of the strong association between electricity and magnetism

electrical insulators   electrical nonconductors, or materials that obstruct the flow of electric current

electrical nonconductors   materials that have electrons that are not moved easily within the material—for example, rubber; electrical nonconductors are also called electrical insulators

electrical resistance   the property of opposing or reducing electric current

electrolyte   water solution of ionic substances that conducts an electric current

electromagnet   a magnet formed by a solenoid that can be turned on and off by turning the current on and off

electromagnetic force   one of four fundamental forces; the force of attraction or repulsion between two charged particles

electromagnetic induction   process in which current is induced by moving a loop of wire in a magnetic field or by changing the magnetic field

electron   subatomic particle that has the smallest negative charge possible, usually found in an orbital of an atom, but gained or lost when atoms become ions

electron configuration   the arrangement of electrons in orbitals and suborbitals about the nucleus of an atom

electron current   opposite to conventional current; that is, considers electric current to consist of a drift of negative charges that flows from the negative terminal to the positive terminal of a battery

electron dot notation   notation made by writing the chemical symbol of an element with dots around the symbol to indicate the number of outer shell electrons

electron pair   a pair of electrons with different spin quantum numbers that may occupy an orbital

electron volt   the energy gained by an electron moving across a potential difference of one volt; equivalent to 1.60×10^{-19} J

electronegativity   the comparative ability of atoms of an element to attract bonding electrons

electrostatic charge   an accumulated electric charge on an object from a surplus or deficiency of electrons; also called "static electricity"

element   a pure chemical substance that cannot be broken down into anything simpler by chemical or physical means; there are over 100 known elements, the fundamental materials of which all matter is made

empirical formula   identifies the elements present in a compound and describes the simplest whole number ratio of atoms of these elements with subscripts

energy   the ability to do work

English system   a system of measurement that originally used sizes of parts of the human body as referents

entropy   the measure of disorder in thermodynamics

eons   major blocks of time in the earth's geologic history

epochs   subdivisions of geologic periods

equation   a statement that describes a relationship in which quantities on one side of the equal sign are identical to quantities on the other side

equation of time   the cumulative variation between the apparent local solar time and the mean solar time

equinoxes   Latin meaning "equal nights"; time when daylight and night are of equal length, which occurs during the spring equinox and the autumnal equinox

eras   the major blocks of time in the earth's geologic history; the Cenozoic, Mesozoic, Paleozoic, and Precambrian

erosion   the process of physically removing weathered materials; for example, rock fragments are physically picked up by an erosion agent such as a stream or a glacier

esters   class of organic compounds with the general structure of RCOOR′, where R is one of the hydrocarbon groups—for example, methyl or ethyl; esters make up fats, oils, and waxes and some give fruit and flowers their taste and odor

ether   class of organic compounds with the general formula ROR′, where R is one of the hydrocarbon groups—for example, methyl or ethyl; mostly used as industrial and laboratory solvents

excited states   as applied to an atom, describes the energy state of an atom that has electrons in a state above the minimum energy state for that atom; as applied to a nucleus, describes the energy state of a nucleus that has particles in a state above the minimum energy state for that nuclear configuration

exfoliation   the fracturing and breaking away of curved, sheetlike plates from bare rock surfaces via physical or chemical weathering, resulting in dome-shaped hills and rounded boulders

exosphere   the outermost layer of the atmosphere where gas molecules merge with the diffuse vacuum of space

experiment   a re-creation of an event in a way that enables a scientist to gain valid and reliable empirical evidence

external energy   the total potential and kinetic energy of an everyday-sized object

extrusive igneous rocks   fine-grained igneous rocks formed as lava cools rapidly on the surface

F

Fahrenheit scale   referent scale that defines numerical values for measuring hotness or coldness, defined as degrees of temperature; based on the reference points of the freezing point of water and the boiling point of water at sea-level pressure, with 180 degrees between the two points

family   vertical columns of the periodic table consisting of elements that have similar properties

fats   organic compounds of esters formed from glycerol and three long-chain carboxylic acids that are also called triglycerides; called fats in animals and oils in plants

fault   a break in the continuity of a rock formation along which relative movement has occurred between the rocks on either side

fault plane   the surface along which relative movement has occurred between the rocks on either side; the surface of the break in continuity of a rock formation

ferromagnesian silicates   silicates that contain iron and magnesium; examples include the dark-colored minerals olivine, augite, hornblende, and biotite

first law of motion   every object remains at rest or in a state of uniform straight-line motion unless acted on by an unbalanced force

first law of thermodynamics   a statement of the law of conservation of energy in the relationship between internal energy, work, and heat

first quarter   the moon phase between the new phase and the full phase when the moon is perpendicular to a line drawn through the earth and the sun; one-half of the lighted moon can be seen from the earth, so this phase is called the first quarter

floodplain   the wide, level floor of a valley built by a stream; the river valley where a stream floods when it spills out of its channel

fluids   matter that has the ability to flow or be poured; the individual molecules of a fluid are able to move, rolling over or by one another

focus   the place beneath the surface where the waves of an earthquake originate

folds   bends in layered bedrock as a result of stress or stresses that occurred when the rock layers were in a ductile condition, probably under considerable confining pressure from deep burial

foliation   the alignment of flat crystal flakes of a rock into parallel sheets

force   a push or pull capable of changing the state of motion of an object; a force has magnitude (strength) as well as direction

force field   a model describing action at a distance by giving the magnitude and direction of force on a unit particle; considers a charge or a mass to alter the space surrounding it and a second charge or mass to interact with the altered space with a force

formula   describes what elements are in a compound and in what proportions

formula weight   the sum of the atomic weights of all the atoms in a chemical formula

fossil   any evidence of former prehistoric life

fossil fuels   organic fuels that contain the stored radiant energy of the sun converted to chemical energy by plants or animals that lived millions of years ago; coal, petroleum, and natural gas are the common fossil fuels

Foucault pendulum   a heavy mass swinging from a long wire that can be used to provide evidence about the rotation of the earth

fracture strain   an adjustment to stress in which materials crack or break as a result of the stress

free fall   when objects fall toward the earth with no forces acting upward; air resistance is neglected when considering an object to be in free fall

freezing point   the temperature at which a phase change of liquid to solid takes place; the same temperature as the melting point for a given substance

frequency   the number of cycles of a vibration or of a wave occurring in one second, measured in units of cycles per second (hertz)

freshwater   water that is not saline and is fit for human consumption

front   the boundary, or thin transition zone, between air masses of different temperatures

frost   ice crystals formed by water vapor condensing directly from the vapor phase; frozen water vapor that forms on objects

frost wedging   the process of freezing and thawing water in small rock pores and cracks that become larger and larger, eventually forcing pieces of rock to break off

fuel rod   long zirconium alloy tubes containing fissionable material for use in a nuclear reactor

full moon   the moon phase when the earth is between the sun and the moon and the entire side of the moon facing the earth is illuminated by sunlight

functional group   the atom or group of atoms in an organic molecule that is responsible for the chemical properties of a particular class or group of organic chemicals

fundamental charge   smallest common charge known; the magnitude of the charge of an electron and a proton, which is 1.60×10^{-19} coulomb

fundamental frequency   the lowest frequency (longest wavelength) that can set up standing waves in an air column or on a string

fundamental properties   a property that cannot be defined in simpler terms other than to describe how it is measured; the fundamental properties are length, mass, time, and charge

G

g symbol representing the acceleration of an object in free fall due to the force of gravity; its magnitude is 9.8 m/s^2 (32 ft/s^2)

galactic clusters gravitationally bound subgroups of as many as 1,000 stars that move together within the Milky Way galaxy

galaxy group of billions and billions of stars that form the basic unit of the universe; for example, Earth is part of the solar system, which is located in the Milky Way galaxy

gamma ray very-short-wavelength electromagnetic radiation emitted by decaying nuclei

gases a phase of matter composed of molecules that are relatively far apart moving freely in a constant, random motion and having weak cohesive forces acting between them, resulting in the characteristic indefinite shape and indefinite volume of a gas

gasohol solution of ethanol and gasoline

Geiger counter a device that indirectly measures ionizing radiation (beta and/or gamma) by detecting "avalanches" of electrons that are able to move because of the ions produced by the passage of ionizing radiation

geologic time scale a "calendar" of geologic history based on the appearance and disappearance of particular fossils in the sedimentary rock record

geomagnetic time scale time scale established from the number and duration of magnetic field reversals during the past 6 million years

giant planets the large outer planets Jupiter, Saturn, Uranus, and Neptune that all have similar densities and compositions

glacier a large mass of ice on land that is formed from compacted snow and slowly moves under its own weight

globular clusters symmetrical and tightly packed clusters of as many as a million stars that move together as subgroups within the Milky Way galaxy

glycerol an alcohol with three hydroxyl groups per molecule; for example, glycerin (1,2,3-propanetriol)

glycogen a highly branched polysaccharide synthesized by the human body and stored in the muscles and liver; serves as a direct reserve source of energy

glycol an alcohol with two hydroxyl groups per molecule; for example, ethylene glycol that is used as an antifreeze

gram-atomic weight the mass in grams of one mole of an element that is numerically equal to its atomic weight

gram-formula weight the mass in grams of one mole of a compound that is numerically equal to its formula weight

gram-molecular weight the gram-formula weight of a molecular compound

granite light-colored, coarse-grained igneous rock common on continents; igneous rocks formed by blends of quartz and feldspars, with small amounts of micas, hornblende, and other minerals

greenhouse effect the process of increasing the temperature of the lower parts of the atmosphere through redirecting energy back toward the surface; the absorption and reemission of infrared radiation by carbon dioxide, water vapor, and a few other gases in the atmosphere

ground state energy state of an atom with electrons at the lowest energy state possible for that atom

groundwater water from a saturated zone beneath the surface; water from beneath the surface that supplies wells and springs

gyre the great circular systems of moving water in each ocean

H

hail a frozen form of precipitation, sometimes with alternating layers of clear and opaque, cloudy ice

hair hygrometer a device that measures relative humidity from changes in the length of hair

half-life the time required for one-half of the unstable nuclei in a radioactive substance to decay into a new element

halogen member of family VIIA of the periodic table, having common properties of very reactive nonmetallic elements common in salt compounds

hard water water that contains relatively high concentrations of dissolved salts of calcium and magnesium

heat total internal energy of molecules, which is increased by gaining energy from a temperature difference (conduction, convection, radiation) or by gaining energy from a form conversion (mechanic, chemical, radiant, electrical, nuclear)

heat of formation energy released in a chemical reaction

Heisenberg uncertainty principle you cannot measure both the exact momentum and the exact position of a subatomic particle at the same time—the more exactly one of the two is known, the less certain you are of the value of the other

hertz unit of frequency; equivalent to one cycle per second

Hertzsprung-Russell diagram diagram to classify stars with a temperature-luminosity graph

high short for high-pressure center (anticyclone), which is associated with clear, fair weather

high latitudes latitudes close to the poles; those that sometimes receive no solar radiation at noon

high-pressure center another term for anticyclone

horsepower measurement of power defined as a power rating of 550 ft·lb/s

hot spots sites on the earth's surface where plumes of hot rock materials rise from deep within the mantle

humid moist climate classification; receives more than 50 cm (20 in) precipitation per year

humidity the amount of water vapor in the air; see *relative humidity*

hurricane a tropical cyclone with heavy rains and winds exceeding 120 km/h

hydration the attraction of water molecules for ions; a reaction that occurs between water and minerals that make up rocks

hydrocarbon an organic compound consisting of only the two elements hydrogen and carbon

hydrocarbon derivatives organic compounds that can be thought of as forming when one or more hydrogen atoms on a hydrocarbon have been replaced by an element or a group of elements other than hydrogen

hydrogen bond a weak to moderate bond between the hydrogen end ($+$) of a polar molecule and the negative end ($-$) of a second polar molecule

hydrologic cycle water vapor cycling into and out of the atmosphere through continuous evaporation of liquid water from the surface and precipitation of water back to the surface

hydronium ion a molecule of water with an attached hydrogen ion, H_3O^+

hypothesis a tentative explanation of a phenomenon that is compatible with the data and provides a framework for understanding and describing that phenomenon

I

ice-crystal process a precipitation-forming process that brings water droplets of a cloud together through the formation of ice crystals

ice-forming nuclei small, solid particles suspended in air; ice can form on the suspended particles

igneous rocks rocks that formed from magma, which is a hot, molten mass of melted rock materials

impulse a change of motion is brought about by an impulse; the product of the size of an applied force and the time the force is applied

incandescent matter emitting visible light as a result of high temperature; for example, a lightbulb, a flame from any burning source, and the sun are all incandescent sources because of high temperature

incident ray line representing the direction of motion of incoming light approaching a boundary

inclination of Earth axis tilt of Earth's axis measured from the plane of the ecliptic (23.5°); considered to be the same throughout the year

index fossils distinctive fossils of organisms that lived only a brief time; used to compare the age of rocks exposed in two different locations

index of refraction the ratio of the speed of light in a vacuum to the speed of light in a material

inertia a property of matter describing the tendency of an object to resist a change in its state of motion; an object will remain in unchanging motion or at rest in the absence of an unbalanced force

infrasonic sound waves having too low a frequency to be heard by the human ear; sound having a frequency of less than 20 Hz

inorganic chemistry the study of all compounds and elements in which carbon is not the principal element

insulators materials that are poor conductors of heat—for example, heat flows slowly through materials with air pockets because the molecules making up air are far apart; also, materials that are poor conductors of electricity, for example, glass or wood

intensity a measure of the energy carried by a wave

interference phenomenon of light whereby the relative phase difference between two light waves produces light or dark spots, a result of light's wavelike nature

intermediate-focus earthquakes earthquakes that occur in the upper part of the mantle, between 70 and 350 km below the surface of the earth

intermolecular forces forces of interaction between molecules

internal energy sum of all the potential energy and all the kinetic energy of all the molecules of an object

international date line the 180° meridian is arbitrarily called the international date line; used to compensate for cumulative time zone changes by adding or subtracting a day when the line is crossed

intertropical convergence zone a part of the lower troposphere in a belt from 10°N to 10°S of the equator where air is heated, expands, and becomes less dense and rises around the belt

intrusive igneous rocks coarse-grained igneous rocks formed as magma cools slowly deep below the surface

inverse proportion the relationship in which the value of one variable increases while the value of the second variable decreases at the same rate (in the same ratio)

inversion a condition of the troposphere when temperature increases with height rather than decreasing with height; a cap of cold air over warmer air that results in increased air pollution

ion an atom or a particle that has a net charge because of the gain or loss of electrons; polyatomic ions are groups of bonded atoms that have a net charge

ion exchange reaction a reaction that takes place when the ions of one compound interact with the ions of another, forming a solid that comes out of solution, a gas, or water

ionic bond chemical bond of electrostatic attraction between negative and positive ions

ionic compounds chemical compounds that are held together by ionic bonds—that is, bonds of electrostatic attraction between negative and positive ions

ionization process of forming ions from molecules

ionization counter a device that measures ionizing radiation (alpha, beta, gamma, etc.) by indirectly counting the ions produced by the radiation

ionized an atom or a particle that has a net charge because it has gained or lost electrons

ionosphere refers to that part of the atmosphere—parts of the thermosphere and upper mesosphere—where free electrons and ions reflect radio waves around the earth and where the northern lights occur

iron meteorites meteorite classification group whose members are composed mainly of iron

island arcs curving chains of volcanic islands that occur over belts of deep-seated earthquakes; for example, the Japanese and Indonesian islands

isomers chemical compounds with the same molecular formula but different molecular structure; compounds that are made from the same numbers of the same elements but have different molecular arrangements

isotope atoms of an element with identical chemical properties but with different masses; isotopes are atoms of the same element with different numbers of neutrons

J

jet stream a powerful, winding belt of wind near the top of the troposphere that tends to extend all the way around the earth, moving generally from the west in both hemispheres at speeds of 160 km/h or more

joint a break in the continuity of a rock formation without a relative movement of the rock on either side of the break

joule metric unit used to measure work and energy; can also be used to measure heat; equivalent to newton-meter

K

Kelvin scale a temperature scale that does not have arbitrarily assigned referent points, and zero means nothing; the zero point on the Kelvin scale (also called absolute scale) is the lowest limit of temperature, where all random kinetic energy of molecules ceases

ketone an organic compound with the general formula RCOR', where R is one of the hydrocarbon groups; for example, methyl or ethyl

kilocalorie the amount of energy required to increase the temperature of 1 kilogram of water 1 degree Celsius: equivalent to 1,000 calories

kilogram the fundamental unit of mass in the metric system of measurement

kinetic energy the energy of motion; can be measured from the work done to put an object in motion, from the mass and velocity of the object while in motion, or from the amount of work the object can do because of its motion

kinetic molecular theory   the collection of assumptions that all matter is made up of tiny atoms and molecules that interact physically, that explain the various states of matter, and that have an average kinetic energy that defines the temperature of a substance

Kuiper Belt   a disk-shaped region of small icy bodies some 30 to 100 AU from the Sun; the source of short-period comets

L

L-wave   seismic waves that move on the solid surface of the earth much as water waves move across the surface of a body of water

laccolith   an intrusive rock feature that formed when magma flowed into the plane of contact between sedimentary rock layers, then raised the overlying rock into a blister-like uplift

lake   a large inland body of standing water

landforms   the features of the surface of Earth such as mountains, valleys, and plains

landslide   general term for rapid movement of any type or mass of materials

last quarter   the moon phase between the full phase and the new phase when the Moon is perpendicular to a line drawn through Earth and the Sun; one-half of the lighted Moon can be seen from Earth, so this phase is called the last quarter

latent heat   refers to the heat "hidden" in phase changes

latent heat of fusion   the heat absorbed when 1 gram of a substance changes from the solid to the liquid phase, or the heat released by 1 gram of a substance when changing from the liquid phase to the solid phase

latent heat of vaporization   the heat absorbed when 1 gram of a substance changes from the liquid phase to the gaseous phase, or the heat released when 1 gram of gas changes from the gaseous phase to the liquid phase

laterites   highly leached soils of tropical climates; usually red with high iron and aluminum oxide content

latitude   the angular distance from the equator to a point on a parallel that tells you how far north or south of the equator the point is located

lava   magma, or molten rock, that is forced to the surface from a volcano or a crack in the earth's surface

law of conservation of energy   energy is never created or destroyed; it can only be converted from one form to another as the total energy remains constant

law of conservation of mass   same as law of conservation of matter; mass, including single atoms, is neither created nor destroyed in a chemical reaction

law of conservation of matter   matter is neither created nor destroyed in a chemical reaction

law of conservation of momentum   the total momentum of a group of interacting objects remains constant in the absence of external forces

light ray   model using lines to show the direction of motion of light to describe the travels of light

light-year   the distance that light travels through empty space in one year, approximately 9.5×10^{12} km $(5.86 \times 10^{12}$ mi)

line spectrum   narrow lines of color in an otherwise dark spectrum; these lines can be used as "fingerprints" to identify gases

linear scale   a scale, generally on a graph, where equal intervals represent equal changes in the value of a variable

lines of force   lines drawn to make an electric field strength map, with each line originating on a positive charge and ending on a negative charge; each line represents a path on which a charge would experience a constant force and lines closer together mean a stronger electric field

liquids   a phase of matter composed of molecules that have interactions stronger than those found in a gas but not strong enough to keep the molecules near the equilibrium positions of a solid, resulting in the characteristic definite volume but indefinite shape of a liquid

liter   a metric system unit of volume usually used for liquids

lithosphere   solid layer of the earth's structure that is above the asthenosphere and includes the entire crust, the Moho, and the upper part of the mantle

loess   very fine dust or silt that has been deposited by the wind over a large area

longitude   angular distance of a point east or west from the prime meridian on a parallel

longitudinal wave   a mechanical disturbance that causes particles to move closer together and farther apart in the same direction that the wave is traveling

longshore current   a current that moves parallel to the shore, pushed along by waves that move accumulated water from breakers

loudness   a subjective interpretation of a sound that is related to the energy of the

vibrating source, related to the condition of the transmitting medium, and related to the distance involved

low latitudes   latitudes close to the equator; those that sometimes receive vertical solar radiation at noon

luminosity   the total amount of energy radiated into space each second from the surface of a star

luminous   an object or objects that produce visible light; for example, the sun, stars, lightbulbs, and burning materials are all luminous

lunar eclipse   occurs when the moon is full and the sun, moon, and earth are lined up so the shadow of earth falls on the moon

lunar highlands   light-colored mountainous regions of the moon

M

macromolecule   very large molecule, with a molecular weight of thousands or millions of atomic mass units, that is made up of a combination of many smaller, similar molecules

magma   a mass of molten rock material either below or on the earth's crust from which igneous rock is formed by cooling and hardening

magnetic domain   tiny physical regions in permanent magnets, approximately 0.01 to 1 mm, that have magnetically aligned atoms, giving the domain an overall polarity

magnetic field   model used to describe how magnetic forces on moving charges act at a distance

magnetic poles   the ends, or sides, of a magnet about which the force of magnetic attraction seems to be concentrated

magnetic quantum number   from the quantum mechanics model of the atom, one of four descriptions of the energy state of an electron wave; this quantum number describes the energy of an electron orbital as the orbital is oriented in space by an external magnetic field, a kind of energy sub-sublevel

magnetic reversal   the flipping of polarity of the earth's magnetic field as the north magnetic pole and the south magnetic pole exchange positions

main sequence stars   normal, mature stars that use their nuclear fuel at a steady rate; stars on the Hertzsprung-Russell diagram in a narrow band that runs from the top left to the lower right

manipulated variable   in an experiment, a quantity that can be controlled or manipulated; also known as the independent variable

mantle   middle part of the earth's interior; a 2,870 km (about 1,780 mile) thick shell between the core and the crust

maria   smooth, dark areas on the moon

marine climate   a climate influenced by air masses from over an ocean, with mild winters and cool summers compared to areas farther inland

maritime air mass   a moist air mass that forms over the ocean

mass   a measure of inertia, which means a resistance to a change of motion

mass defect   the difference between the sum of the masses of the individual nucleons forming a nucleus and the actual mass of that nucleus

mass movement   erosion caused by the direct action of gravity

mass number   the sum of the number of protons and neutrons in a nucleus defines the mass number of an atom; used to identify isotopes; for example, uranium-238

matter   anything that occupies space and has mass

matter waves   any moving object has wave properties, but at ordinary velocities these properties are observed only for objects with a tiny mass; term for the wavelike properties of subatomic particles

mean solar day   is 24 hours long and is averaged from the mean solar time

mean solar time   a uniform time averaged from the apparent solar time

meanders   winding, circuitous turns or bends of a stream

measurement   the process of comparing a property of an object to a well-defined and agreed-upon referent

mechanical energy   the form of energy associated with machines, objects in motion, and objects having potential energy that results from gravity

mechanical weathering   the physical breaking up of rocks without any changes in their chemical composition

melting point   the temperature at which a phase change of solid to liquid takes place; the same temperature as the freezing point for a given substance

Mercalli scale   expresses the relative intensity of an earthquake in terms of effects on people and buildings using Roman numerals that range from I to XII

meridians   north-south running arcs that intersect at both poles and are perpendicular to the parallels

mesosphere   the term means "middle layer"—the solid, dense layer of the earth's structure below the asthenosphere but above the core; also the layer of the atmosphere below the thermosphere and above the stratosphere

Mesozoic   one of four geologic eras; the time of middle life, meaning some of the fossils for this time period are similar to the life found on the earth today, but many are different from anything living today

metal   matter having the physical properties of conductivity, malleability, ductility, and luster

metamorphic rocks   previously existing rocks that have been changed into a distinctly different rock by heat, pressure, or hot solutions

meteor   the streak of light and smoke that appears in the sky when a meteoroid is made incandescent by friction with the earth's atmosphere

meteor shower   event when many meteorites fall in a short period of time

meteorite   the solid iron or stony material of a meteoroid that survives passage through the earth's atmosphere and reaches the surface

meteoroids   remnants of comets and asteroids in space

meteorology   the science of understanding and predicting weather

meter   the fundamental metric unit of length

metric system   a system of referent units based on invariable referents of nature that have been defined as standards

microclimate   a local, small-scale pattern of climate; for example, the north side of a house has a different microclimate than the south side

middle latitudes   latitudes equally far from the poles and equator; between the high and low latitudes

mineral   a naturally occurring, inorganic solid element or chemical compound with a crystalline structure

miscible fluids   fluids that can mix in any proportion

mixture   matter made of unlike parts that have a variable composition and can be separated into their component parts by physical means

model   a mental or physical representation of something that cannot be observed directly that is usually used as an aid to understanding

moderator   a substance in a nuclear reactor that slows fast neutrons so the neutrons can participate in nuclear reactions

Mohorovicic discontinuity   boundary between the crust and mantle that is marked by a sharp increase in the velocity of seismic waves as they pass from the crust to the mantle

molarity   a measure of the concentration of a solution—number of moles of a solute dissolved in one liter of solution

mold   the preservation of the shape of an organism by the dissolution of the remains of a buried organism, leaving an empty space where the remains were

mole   an amount of a substance that contains Avogadro's number of atoms, ions, molecules, or any other chemical unit; a mole is thus 6.02×10^{23} atoms, ions, or other chemical units

molecular formula   a chemical formula that identifies the actual numbers of atoms in a molecule

molecular weight   the formula weight of a molecular substance

molecule   from the chemical point of view: a particle composed of two or more atoms held together by an attractive force called a chemical bond; from the kinetic theory point of view: smallest particle of a compound or gaseous element that can exist and still retain the characteristic properties of a substance

momentum   the product of the mass of an object times its velocity

monadnocks   hills of resistant rock that are found on peneplains

monosaccharides   simple sugars that are mostly 6-carbon molecules such as glucose and fructose

moraines   deposits of bulldozed rocks and other mounded materials left behind by a melted glacier

mountain   a natural elevation of the earth's crust that rises above the surrounding surface

mudflow   a mass movement of a slurry of debris and water with the consistency of a thick milkshake

N

natural frequency   the frequency of vibration of an elastic object that depends on the size, composition, and shape of the object

neap tide   period of less-pronounced high and low tides: occurs when the sun and moon are at right angles to one another

nebula   a diffuse mass of interstellar clouds of hydrogen gas or dust

negative electric charge   one of the two types of electric charge; repels other negative charges and attracts positive charges

negative ion   atom or particle that has a surplus, or imbalance, of electrons and, thus, a negative charge

net force   the resulting force after all vector forces have been added; if a net force is zero, all the forces have canceled each other and there is not an unbalanced force

neutralized   acid or base properties have been lost through a chemical reaction

neutron   neutral subatomic particle usually found in the nucleus of an atom

neutron star   very small superdense remains of a supernova with a center core of pure neutrons

new crust zone   zone of a divergent boundary where new crust is formed by magma upwelling at the boundary

new moon   the moon phase when the Moon is between Earth and the Sun and the entire side of the Moon facing Earth is dark

newton   a unit of force defined as kg·m/s²; that is, a 1 newton force is needed to accelerate a 1 kg mass 1 m/s²

noble gas   members of family VIII of the periodic table, having common properties of colorless, odorless, chemically inert gases; also known as rare gases or inert gases

node   regions on a standing wave that do not oscillate

noise   sounds made up of groups of waves of random frequency and intensity

nonelectrolytes   water solutions that do not conduct an electric current; covalent compounds that form molecular solutions and cannot conduct an electric current

nonferromagnesian silicates   silicates that do not contain iron or magnesium ions; examples include the minerals of muscovite (white mica), the feldspars, and quartz

nonmetal   an element that is brittle (when a solid), does not have a metallic luster, is a poor conductor of heat and electricity, and is not malleable or ductile

nonsilicates   minerals that do not have the silicon-oxygen tetrahedra in their crystal structure

noon   the event of time when the sun moves across the celestial meridian

normal   a line perpendicular to the surface of a boundary

normal fault   a fault where the hanging wall has moved downward with respect to the foot wall

north celestial pole   a point directly above the north pole of the earth; the point above the north pole on the celestial sphere

north pole   the north pole of a magnet or lodestone is "north seeking," meaning that the pole of a magnet points northward when the magnet is free to turn

nova   a star that explodes or suddenly erupts and increases in brightness

nuclear energy   the form of energy from reactions involving the nucleus, the innermost part of an atom

nuclear fission   nuclear reaction of splitting a massive nucleus into more stable, less-massive nuclei with an accompanying release of energy

nuclear force   one of four fundamental forces, a strong force of attraction that operates over very short distances between subatomic particles; this force overcomes the electric repulsion of protons in a nucleus and binds the nucleus together

nuclear fusion   nuclear reaction of low-mass nuclei fusing together to form more stable and more massive nuclei with an accompanying release of energy

nuclear reactor   steel vessel in which a controlled chain reaction of fissionable materials releases energy

nucleons   name used to refer to both the protons and neutrons in the nucleus of an atom

nucleus   tiny, relatively massive and positively charged center of an atom containing protons and neutrons; the small, dense center of an atom

numerical constant   a constant without units; a number

oblate spheroid   the shape of the earth—a somewhat squashed spherical shape

observed lapse rate   the rate of change in temperature compared to change in altitude

occluded front   a front that has been lifted completely off the ground into the atmosphere, forming a cyclonic storm

ocean   the single, continuous body of salt water on the surface of the earth

ocean basin   the deep bottom of the ocean floor, which starts beyond the continental slope

ocean currents   streams of water within the ocean that stay in about the same path as they move over large distances; steady and continuous onward movement of a channel of water in the ocean

ocean wave   a moving disturbance that travels across the surface of the ocean

oceanic ridges   long, high, continuous, suboceanic mountain chains; for example, the Mid-Atlantic Ridge in the center of the Atlantic Ocean Basin

oceanic trenches   long, narrow, deep troughs with steep sides that run parallel to the edges of continents

octet rule   a generalization that helps keep track of the valence electrons in most representative elements; atoms of the representative elements (A families) attempt to acquire an outer orbital with eight electrons through chemical reactions

ohm   unit of resistance; equivalent to volts/amps

Ohm's law   the electric potential difference is directly proportional to the product of the current times the resistance

oil field   petroleum accumulated and trapped in extensive porous rock structure or structures

oils   organic compounds of esters formed from glycerol and three long-chain carboxylic acids that are also called triglycerides; called fats in animals and oils in plants

Oort cloud   a spherical "cloud" of small, icy bodies from 30,000 AU out to a light-year from the sun; the source of long-period comets

opaque   materials that do not allow the transmission of any light

orbital   the region of space around the nucleus of an atom where an electron is likely to be found

ore mineral   mineral deposits with an economic value

organic acids   acids derived from organisms; organic compounds with a general formula of RCOOH, where R is one of the hydrocarbon groups; for example, methyl or ethyl

organic chemistry   the study of compounds in which carbon is the principal element

orientation of the earth's axis   direction that the earth's axis points; considered to be the same throughout the year

origin   the only point on a graph where both the x and y variables have a value of zero at the same time

overtones   higher resonant frequencies that occur at the same time as the fundamental frequency, giving a musical instrument its characteristic sound quality

oxbow lake a small body of water, or lake, that formed when two bends of a stream came together and cut off a meander

oxidation the process of a substance losing electrons during a chemical reaction; a reaction between oxygen and the minerals making up rocks

oxidation-reduction reaction a chemical reaction in which electrons are transferred from one atom to another; sometimes called "redox" for short

oxidizing agents substances that take electrons from other substances

ozone shield concentration of ozone in the upper portions of the stratosphere that absorbs potentially damaging ultraviolet radiation, preventing it from reaching the surface of the earth

P

P-wave a pressure, or compressional wave in which a disturbance vibrates materials back and forth in the same direction as the direction of wave movement

P-wave shadow zone a region on the earth between 103° and 142° of arc from an earthquake where no P-waves are received; believed to be explained by P-waves being refracted by the core

Paleozoic one of four geologic eras; time of ancient life, meaning the fossils from this time period are very different from anything living on the earth today

parallels reference lines on the earth used to identify where in the world you are northward or southward from the equator; east and west running circles that are parallel to the equator on a globe with the distance from the equator called the latitude

parts per billion concentration ratio of parts of solute in every one billion parts of solution (ppb); could be expressed as ppb by volume or as ppb by weight

parts per million concentration ratio of parts of solute in every one million parts of solution (ppm); could be expressed as ppm by volume or as ppm by weight

Pauli exclusion principle no two electrons in an atom can have the same four quantum numbers; thus, a maximum of two electrons can occupy a given orbital

peneplain a nearly flat landform that is the end result of the weathering and erosion of the land surface

penumbra the zone of partial darkness in a shadow

percent by volume the volume of solute in 100 volumes of solution

percent by weight the weight of solute in 100 weight units of solution

perigee when the Moon's elliptical orbit brings the Moon closest to Earth

period (geologic time) subdivisions of geologic eras

period (periodic table) horizontal rows of elements with increasing atomic numbers; runs from left to right on the element table

period (wave) the time required for one complete cycle of a wave

periodic law similar physical and chemical properties recur periodically when the elements are listed in order of increasing atomic number

permeability the ability to transmit fluids through openings, small passageways, or gaps

permineralization the process that forms a fossil by alteration of an organism's buried remains by circulating groundwater depositing calcium carbonate, silica, or pyrite

petroleum oil that comes from oil-bearing rock, a mixture of hydrocarbons that is believed to have formed from ancient accumulations of buried organic materials such as remains of algae

pH scale scale that measures the acidity of a solution with numbers below 7 representing acids, 7 representing neutral, and numbers above 7 representing bases

Phanerozoic the eon of an abundant fossil record and living organisms

phase change the action of a substance changing from one state of matter to another; a phase change always absorbs or releases internal potential energy that is not associated with a temperature change

phases of matter the different physical forms that matter can take as a result of different molecular arrangements, resulting in characteristics of the common phases of a solid, liquid, or gas

photoelectric effect the movement of electrons in some materials as a result of energy acquired from absorbed light

photon a quanta of energy in a light wave; the particle associated with light

physical change a change of the state of a substance but not the identity of the substance

pitch the frequency of a sound wave

Planck's constant proportionality constant in the relationship between the energy of vibrating molecules and their frequency of vibration; a value of 6.63×10^{-34} Js

plane of the ecliptic the plane of Earth's orbit

plasma a phase of matter; a very hot gas consisting of electrons and atoms that have been stripped of their electrons because of high kinetic energies

plastic strain an adjustment to stress in which materials become molded or bent out of shape under stress and do not return to their original shape after the stress is released

plate tectonics the theory that the earth's crust is made of rigid plates that float on the upper mantle

plunging folds synclines and anticlines that are not parallel to the surface of the earth

polar air mass cold air mass that forms in cold regions

polar climate zone climate zone of the high latitudes; average monthly temperatures stay below 10°C (50°F), even during the warmest month of the year

polar covalent bond a covalent bond in which there is an unequal sharing of bonding electrons

polarized light whose constituent transverse waves are all vibrating in the same plane; also known as plane-polarized light

Polaroid a film that transmits only polarized light

polyatomic ion ion made up of many atoms

polymers huge, chainlike molecules made of hundreds or thousands of smaller repeating molecular units called monomers

polysaccharides polymers consisting of monosaccharide units joined together in straight or branched chains; starches, glycogen, or cellulose

pond a small body of standing water, smaller than a lake

porosity the ratio of pore space to the total volume of a rock or soil sample, expressed as a percentage; freely admitting the passage of fluids through pores or small spaces between parts of the rock or soil

positive electric charge one of the two types of electric charge; repels other positive charges and attracts negative charges

positive ion atom or particle that has a net positive charge due to an electron or electrons being torn away

potential energy energy due to position; energy associated with changes in position (e.g., gravitational potential energy) or changes in shape (e.g., compressed or stretched spring)

power the rate at which energy is transferred or the rate at which work is performed; defined as work per unit of time

Prearchean the earliest of the eons before life

Precambrian one of four geologic eras; the time before the time of ancient life, meaning the rocks for this time period contain very few fossils

precession the slow wobble of the axis of the earth similar to the wobble of a spinning top

precipitation water that falls to the surface of the earth in the solid or liquid form

pressure defined as force per unit area; for example, pounds per square inch (lb/in^2)

primary coil part of a transformer; a coil of wire that is connected to a source of alternating current

primary loop part of the energy-converting system of a nuclear power plant; the closed pipe system that carries heated water from the nuclear reactor to a steam generator

prime meridian the referent meridian (0°) that passes through the Greenwich Observatory in England

principal quantum number from the quantum mechanics model of the atom, one of four descriptions of the energy state of an electron wave; this quantum number describes the main energy level of an electron in terms of its most probable distance from the nucleus

principle of crosscutting relationships a frame of reference based on the understanding that any geologic feature that cuts across or is intruded into a rock mass must be younger than the rock mass

principle of faunal succession a frame of reference based on the understanding that life forms have changed through time as old life forms disappear from the fossil record and new ones appear, but the same form is never exactly duplicated independently at two different times in history

principle of original horizontality a frame of reference based on the understanding that on a large scale sediments are deposited in flat-lying layers, so any layers of sedimentary rocks that are not horizontal have been subjected to forces that have deformed the earth's surface

principle of superposition a frame of reference based on the understanding that an undisturbed sequence of horizontal rock layers is arranged in chronological order with the oldest layers at the bottom, and each consecutive layer will be younger than the one below it

principle of uniformity a frame of reference of slow, uniform changes in the earth's history; the processes changing rocks today are the processes that changed them in the past, or "the present is the key to the past"

proof a measure of ethanol concentration of an alcoholic beverage; proof is double the concentration by volume; for example, 50 percent by volume is 100 proof

properties qualities or attributes that, taken together, are usually unique to an object; for example, color, texture, and size

proportionality constant a constant applied to a proportionality statement that transforms the statement into an equation

proteins macromolecular polymers made of smaller molecules of amino acids, with molecular weight from about six thousand to fifty million; proteins are amino acid polymers with roles in biological structures or functions; without such a function, they are known as polypeptides

Proterozoic the eon before the Phanerozoic, meaning "beginning life"

protogalaxy collection of gas, dust, and young stars in the process of forming a galaxy

proton subatomic particle that has the smallest possible positive charge, usually found in the nucleus of an atom

protoplanet nebular model a model of the formation of the solar system that states that the planets formed from gas and dust left over from the formation of the sun

protostar an accumulation of gases that will become a star

psychrometer a two-thermometer device used to measure the relative humidity

Ptolemaic system geocentric model of the structure of the solar system that uses epicycles to explain retrograde motion

pulsars the source of regular, equally spaced pulsating radio signals believed to be the result of the magnetic field of a rotating neutron star

pure substance materials that are the same throughout and have a fixed definite composition

pure tone sound made by very regular intensities and very regular frequencies from regular repeating vibrations

quad one quadrillion Btu (10^{15} Btu); used to describe very large amounts of energy

quanta fixed amounts; usually referring to fixed amounts of energy absorbed or emitted by matter ("quanta" is plural, and "quantum" is singular)

quantities measured properties; includes the numerical value of the measurement and the unit used in the measurement

quantum mechanics model of the atom based on the wave nature of subatomic particles, the mechanics of electron waves; also called wave mechanics

quantum numbers numbers that describe energy states of an electron; in the Bohr model of the atom, the orbit quantum numbers could be any whole number 1, 2, 3, and so on out from the nucleus; in the quantum mechanics model of the atom, four quantum numbers are used to describe the energy state of an electron wave

R

rad a measure of radiation received by a material (radiation absorbed dose)

radiant energy the form of energy that can travel through space; for example, visible light and other parts of the electromagnetic spectrum

radiation the transfer of heat from a region of higher temperature to a region of lower temperature by greater emission of radiant energy from the region of higher temperature

radiation zone part of the interior of a star according to a model; the region directly above the core where gamma and X rays from the core are absorbed and reemitted, with the radiation slowly working its way outward

radioactive decay the natural spontaneous disintegration or decomposition of a nucleus

radioactive decay constant a specific constant for a particular isotope that is the ratio of the rate of nuclear disintegration per unit of time to the total number of radioactive nuclei

radioactive decay series series of decay reactions that begins with one radioactive nucleus that decays to a second nucleus that decays to a third nucleus and so on until a stable nucleus is reached

radioactivity spontaneous emission of particles or energy from an atomic nucleus as it disintegrates

radiometric age age of rocks determined by measuring the radioactive decay of unstable elements within the crystals of certain minerals in the rocks

rarefaction a thinning or pulse of decreased density and pressure of gas molecules

ratio a relationship between two numbers, one divided by the other; the ratio of distance per time is speed

real image an image generated by a lens or mirror that can be projected onto a screen

red giant stars one of two groups of stars on the Hertzsprung-Russell diagram that have a different set of properties than the main sequence stars; bright, low temperature giant stars that are enormously bright for their temperature

redox reaction short name for oxidation-reduction reaction

reducing agent supplies electrons to the substance being reduced in a chemical reaction

referent referring to or thinking of a property in terms of another, more familiar object

reflected ray a line representing direction of motion of light reflected from a boundary

reflection the change when light, sound, or other waves bounce backward off a boundary

refraction a change in the direction of travel of light, sound, or other waves crossing a boundary

rejuvenation process of uplifting land that renews the effectiveness of weathering and erosion processes

relative dating dating the age of a rock unit or geologic event relative to some other unit or event

relative humidity ratio (times 100%) of how much water vapor is in the air to the maximum amount of water vapor that could be in the air at a given temperature

rem measure of radiation that considers the biological effects of different kinds of ionizing radiation

replacement (chemical reaction) reaction in which an atom or polyatomic ion is replaced in a compound by a different atom or polyatomic ion

replacement (fossil formation) process in which an organism's buried remains are altered by circulating groundwaters carrying elements in solution; the removal of original materials by dissolutions and the replacement of new materials an atom or molecule at a time

representative elements name given to the members of the A-group families of the periodic table; also called the main-group elements

reservoir a natural or artificial pond or lake used to store water, control floods, or generate electricity; a body of water stored for public use

resonance when the frequency of an external force matches the natural frequency and standing waves are set up

responding variable the variable that responds to changes in the manipulated variable; also known as the dependent variable because its value depends on the value of the manipulated variable

reverberation apparent increase in volume caused by reflections, usually arriving within 0.1 second after the original sound

reverse fault a fault where the hanging wall has moved upward with respect to the foot wall

revolution the motion of a planet as it orbits the sun

Richter scale expresses the intensity of an earthquake in terms of a scale with each higher number indicating 10 times more ground movement and about 30 times more energy released than the preceding number

ridges long, rugged mountain chains rising thousands of meters above the abyssal plains of the ocean basin

rift a split or fracture in a rock formation, in a land formation, or in the crust of the earth

rip current strong, brief current that runs against the surf and out to sea

rock a solid aggregation of minerals or mineral materials that have been brought together into a cohesive solid

rock cycle understanding of igneous, sedimentary, or metamorphic rock as a temporary state in an ongoing transformation of rocks into new types; the process of rocks continually changing from one type to another

rock flour rock pulverized by a glacier into powdery, silt-sized sediment

rockfall the rapid tumbling, bouncing, or free fall of rock fragments from a cliff or steep slope

rockslide a sudden, rapid movement of a coherent unit of rock along a clearly defined surface or plane

rotation the spinning of a planet on its axis

runoff water moving across the surface of the earth as opposed to soaking into the ground

S

S-wave a sideways, or shear wave in which a disturbance vibrates materials from side to side, perpendicular to the direction of wave movement

S-wave shadow zone a region of the earth more than 103° of arc away from the epicenter of an earthquake where S-waves

are not recorded; believed to be the result of the core of the earth being a liquid, or at least acting like a liquid

salinity a measure of dissolved salts in seawater, defined as the mass of salts dissolved in 1,000 g of solution

salt any ionic compound except one with hydroxide or oxide ions

San Andreas fault in California, the boundary between the North American Plate and the Pacific Plate that runs north-south for some 1,300 km (800 miles) with the Pacific Plate moving northwest and the North American Plate moving southeast

saturated air air in which an equilibrium exists between evaporation and condensation; the relative humidity is 100 percent

saturated molecule an organic molecule that has the maximum number of hydrogen atoms possible

saturated solution the apparent limit to dissolving a given solid in a specified amount of water at a given temperature; a state of equilibrium that exists between dissolving solute and solute coming out of solution

scientific law a relationship between quantities, usually described by an equation in the physical sciences; is more important and describes a wider range of phenomena than a scientific principle

scientific principle a relationship between quantities concerned with a specific, or narrow range of observations and behavior

scintillation counter a device that indirectly measures ionizing radiation (alpha, beta, gamma, etc.) by measuring the flashes of light produced when the radiation strikes a phosphor

sea a smaller part of the ocean with characteristics that distinguish it from the larger ocean

sea breeze cool, dense air from over water moving over land as part of convective circulation

seafloor spreading the process by which hot, molten rock moves up from the interior of the earth to emerge along mid-oceanic rifts, flowing out in both directions to create new rocks

seamounts steep, submerged volcanic peaks on the abyssal plain

second the standard unit of time in both the metric and English systems of measurement

second law of motion the acceleration of an object is directly proportional to the net force acting on that object and inversely proportional to the mass of the object

second law of thermodynamics a statement that the natural process proceeds from a state of higher order to a state of greater disorder

secondary coil part of a transformer, a coil of wire in which the voltage of the original alternating current in the primary coil is stepped up or down by way of electromagnetic induction

secondary loop part of nuclear power plant; the closed-pipe system that carries steam from a steam generator to the turbines, then back to the steam generator as feedwater

sedimentary rocks rocks formed from particles or dissolved minerals from previously existing rocks

sediments accumulations of silt, sand, or gravel that settled out of the atmosphere or out of water

seismic waves vibrations that move as waves through any part of the earth, usually associated with earthquakes, volcanoes, or large explosions

seismograph an instrument that measures and records seismic wave data

semiarid climate classification between arid and humid; receives between 25 and 50 cm (10 and 20 in) precipitation per year

semiconductors elements that have properties between those of a metal and those of a nonmetal, sometimes conducting an electric current and sometimes acting like an electrical insulator depending on the conditions and their purity; also called metalloids

shallow-focus earthquakes earthquakes that occur from the surface down to 70 km deep

shear stress produced when two plates slide past one another or by one plate sliding past another plate that is not moving

shell model of the nucleus model of the nucleus that has protons and neutrons moving in energy levels or shells in the nucleus (similar to the shell structure of electrons in an atom)

shield volcano a broad, gently sloping volcanic cone constructed of solidified lava flows

shock wave a large, intense wave disturbance of very high pressure; the pressure wave created by an explosion, for example

sidereal day the interval between two consecutive crossings of the celestial meridian by a particular star

sidereal month the time interval between two consecutive crossings of the moon across any star

sidereal year the time interval required for the earth to move around its orbit so that the sun is again in the same position against the stars

silicates minerals that contain silicon-oxygen tetrahedra either isolated or joined together in a crystal structure

sill a tabular-shaped intrusive rock that formed when magma moved into the plane of contact between sedimentary rock layers

simple harmonic motion the vibratory motion that occurs when there is a restoring force opposite to and proportional to a displacement

single bond covalent bond in which a single pair of electrons is shared by two atoms

slope the ratio of changes in the y variable to changes in the x variable or how fast the y-value increases as the x-value increases

soil a mixture of unconsolidated weathered earth materials and humus, which is altered, decay-resistant organic matter

solar constant the averaged solar power received by the outermost part of the earth's atmosphere when the sunlight is perpendicular to the outer edge and the earth is at an average distance from the sun; about 1,370 watts per square meter

solenoid a cylindrical coil of wire that becomes electromagnetic when a current runs through it

solids a phase of matter with molecules that remain close to fixed equilibrium positions due to strong interactions between the molecules, resulting in the characteristic definite shape and definite volume of a solid

solstices time when the sun is at its maximum or minimum altitude in the sky, known as the summer solstice and the winter solstice

solubility dissolving ability of a given solute in a specified amount of solvent, the concentration that is reached as a saturated solution is achieved at a particular temperature

solute the component of a solution that dissolves in the other component; the solvent

solution a homogeneous mixture of ions or molecules of two or more substances

solvent the component of a solution present in the larger amount; the solute dissolves in the solvent to make a solution

sonic boom sound waves that pile up into a shock wave when a source is traveling at or faster than the speed of sound

sound quality characteristic of the sound produced by a musical instrument; determined by the presence and relative strengths of the overtones produced by the instrument

south celestial pole a point directly above the south pole of earth; the point above the south pole on the celestial sphere

south pole short for "south seeking"; the pole of a magnet that points southward when it is free to turn

specific heat each substance has its own specific heat, which is defined as the amount of energy (or heat) needed to increase the temperature of 1 gram of a substance 1 degree Celsius

speed a measure of how fast an object is moving—the rate of change of position per change in time; speed has magnitude only and does not include the direction of change

spin quantum number from the quantum mechanics model of the atom, one of four descriptions of the energy state of an electron wave; this quantum number describes the spin orientation of an electron relative to an external magnetic field

spring equinox one of two times a year that daylight and night are of equal length; occurs on or about March 21 and identifies the beginning of the spring season

spring tides unusually high and low tides that occur every two weeks because of the relative positions of the earth, moon, and sun

standard atmospheric pressure the average atmospheric pressure at sea level, which is also known as normal pressure; the standard pressure is 29.92 inches or 760.0 mm of mercury (1,013.25 millibar)

standard time zones 15° wide zones defined to have the same time throughout the zone, defined as the mean solar time at the middle of each zone

standard unit a measurement unit established as the standard upon which the value of the other referent units of the same type are based

standing waves condition where two waves of equal frequency traveling in opposite directions meet and form stationary regions of maximum displacement due to constructive interference and stationary regions of zero displacement due to destructive interference

starch group of complex carbohydrates (polysaccharides) that plants use as a stored food source and that serves as an important source of food for animals

stationary front occurs when the edge of a front is not advancing

steam generator   part of nuclear power plant; the heat exchanger that heats feedwater from the secondary loop to steam with the very hot water from the primary loop

step-down transformer   a transformer that decreases the voltage of a current

step-up transformer   a transformer that increases the voltage of a current

stony-iron meteorites   meteorites composed of silicate minerals and metallic iron

stony meteorites   meteorites composed mostly of silicate minerals that usually make up rocks on the earth

storm   a rapid and violent weather change with strong winds, heavy rain, snow, or hail

strain   adjustment to stress; a rock unit might respond to stress by changes in volume, changes in shape, or breaking

stratopause   the upper boundary of the stratosphere

stratosphere   the layer of the atmosphere above the troposphere where temperature increases with height

stream   a large or small body of running water

stress   a force that tends to compress, pull apart, or deform rock; stress on rocks in the earth's solid outer crust results as the earth's plates move into, away from, or alongside each other

strong acid   acid that ionizes completely in water, with all molecules dissociating into ions

strong base   base that is completely ionic in solution and has hydroxide ions

subduction zone   the region of a convergent boundary where the crust of one plate is forced under the crust of another plate into the interior of the earth

sublimation   the phase change of a solid directly into a vapor or gas

submarine canyons   a feature of the ocean basin; deep, steep-sided canyons that cut through the continental slopes

summer solstice   in the Northern Hemisphere, the time when the sun reaches its maximum altitude in the sky, which occurs on or about June 22 and identifies the beginning of the summer season

superconductors   some materials in which, under certain conditions, the electrical resistance approaches zero

supercooled   water in the liquid phase when the temperature is below the freezing point

supernova   a rare catastrophic explosion of a star into an extremely bright, but short-lived phenomenon

supersaturated   containing more than the normal saturation amount of a solute at a given temperature

surf   the zone where breakers occur; the water zone between the shoreline and the outermost boundary of the breakers

surface wave   a seismic wave that moves across the earth's surface, spreading across the surface as water waves spread on the surface of a pond from a disturbance

swell   regular groups of low-profile, long-wavelength waves that move continuously

syncline   a trough-shaped fold in layered bedrock

synodic month   the interval of time from new moon to new moon (or any two consecutive identical phases)

T

talus   steep, conical or apron-like accumulations of rock fragments at the base of a slope

temperate climate zone   climate zone of the middle latitudes; average monthly temperatures stay between 10°C and 18°C (50°F and 64°F) throughout the year

temperature   how hot or how cold something is; a measure of the average kinetic energy of the molecules making up a substance

tensional stress   the opposite of compressional stress; occurs when one part of a plate moves away from another part that does not move

terrestrial planets   planets Mercury, Venus, Earth, and Mars that have similar densities and compositions as compared to the outer giant planets

theory   a broad, detailed explanation that guides the development of hypotheses and interpretations of experiments in a field of study

thermometer   a device used to measure the hotness or coldness of a substance

thermosphere   thin, high, outer atmospheric layer of the earth where the molecules are far apart and have a high kinetic energy

third law of motion   whenever two objects interact, the force exerted on one object is equal in size and opposite in direction to the force exerted on the other object; forces always occur in matched pairs that are equal and opposite

thrust fault   a reverse fault with a low-angle fault plane

thunderstorm   a brief, intense electrical storm with rain, lightning, thunder, strong winds, and sometimes hail

tidal bore   a strong tidal current, sometimes resembling a wave, produced in very long, very narrow bays as the tide rises

tidal currents   a steady and continuous onward movement of water produced in narrow bays by the tides

tides   periodic rise and fall of the level of the sea from the gravitational attraction of the moon and sun

tornado   a long, narrow, funnel-shaped column of violently whirling air from a thundercloud that moves destructively over a narrow path when it touches the ground

total internal reflection   condition where all light is reflected back from a boundary between materials; occurs when light arrives at a boundary at the critical angle or beyond

total solar eclipse   eclipse that occurs when the earth, the moon, and the sun are lined up so the new moon completely covers the disk of the sun; the umbra of the moon's shadow falls on the surface of the earth

transform boundaries   in plate tectonics, boundaries that occur between two plates sliding horizontally by each other along a long, vertical fault; sudden jerks along the boundary result in the vibrations of earthquakes

transformer   a device consisting of a primary coil of wire connected to a source of alternating current and a secondary coil of wire in which electromagnetic induction increases or decreases the voltage of the source

transition elements   members of the B-group families of the periodic table

transparent   term describing materials that allow the transmission of light; for example, glass and clear water are transparent materials

transportation   the movement of eroded materials by agents such as rivers, glaciers, wind, or waves

transverse wave   a mechanical disturbance that causes particles to move perpendicular to the direction that the wave is traveling

trenches   a long, relatively narrow, steep-sided trough that occurs along the edges of the ocean basins

triglyceride   organic compound of esters formed from glycerol and three long-chain carboxylic acids; also called fats in animals and oil in plants

triple bond   covalent bond formed when three pairs of electrons are shared by two atoms

tropic of Cancer parallel identifying the northern limit where the sun appears directly overhead; located at 23.5°N latitude

tropic of Capricorn parallel identifying the southern limit where the sun appears directly overhead; located at 23.5°S latitude

tropical air mass a warm air mass from warm regions

tropical climate zone climate zone of the low latitudes; average monthly temperatures stay above 18°C (64°F), even during the coldest month of the year

tropical cyclone a large, violent circular storm that is born over the warm, tropical ocean near the equator; also called hurricane (Atlantic and eastern Pacific) and typhoon (in western Pacific)

tropical year the time interval between two consecutive spring equinoxes; used as standard for the common calendar year

tropopause the upper boundary of the troposphere, identified by the altitude where the temperature stops decreasing and remains constant with increasing altitude

troposphere layer of the atmosphere from the surface to where the temperature stops decreasing with height

trough the low mound of water that is part of a wave; also refers to the rarefaction, or low-pressure part of a sound wave

tsunami very large, fast, and destructive ocean wave created by an undersea earthquake, landslide, or volcanic explosion; a seismic sea wave

turbidity current a muddy current produced by underwater landslides

typhoon the name for hurricanes in the western Pacific

U

ultrasonic sound waves too high in frequency to be heard by the human ear; frequencies above 20,000 Hz

umbra the inner core of a complete shadow

unconformity a time break in the rock record

undertow a current beneath the surface of the water produced by the return of water from the shore to the sea

unit in measurement, a well-defined and agreed-upon referent

universal law of gravitation every object in the universe is attracted to every other object with a force directly proportional to the product of their masses and inversely

proportional to the square of the distance between the centers of the two masses

unpolarized light light consisting of transverse waves vibrating in all conceivable random directions

unsaturated molecule an organic molecule that does not contain the maximum number of hydrogen atoms; a molecule that can add more hydrogen atoms because of the presence of double or triple bonds

V

valence the number of covalent bonds an atom can form

valence electrons electrons of the outermost shell; the electrons that determine the chemical properties of an atom and the electrons that participate in chemical bonding

Van Allen belts belts of radiation caused by cosmic-ray particles becoming trapped and following the earth's magnetic field lines between the poles

vapor the gaseous state of a substance that is normally in the liquid state

variable changing quantity usually represented by a letter or symbol

velocity describes both the speed and direction of a moving object; a change in velocity is a change in speed, in direction of travel, or both

ventifacts rocks sculpted by wind abrasion

vernal equinox another name for the spring equinox, which occurs on or about March 21 and marks the beginning of the spring season

vibration a back-and-forth motion that repeats itself

virtual image an image where light rays appear to originate from a mirror or lens; this image cannot be projected on a screen

volcanism volcanic activity; the movement of magma

volcano a hill or mountain formed by the extrusion of lava or rock fragments from a mass of magma below

volt unit of potential difference equivalent to J/C

voltage drop the electric potential difference across a resistor or other part of a circuit that consumes power

voltage source source of electric power in an electric circuit that maintains a constant voltage supply to the circuit

volume how much space something occupies

vulcanism volcanic activity; the movement of magma

W

warm front the front that forms when a warm air mass advances against a cool air mass

water table the boundary below which the ground is saturated with water

watershed the region or land area drained by a stream; a stream drainage basin

watt metric unit for power; equivalent to J/s

wave a disturbance or oscillation that moves through a medium

wave equation the relationship of the velocity of a wave to the product of the wavelength and frequency of the wave

wave front a region of maximum displacement in a wave; a condensation in a sound wave

wave height the vertical distance of an ocean wave between the top of the wave crest and the bottom of the next trough

wave mechanics alternate name for quantum mechanics derived from the wavelike properties of subatomic particles

wave period the time required for two successive crests or other successive parts of the wave to pass a given point

wavelength the horizontal distance between successive wave crests or other successive parts of the wave

weak acid acids that only partially ionize because of an equilibrium reaction with water

weak base a base only partially ionized because of an equilibrium reaction with water

weathering slow changes that result in the breaking up, crumbling, and destruction of any kind of solid rock

white dwarf stars one of two groups of stars on the Hertzsprung-Russell diagram that have a different set of properties than the main sequence stars; faint, white-hot stars that are very small and dense

wind a horizontal movement of air that moves along or parallel to the ground, sometimes in currents or streams

wind abrasion the natural sand-blasting process that occurs when wind particles break off small particles of rock and polish the rock they strike

wind chill factor the cooling equivalent temperature that results from the wind making the air temperature seem much lower; the cooling power of wind

winter solstice in the Northern Hemisphere, the time when the sun reaches its minimum altitude, which occurs on or about December 22 and identifies the beginning of the winter season

work the magnitude of applied force times the distance through which the force acts; can be thought of as the process by which one form of energy is transformed to another

Z

zone of saturation zone of sediments beneath the surface in which water has collected in all available spaces

Credits

Photographs

Table of Contents **Ch 1:** © Werner H. Muller/Peter Arnold, Inc.; **Ch 2:** © Douglas Faulkner/Photo Researchers; **Ch 3:** © Hank Morgan/Science Source/Photo Researchers; **Ch 4:** © Lawrence Livermore/National Library/SPL/Photo Researchers; **Ch 5:** © Clyde H. Smith/Peter Arnold, Inc.; **Ch 6:** © Keith Kent/Peter Arnold, Inc.; **Ch 7:** © James L. Amos/Peter Arnold, Inc.; **Ch 8:** © Astrid & Hanns-Frieder Michler/ SPL/Photo Researchers, Inc.; **Ch 9:** © David Weintraub/Photo Researchers, Inc.; **Ch 10:** © Fundamental Photographs; **Ch 11:** © Blair Seitz/ Photo Researchers; **Ch 12:** © Kenneth Eward/Bio Grafix-Science Source/Photo Researchers Inc.; **Ch 13:** © Yann Arthus-Bertrand/ Peter Arnold, Inc; **Ch 14, Ch 15:** © NASA; **Ch 16:** © Science VU/Visuals Unlimited; **Ch 17:** © Brian Parker/Tom Stack & Associates; **Ch 18:** © Jeff Foot/Tom Stack & Associates; **Ch 19:** © Cheeseman/ Peter Arnold, Inc; **Ch 20:** © G. Ziesler/ Peter Arnold, Inc; **Ch 21:** © Alfred Pasieka/ SPL/Photo Researchers, Inc; **Ch 22:** © Galen Rowell/ Mountain Light Photography; **Ch 23:** © Charles Mayer/Science Source/Photo Researchers; **Ch 24:** © James H. Karales/Peter Arnold, Inc; **Page xiii:** © Vol. 223/ Corbis RF.

Chapter 1 **Opener:** © Werner H. Muller/ Peter Arnold, Inc.; **1.1:** © Spencer Grant/Photo Researchers, Inc; **1.2:** © Rafael Macia/Photo Researchers, Inc; **1.3:** © Bill W. Tillery; **1.12:** © Chad Slattery; **1.15:** © David Frazier/Photo Researchers, Inc.

Chapter 2 **Opener:** © Douglas Faulkner/ Photo Researchers; **2.1:** © Darrell Wong/Tony Stone/Getty; **2.7:** © Bill W. Tillery; **Box Figure 2.1:** © Eunice Harris/Photo Researchers; **2.14:** © Richard Megna/Fundamental Photographs; **2.18:** © Mark Wagner/Tony Stone/Getty; **2.23:** © Keith Jennings, Arizona State University Media Relations Office.

Chapter 3 **Opener:** © Hank Morgan/ Science Source/Photo Researchers; **3.1, 3.10, 3.11A:** © Bill W. Tillery; **3.11B:** © Alan Oddie/ PhotoEdit, Inc; **3.12, 3.14:** Courtesy of Arizona Public Service Company; **Box Figure 3.3:** © Bill W. Tillery; **p. 79:** © Archive Photos/Getty.

Chapter 4 **Opener:** © Lawrence Livermore/National Library/SPL/Photo Researchers; **4.1:** Courtesy of International Steel Group Inc; **4.2:** Courtesy of John C. Wheatly, Center for High Resolution Electron Microscopy; **4.6:** Courtesy of Honeywell, Inc; **4.14:** Manville Company; **4.22:** © Bill W. Tillery; **Box Figure 4.4:** Dippy Bird™ Courtesy of Bits & Pieces, Inc.; **Page 108:** North Wind Picture Archives.

Chapter 5 **Opener:** © Clyde H. Smith/Peter Arnold, Inc.; **5.1:** © Amy Etra/PhotoEdit, Inc.; **Box Figure 5.1:** Jones & Childers; **5.13:** © The McGraw-Hill Company, Inc. Bob Coyle, Photographer; **Page 133:** Johann Christian Doppler Stamp; **5.26:** © NASA.

Chapter 6 **Opener:** © Keith Kent/Peter Arnold, Inc.; **6.1:** © Bill W. Tillery; **6.13:** Courtesy of Arizona Public Service Company; **Box Figure 6.1, 6.17A-C:** © Bill W. Tillery; **6.18:** © Steve Cole/ PhotoDisc/Getty; **6.36A-B, Box Figure 6.4A-B:** © Bill W. Tillery; **Page 172:** © Hulton-Deutsch Collection/CORBIS.

Chapter 7 **Opener:** © James L. Amos/Peter Arnold, Inc.; **7.1:** © John Kieffer/Peter Arnold, Inc.; **7.8:** © Bill W. Tillery; **7.9A-B:** 1987 Libbey-Owens-Ford Co.; **7.17:** © Bill W. Tillery; **Page 200:** © Modern Technologies, Photo Disc.

Chapter 8 **Opener:** © Astrid & Hanns-Frieder Michler/SPL/Photo Researchers, Inc.; **8.1:** © Ken Eward/Science Source/Photo Researchers, Inc.; **8.12:** © Weiss/Jerrican/Photo Researchers.

Chapter 9 **Opener:** © David Weintraub/ Photo Researchers, Inc.; **9.1:** © Bill W. Tillery; **9.3:** © Richard Megna/Fundamental Photographs; **9.7:** © Charles M. Falco/Photo Researchers, Inc.; **9.9:** © Arthur S. Aubry/Photodisc/Getty; **9.12:** © Bill W. Tillery.

Chapter 10 **Opener:** © Fundamental Photographs; **10.1:** © Visuals Unlimited; **10.3A-D:** © Bill W. Tillery; **10.4:** © Paul Sisul/ Tony Stone/ Getty; **10.7:** © Donovan Reese/Tony Stone/Getty; **10.8, 10.9:** © Bill W. Tillery; **10.10:** © Paul Silverman/Fundamental Photographs; **10.11, 10.13:** © Bill W. Tillery; **Box Figure 10.1:** © Jon Brennis.

Chapter 11 **Opener:** © Blair Seitz/Photo Researchers; **11.1:** © Craig Tuttle/Corbis; **11.14A-B, 11.15, 11.16, Box Figure 11.1:** © Bill W. Tillery.

Chapter 12 **Opener:** © Kenneth Eward/Bio Grafix-Science Source/Photo Researchers Inc.; **12.1:** © American Petroleum Institute Photograph and Film Collection, Archives Center, National Museum of American History, Behring Center, Smithsonian Institution; **12.7, 12.15:** © Bill W. Tillery; **12.18:** © R. J. Erwin 1988/Photo Researchers, Inc; **12.21:** © Bill W. Tillery.

Chapter 13 **Opener:** © Yann Arthus-Bertrand/Peter Arnold, Inc; **13.1:** © Bill W. Tillery; **13.2A-B:** © Fundamental Photographs; **13.3:** © Bill W. Tillery; **13.10:** Arizona Public Service Company; **13.2A:** © SPL/Photo Researchers; **13.2B:** © SIU/ Photo Researchers; **13.17A-B, 13.19, 13.20:** Arizona Public Service Company; **Page 348:** © Bettmann/ Corbis.

Chapter 14 **Opener:** © NASA; **14.1A:** Lick Observatory, National Optical Astronomy Observatories; **14.2:** © J. Fuller/Visuals Unlimited; **14.12:** Lick Observatory; **14.14:** © National Optical Astronomy Observatory/ Association of Universities for Research in Astronomy/National Science Foundation; **Box Figure 14.1:** © Hencoup Enterprises/SPL/Photo Researchers; **14.16:** Lick Observatory.

Chapter 15 **Opener, 15.1, 15.4, 15.5, 15.6, 15.7, 15.9A-B, 15.10:** © NASA; **15.11A-B:** © Space Telescope Science Institute/NASA/SPL/ Photo Researchers, Inc.; **15.12, 15.13:** © NASA; **15.17:** Lick Observatory; **15.19A-B:** Center for Meteorite Studies, Arizona State University.

Chapter 16 **Opener:** © Science VU/Visuals Unlimited; **16.3:** © NASA; **16.8:** Public Domain; **16.19:** © Bill W. Tillery; **16.27:** Lick Observatory.

Chapter 17 **Opener:** © Brian Parker/Tom Stack & Associates; **17.1:** © Bill W. Tillery; **17.4:** © Charles Falco/Photo Researchers; **17.5, 17.9, 17.10:** © Bill W. Tillery; **17.11A-B, 17.13, Box Figure 17.1, 17.15:** © C.C. Plummer; **17.16, 17.17A, 17.18:** © David McGeary; **17.21:** © C.C. Plummer; **Page 439:** Courtesy of the Smithsonian Institution; **17.22:** © C.C. Plummer.

Chapter 18 Opener: © Jeff Foot/Tom Stack & Associates; **18.1:** © USGS Photo Library, Denver, CO; **18.7:** © NASA; **Page 455:** © 2000 by the Trustees of Princeton University; **Page 457:** © Sheila Davies.

Chapter 19 Opener: © Douglas Cheeseman/ Peter Arnold, Inc.; **19.1:** © A. Post, USGS Photo Library, Denver, CO; **19.2:** © Bill W. Tillery; **19.4A:** © Robert W. Northrop, Photographer/ Illustrator; **19.4B:** © Bill W. Tillery; **19.5:** C.C. Plummer; **19.6:** © John S. Shelton; **19.9A:** © National Park Service/Photo by Cecil W. Stoughton; **19.9B:** © D.E. Trimble, USGS Photo Library, Denver, CO; **19.10:** © Frank M. Hanna; **19.14:** © University of Colorado, Courtesy National Geophysical Data Center, Boulder, CO; **19.18:** © NASA; **19.21:** © Bill W. Tillery; **19.22:** © D.W. Peterson, USGS; **19.23:** © B. Amundson; **Page 478:** Edinburgh University Library.

Chapter 20 Opener: © G Ziesler/Peter Arnold, Inc; **20.1:** © W.R. Hansen, USGS Photo Library, Denver, CO; **20.2:** National Park Service, Photo by Wm. Belnap, Jr.; **20.4A:** © A. J. Copley/Visuals Unlimited; **20.4B:** © L. Linkhart/ Visuals Unlimited; **20.5:** © Ken Wagner/Visuals Unlimited; **20.6A-B:** © Bill W. Tillery; **20.8:** © William J. Weber/Visuals Unlimited; **20.9:** © Doug Sherman/ Geofile; **20.11:** © B. Amundson; **20.12A:** © D.A. Rahm, photo courtesy of Rahn Memorial Collections, Western Washington University; **20.13:** © C.C. Plummer; **20.16A-B:** © Bill W. Tillery; **Page 496:** U.S. Geological Survey.

Chapter 21 Opener: © Alfred Pasieka/SPL/ Photo Researchers, Inc; **21.1, 21.2:** © Robert W. Northrop, Photographer/Illustrator; **21.7:** © Bob Wallen; **21.10:** © Frank M. Hanna; **21.13:** U.S. Geological Survey; **Page 513:** Library of the Geological Survey of Austria.

Chapter 22 Opener: © Galen Rowell/ Mountain Light Photography; **22.1, 22.16, 22.18A-B, 22.19A-F:** © Bill W. Tillery.

Chapter 23 Opener: © Charles Mayer/ Science Source/Photo Researchers; **23.1:** © Peter Arnold/Peter Arnold, Inc; **23.5, 23.7:** © NOAA; **23.8:** © Rachel Epstein/PhotoEdit, Inc; **23.17:** © Telegraph Herald/Photo by Patti Carr; **23.18, 23.19:** © NOAA; **23.21:** © David Parker/SPL/ Photo Researchers; **23.22:** © Bill W. Tillery; **23.25:** © Dr. Charles Hogue, Curator of Entomology, Los Angeles County Museum of Natural History; **23.26:** © Bob Wallen; **23.27:** © Elizabeth Wallen; **Page 563:** Public Domain; **Page 564:** Courtesy of the National Portrait Gallery London.

Chapter 24 Opener: © James H. Karales/ Peter Arnold, Inc; **24.1:** Salt River Project; **24.9:** City of Tempe, AZ; **24.10:** Salt River Project; **24.11, 24.14, 24.15:** © Bill W. Tillery; **24.19:** © John S. Shelton; **Page 588:** © Bodleian Library, University of Oxford (MS. Eng. Misc. c. 1103, F2).

Line Art/Text

Chapter 3 **3.19:** Source: Energy Information Administration www.eia.doe.gov/emen/sep/us/ frame.html.

Chapter 9 **Box Figure 9.3:** Source: www.accessscience.com/server-ava/arknoid/ science/AS/Biographies/B (see Bunsen). Reprinted by permission of The McGraw-Hill Companies.

Chapter 16 **Table 16.1:** Source: Data from NASA.

Chapter 17 **17.23:** From Carla W. Montgomery, *Physical Geology*, 3rd edition. Copyright © 1993. Reprinted by permission of The McGraw-Hill Companies.

Chapter 18 **18.A-B:** From Charles C. Plummer and David McGeary, *Physical Geology*, 6th edition. Copyright © 1993. Reprinted by permission of The McGraw-Hill Companies. **18.5:** From Charles C. Plummer and David McGeary, *Physical Geology*, 6th edition. Copyright © 1993. Reprinted by permission of The McGraw-Hill Companies. **18.6:** From Charles C. Plummer and David McGeary, *Physical Geology*, 6th edition. Copyright © 1993. Reprinted by permission of The McGraw-Hill Companies. **18.11:** Pitman, W.C., III, Larson, R.L., and Herron, E.M., compilers, 1974, The age of the ocean basin: Boulder, Colorado, Geological Society of America Maps and Charts 6, 2 sheets. Reprinted by permission of The Geological Society of America, Boulder, CO. **18.12:** From Carla W. Montgomery, *Physical Geology*, 3rd edition. Copyright © 1993. Reprinted by permission of The McGraw-Hill Companies. **18.13:** Source: After W. Hamilton, U.S. Geological Survey. **18.18:** From Charles C. Plummer and David McGeary, *Physical Geology*, 6th edition. Copyright © 1993. Reprinted by permission of The McGraw-Hill Companies.

Chapter 19 **Table 19.1:** Source: National Oceanic and Atmospheric Administration. **19.3:** From Carla W. Montgomery, *Physical Geology*, 3rd edition. Copyright © 1993. Reprinted by permission of The McGraw-Hill Companies. **19.6A:** From David McGeary and Charles C. Plummer, *Physical Geology, Earth Revealed*, 2nd Edition. Copyright © 1994. Reprinted by permission of The McGraw-Hill Companies. **19.7:** From Carla W. Montgomery, *Physical Geology*, 3rd edition. Copyright © 1993. Reprinted by permission of The McGraw-Hill Companies. **19.8:** From Carla W. Montgomery, *Physical Geology*, 3rd edition. Copyright © 1993. Reprinted by permission of The McGraw-Hill Companies. **19.10A:** From David McGeary and Charles C. Plummer, *Physical Geology, Earth Revealed*, 2nd Edition. Copyright © 1994. Reprinted by permission of The McGraw-Hill Companies. **19.11A-C:** From Charles C. Plummer and David McGeary, *Physical Geology*, 6th edition. Copyright © 1993. Reprinted by permission of The McGraw-Hill Companies. **19.12A-B:** From Charles C. Plummer and David McGeary, *Physical Geology*, 6th edition. Copyright © 1993. Reprinted by permission of The McGraw-Hill Companies. **19.14A-C:** From Charles C. Plummer and David McGeary, *Physical Geology*, 6th edition. Copyright © 1993. Reprinted by permission of The McGraw-Hill Companies. **19.15:** From Carla W. Montgomery, *Physical Geology*, 3rd edition. Copyright © 1993. Reprinted by permission of The McGraw-Hill Companies. **19.17A-B:** From Carla W. Montgomery, *Physical Geology*, 3rd edition. Copyright ©1993. Reprinted by permission of The McGraw-Hill Companies. **19.20:** From Carla W. Montgomery, Physical Geology, 3rd edition. Copyright © 1993. Reprinted by permission of The McGraw-Hill Companies. 19.25: From Charles C. Plummer and David McGeary, *Physical Geology*, 5th edition. Copyright © 1991. Reprinted by permission of The McGraw-Hill Companies.

Chapter 20 **20.3A-B:** From Carla W. Montgomery, *Physical Geology*, 3rd edition. Copyright © 1993. Reprinted by permission of The McGraw-Hill Companies. **20.10A-C:** From Charles C. Plummer and David McGeary, *Physical Geology*, 6th edition. Copyright © 1993. Reprinted by permission of The McGraw-Hill Companies. **20.14:** From Charles C. Plummer and David McGeary, *Physical Geology*, 5th edition. Copyright © 1991. Reprinted by permission of The McGraw-Hill Companies.

Chapter 21 **21.3:** From Carla W. Montgomery and David Dathe, *Earth: Then and Now*, 2nd Edition. Copyright © 1994. Reprinted by permission of The McGraw-Hill Companies. **21.5A-B:** From Carla W. Montgomery, *Physical Geology*, 3rd edition. Copyright ©1993. Reprinted by permission of The McGraw-Hill Companies. **21.6:** From Carla W. Montgomery, *Physical Geology*, 3rd edition. Copyright © 1993. Reprinted by permission of The McGraw-Hill Companies. **21.8:** From Carla W. Montgomery, *Physical Geology*, 3rd edition. Copyright © 1993. Reprinted by permission of The McGraw-Hill Companies. **21.9:** From Carla W. Montgomery, *Physical Geology*, 3rd edition. Copyright © 1993. Reprinted by permission of The McGraw-Hill Companies. **21.11:** From Carla W. Montgomery, *Physical Geology*, 3rd edition. Copyright © 1993. Reprinted by permission of The McGraw-Hill Companies. **21.12:** From Carla W. Montgomery, *Physical Geology*, 3rd edition. Copyright © 1993. Reprinted by permission of The McGraw-Hill Companies. **21.15:** Modified from "Decade of North American Geology," 1983 Geologic Time Scale—Geological Society of America.

Chapter 24 **Pages 575–577:** Drawings, and some text, from *How Wastewater Treatment Works . . . The Basics*, The Environmental Protection Agency, Office of Water, www.epa.gov/owmitnet/ basics.html.

Index

Fluorine, 237, 238, 239
Fluorite, 432
Fly ash, 78
Foci of earthquakes, 469–70
Fog, 532–34
Folding
  causes, 464–66, 478
  mountains created by, 472–73
Foliation, 438
Follow-through, 49
Food preservation, 337
Foot-pounds, 62–63
Footprints, fossil, 503
Footwalls, 467
Force arrows, 34
Force fields. *See* Fields of force
Forces
  on elastic materials, 116–18
  measuring, 43
  overview, 32–34
  relation to change of motion, 41
  relation to work, 62
Formaldehyde, 311
Formic acid, 286, 287, 312
Forms of energy, 71–73
Formulas
  covalent compounds, 240, 247
  empirical and molecular, 254–56
  ionic compounds, 237, 244–45
Formula weight, 255, 271
Fossil fuels, 76. *See also* Petroleum
Fossils
  dating rocks from, 505, *507*, 513
  defined, 500, 513
  early concepts of, 500–501
  incomplete record, 509
  interpreting, 511
  types, 501–3
Foucault, J. L., 195, 402
Foucault pendulum, 402
Fractions, 593
Fracture of minerals, 430–31
Free fall, 35, 36–37, 38
Freezing
  fossil preservation by, 502
  of seawater, 286, 579, 585
  weathering effects, 483
Freezing point, 99, 286, 294
Freon, 523, 524
Frequency. *See also* Spectra
  defined, 117, 135
  interference, 126–27
  motion and, 132–35
  natural, 129, 135
  of waves, 123, 180
Freshwater distribution, 568–74
Fresnel, A. J., 192, 200
Friction
  electrostatic charge from, 144
  heat from, 92
  resistance to moving objects from, 34, 36
  work against, 71
Fronts, 548–50, 563, 564
Frost formation, 531–32
Frost wedging, 483, *485*
Fructose, 314

Fuel rods, 341, 342, *345*
Fullerenes, 270
Full moon, 415
Functional groups, 310
Fundamental charge of electrons, 145, 174
Fundamental frequency, 131–32, 135
Fundamental properties, 6–7
Fuses, 154
Fusion, latent heat of, 100
Fusion (nuclear)
  overview, 345–46
  proposed research facility, 11
  in stars, 345, 357, 360

G

Gabbro, 435
Gaia hypothesis, 535
Galactic clusters, 363
Galaxies, 363–68, 372
Galena, 432
Galilean moons, 383, *385*
Galileo
  motion theory, 34, 35–37, 55
  use of scientific method, 15–16
Galileo spacecraft, 380
Gallium scans, 338
Galvanometers, 162–63
Gamma cameras, 338
Gamma photons, 328
Gamma radiation
  basic features, 330–31
  defined, 327, 349
  particle properties, 197
Ganymede, 383, *385*
Gases
  as compounds, 232
  convection in, 97
  evaporation and condensation, 103–4, 110
  line spectra, 210–11
  molecular arrangement, 88, 109
  nebulae, *353, 356*
  origins of stars from, 356–57
  sound wave transmission in, 123
Gasohol, *311*
Gasoline, 10, 308–9, *311*
Gastric solution, 290
Gay-Lussac, Joseph, 266
Geiger counters, 333, *334*
Generalization, 9–10, 16, 261–62
Generators, 166, 174
Geologic events, 503–7
Geologic time, 507–11
Geomagnetic time scale, 508
German Plankton Expedition, 588
Giant planets, 376
Giotto spacecraft, 388
Glaciers, 404, 490–91, *492*
Glass
  formation of, 431, 433, 434
  rate of conduction, 97
  refraction index, 187
  specific heat, 95
  speed of sound in, 124
Glass wool, 97

Glen Canyon Dam, *63*
Global Surveyor spacecraft, 381
Global wind patterns, 527–28
Globular clusters, 363
Glucose
  basic features, 314–15
  molecular formula, 254, 314
  molecular structure, 314
  oxidation equation, 261–62
Glycerol, 311
Glycogen, 315
Glycols, 311
Gneiss, 438
Gnomons, 407
Gold, 95, 238
Goldschmidt, Victor Moritz, 439
Grabens, 467
Gram-atomic weight, 268, 271
Gram-formula weight, 268, 271
Gram-molecular weight, 268, 271
Granite
  basic features, 434
  component minerals, 432, *433*
  weathering of, 484
Grapefruit pH, 290
Graphs, 599–600
Gravel, 435
Gravitational fields, 147
Gravitational forces
  Jupiter's influence on asteroid belt, 394
  Newton's law, 50–53, 54, 56
  role in star formation, 356–57, 360
  as source of nuclear energy, 347
  tides and, 417–19
Gravitational potential energy, 67
Gravitons, 148
Gravity. *See also* Acceleration due to gravity; Gravitational forces
  erosion due to, 487
  role in weight, 44, 53
  work against, 71
Grays, 335
Great Dark Spot (Neptune), *375*
Great Red Spot (Jupiter), 383, *384*
Greek letters, 12
Greenhouse effect
  on Earth, 522, 533
  in passive solar applications, 101
  on Venus, 379
Gregorian calendar, 410
Grit chambers, 575–76
Ground-fault interrupters, 155
Ground state of electrons, 212–13
Groundwater, 569, 570–74, 589
Gulf of Mexico, 417
Gulf Stream, 585
Guyots, 455
Gypsum, *430*
Gyres, 585

H

Hail, 552–53
Hair hygrometer, 530
Half-life, 333
Halley, Edmund, 54

Halley's comet, 390
Halogens, 220, 239, 309
Hanging walls, 467
Hardness of minerals, 429–30
Hard water, 290–91
Hardy, Alistair Clavering, 588–89
Hardy Plankton Continuous Recorder, 588–89
Harmonic instruments, 132
Hawaiian volcanoes, 474
Hawking, Stephen William, 370
Health
  effects of radiation on, 335–36
  nuclear medicine technology, 338–39
Hearing, 120–21
Heat. *See also* Temperature
  basic attributes, 92–93
  evaporation and condensation, 103–4, 110
  flow of, 96–99
  measuring, 93–94, 110
  mechanical equivalent, 79
  phase changes and, 99–103
  specific heat, 94–96
  thermodynamics, 104–9
Heat death of the universe, 109
Heat engines, 104–5
Heat islands, 561, 562
Heat of formation, 236
Heat pumps, 106, *107*
Heisenberg uncertainty principle, 216
Helium
  fusion in stars, 361
  as nuclear fusion product, 345, 357, 360
  sound transmission in, 123, 124
Hematite, 429, 484
Henry, Joseph, 165, 173
Heptane, 303
Herodotus, 500
Hertz, 117
Hertzsprung-Russell diagram, 359–60, 372
Hess, Harry Hammond, 455
Hexane, 303
High latitudes, 556–57
High-level nuclear wastes, 347
High-performance alloys, 223
High-pressure centers, 550
High-quality energy, 107
Hipparchus, 357
Historical geology, 500
Hogbacks, 475, *476*
"Hole in the wall" sun calendar, 407
Horizontal motion, 34, 39–40
Hornblende, *429*
Horsepower, 66, 156
Horsts, 467
Household circuits, 154–55
Howard, Luke, 534
Hubble, Edwin, 360, 365, 366
Hubble's law, 360
Hubble Space Telescope, 363, 366
Huggins, William, 133
Humid climates, 560

Table of Atomic Weights (Based on Carbon-12)

Name	Symbol	Atomic Number	Atomic Weight	Name	Symbol	Atomic Number	Atomic Weight
Actinium	Ac	89	(227)	Meitnerium	Mt	109	(268)
Aluminum	Al	13	26.9815	Mendelevium	Md	101	258.10
Americium	Am	95	(243)	Mercury	Hg	80	200.59
Antimony	Sb	51	121.75	Molybdenum	Mo	42	95.94
Argon	Ar	18	39.948	Neodymium	Nd	60	144.24
Arsenic	As	33	74.922	Neon	Ne	10	20.179
Astatine	At	85	(210)	Neptunium	Np	93	(237)
Barium	Ba	56	137.34	Nickel	Ni	28	58.71
Berkelium	Bk	97	(247)	Niobium	Nb	41	92.906
Beryllium	Be	4	9.0122	Nitrogen	N	7	14.0067
Bismuth	Bi	83	208.980	Nobelium	No	102	259.101
Bohrium	Bh	107	(264)	Osmium	Os	76	190.2
Boron	B	5	10.811	Oxygen	O	8	15.9994
Bromine	Br	35	79.904	Palladium	Pd	46	106.4
Cadmium	Cd	48	112.40	Phosphorus	P	15	30.9738
Calcium	Ca	20	40.08	Platinum	Pt	78	195.09
Californium	Cf	98	242.058	Plutonium	Pu	94	244.064
Carbon	C	6	12.0112	Polonium	Po	84	(209)
Cerium	Ce	58	140.12	Potassium	K	19	39.098
Cesium	Cs	55	132.905	Praseodymium	Pr	59	140.907
Chlorine	Cl	17	35.453	Promethium	Pm	61	144.913
Chromium	Cr	24	51.996	Protactinium	Pa	91	(231)
Cobalt	Co	27	58.933	Radium	Ra	88	(226)
Copper	Cu	29	63.546	Radon	Rn	86	(222)
Curium	Cm	96	(247)	Rhenium	Re	75	186.2
Darmstadtium	Ds	110	(281)	Rhodium	Rh	45	102.905
Dubnium	Db	105	(262)	Rubidium	Rb	37	85.468
Dysprosium	Dy	66	162.50	Ruthenium	Ru	44	101.07
Einsteinium	Es	99	(254)	Rutherfordium	Rf	104	(261)
Erbium	Er	68	167.26	Samarium	Sm	62	150.35
Europium	Eu	63	151.96	Scandium	Sc	21	44.956
Fermium	Fm	100	257.095	Seaborgium	Sg	106	(266)
Fluorine	F	9	18.9984	Selenium	Se	34	78.96
Francium	Fr	87	(223)	Silicon	Si	14	28.086
Gadolinium	Gd	64	157.25	Silver	Ag	47	107.868
Gallium	Ga	31	69.723	Sodium	Na	11	22.989
Germanium	Ge	32	72.59	Strontium	Sr	38	87.62
Gold	Au	79	196.967	Sulfur	S	16	32.064
Hafnium	Hf	72	178.49	Tantalum	Ta	73	180.948
Hassium	Hs	108	(269)	Technetium	Tc	43	(99)
Helium	He	2	4.0026	Tellurium	Te	52	127.60
Holmium	Ho	67	164.930	Terbium	Tb	65	158.925
Hydrogen	H	1	1.0079	Thallium	Tl	81	204.37
Indium	In	49	114.82	Thorium	Th	90	232.038
Iodine	I	53	126.904	Thulium	Tm	69	168.934
Iridium	Ir	77	192.2	Tin	Sn	50	118.69
Iron	Fe	26	55.847	Titanium	Ti	22	47.90
Krypton	Kr	36	83.80	Tungsten	W	74	183.85
Lanthanum	La	57	138.91	Uranium	U	92	238.03
Lawrencium	Lr	103	260.105	Vanadium	V	23	50.942
Lead	Pb	82	207.19	Xenon	Xe	54	131.30
Lithium	Li	3	6.941	Ytterbium	Yb	70	173.04
Lutetium	Lu	71	174.97	Yttrium	Y	39	88.905
Magnesium	Mg	12	24.305	Zinc	Zn	30	65.38
Manganese	Mn	25	54.938	Zirconium	Zr	40	91.22

Note: A value given in parentheses denotes the number of the longest-lived or best-known isotope.